Chloride Transport
in Biological Membranes

Academic Press Rapid Manuscript Reproduction

Based on the Symposium
Chloride Transport in Biological Membranes
Sponsored by the American Physiological Society

Chloride Transport in Biological Membranes

Edited by

José A. Zadunaisky
Department of Physiology and Biophysics
New York University Medical Center
School of Medicine
New York, New York

1982

ACADEMIC PRESS
A Subsidiary of Harcourt Brace Jovanovich, Publishers

New York   London
Paris   San Diego   San Francisco   São Paulo   Sydney   Tokyo   Toronto

ACADEMIC PRESS, INC.
111 Fifth Avenue, New York, New York 10003

United Kingdom Edition published by
ACADEMIC PRESS, INC. (LONDON) LTD.
24/28 Oval Road, London NW1   7DX

Library of Congress Cataloging in Publication Data
Main entry under title:

Chloride transport in biological membranes.

   Proceedings of a symposium, sponsored by the American
Physiological Society, held at the New York University
Medical Center in April 1979.
   Includes index.
   1. Chlorides in the body--Congresses. 2. Biological
transport, Active--Congresses. 3. Plasma membranes--
Congresses. I. Zadunaisky, José A.   II. American
Physiological Society (1887-   ) [DNLM: 1. Biological
transport--Congresses. 2. Chlorides--Metabolism--
Congresses. 3. Membranes--Metabolism--Congresses.
QU 130 C544 1979]
QP535.C5C55   574.19'214   81-8107
ISBN 0-12-775280-3   AACR2

PRINTED IN THE UNITED STATES OF AMERICA

82 83 84 85   9 8 7 6 5 4 3 2 1

Contents

Contributors

Numbers in parentheses indicate the pages on which the authors' contributions begin.

Haim Breitbart (151), *Department of Physiology-Anatomy, University of California, Berkeley, California 94720*

William A. Brodsky (243), *Department of Physiology and Biophysics, Mount Sinai School of Medicine, New York, New York 10029*

Maurice B. Burg (111), *Laboratory of Kidney and Electrolyte Metabolism, National Heart, Lung, and Blood Institute, National Institute of Health, Bethesda, Maryland 20205*

Oscar A. Candia (223), *Departments of Ophthalmology, and Physiology and Biophysics, Mount Sinai School of Medicine of the City University of New York, New York, New York 10029*

Kevin J. Degnan[1] (295), *Department of Physiology and Biophysics, New York University Medical Center, New York, New York 10016; Mount Desert Island Biological Laboratory, Salsbury Cove, Maine 04672*

John H. Durham (243), *Department of Physiology and Biophysics, Mount Sinai School of Medicine, New York, New York 10029*

Gerhardt Ehrenspeck (243), *Department of Zoology, Ohio University, Athens, Ohio 45701*

Franklin H. Epstein (261, 277), *Department of Medicine and Thorndike Laboratory, Harvard Medical School and Beth Israel Hospital, Boston, Massachusetts 02215; Mount Desert Island Biological Laboratory, Salsbury Cove, Maine 04672*

H. Fasold (1), *Biochemisches Institut der Universität, Frankfurt/Main, West Germany*

---

[1]Present address: *Physiology Section, Division of Biology and Medicine, Brown University, Providence, Rhode Island 02912.*

John G. Forte (151), *Department of Physiology-Anatomy, University of California, Berkeley, California 94720*

O. Fröhlich (33), *Department of Pharmacological and Physiological Sciences, University of Chicago, Chicago, Illinois 60601*

Jared Grantham (135), *Departments of Medicine and Physiology, University of Kansas School of Medicine, Kansas City, Kansas 66103*

Sergio Grinstein (61), *Research Institute, The Hospital for Sick Children, Toronto, Ontario, Canada M5G 1X8*

R. B. Gunn (33), *Department of Pharmacological and Physiological Sciences, University of Chicago, Chicago, Illinois 60601*

John Gutknecht (91), *Duke University, Marine Laboratory, Beaufort, North Carolina 28516*

James Irish III (135), *Departments of Medicine and Physiology, University of Kansas School of Medicine, Kansas City, Kansas 66103*

M. L. Jennings[2] (1), *Max-Planck-Institut für Biophysik, and Biochemisches Institut der Universität, Frankfurt/Main, West Germany*

Rolf Kinne (173), *Department of Physiology, Albert Einstein College of Medicine, Bronx, New York 10461*

Stephen D. Klyce[3] (199), *Department of Surgery, Division of Ophthalmology, Stanford University of Medicine, Stanford, California 94305*

Philip A. Knauf (61), *Department of Radiation Biology and Biophysics, University of Rochester School of Medicine and Dentistry, Rochester, New York 14642*

Juha P. Kokko (123), *Department of Internal Medicine, University of Texas Health Sciences Center, Southwestern Medical School, Dallas, Texas 75235*

Poul Kristensen (319), *Institute of Biological Chemistry A, August Krogh Institute, Universitetsparken 13, DK 2100 Copenhagen Ø, Denmark*

Erik Hviid Larsen (333), *Zoophysiological Laboratory A, August Krogh Institute, University of Copenhagen, DK 2100 Copenhagen Ø, Denmark*

Hon Cheung Lee (151), *Department of Physiology-Anatomy, University of California, Berkeley, California 94720*

S. Lepke (1), *Max-Planck-Institut für Biophysik, and Biochemisches Institut der Universität, Frankfurt/Main, West Germany*

Charles Levinson (383), *Department of Physiology, University of Texas Health Science Center, San Antonio, Texas 78284*

Christina Matons (243), *Departments of Physiology and Biophysics, Mount Sinai School of Medicine, New York, New York 10029*

Heini Murer (173), *Physiologisches Institut der Universität, CH-8028 Zürich, Switzerland*

[2]Present address: *Department of Physiology and Biophysics, The University of Iowa, Iowa City, Iowa 52242.*
[3]Present address: *LSU Eye Center, LSU Medical Center School of Medicine, New Orleans, Louisiana 70112.*

H. Passow (1), *Max-Planck-Institut für Biophysik, and Biochemisches Institut der Universität, Frankfurt/Main, West Germany*

Patricio Silva (261, 277), *Department of Medicine and Thorndike Laboratory, Harvard Medical School and Beth Israel Hospital, Boston, Massachusetts 02215; Mount Desert Island Biological Laboratory, Salsbury Cove, Maine 04672*

Jeffrey S. Stoff (277), *Department of Medicine and Thorndike Laboratory, Harvard Medical School and Beth Israel Hospital, Boston, Massachusetts 02215*

Daniel Terreros (135), *Department of Physiology, School of Medicine, West Virginia University, Morgantown, West Virginia 26505*

Anne Walter (91), *Duke University, Marine Laboratory, Beaufort, North Carolina 28516*

Charles O. Watlington (365), *Department of Medicine, Division of Endocrinology, Medical College of Virginia, Virginia Commonwealth University, Richmond, Virginia 23298*

José A. Zadunaisky (295), *Department of Physiology and Biophysics, and Department of Ophthalmology , New York University Medical Center, New York, New York; Mount Desert Island Biological Laboratory, Salsbury Cove, Maine 04672*

Preface

The purpose of this volume is to present the most recent advances and "state of knowledge" in the field of chloride and other anion transport across biological membranes. The systems examined here by the different authors cover most of the areas where very rapid and active research is being performed. However, it is impossible to cover all those areas where the translocation of chloride by means of a carrier-mediated event in the cell membrane has been found. Therefore the different chapters are representative of the basic phenomena that are found in all biological systems.

The concept of a mediated translocation or transport of chloride across the cell membrane has developed into an established field of research affecting the interpretation of the anion exchanges in the red cell membrane, the mechanisms of reabsorbtion in kidney tubules, the understanding of gastric secretion, the control of fluid movements and transparency of the cornea in the eye, and the mechanisms of osmoregulation in fishes and amphibians. These areas are covered in this volume, but advances in the nervous system, the intestine, muscle, and plant cells could not be included in a volume of this size.

The molecular mechanisms are presented first in studies of red cell membranes where the discovery of the relationship of the most abundant protein of the cell membrane, band 3 protein, to anion movements has permitted an interpretation of the intimate kinetics and chemical events for anion exchange. The section on the kidney has contributions on the renal tubules in general and on the collecting ducts in particular. Chloride or other anion translocation in vesicle preparations of cell membranes from the gastric mucosa, the kidney, and small intestine follows. Chloride secretion in the corneal epithelium is presented in two chapters, emphasizing the importance of these processes in the eye. As examples of chloride transport in tissues of amphibians and fishes, the turtle bladder, the shark rectal gland, the opercular epithelium of teleosts, and the isolated frog skin are discussed as models or because of their specific function in the organ or tissue.

Finally, chloride transport in the mouse ascitis tumor cell is discussed.

Because of the rapid expansion of knowledge in this aspect of cellular and molecular biology, it is not likely that several chloride transporting systems will be encompassed in one volume such as this in the future. On the basis of the information presented here, the reader should, in the future, look for new information in specific reviews or books on red cells, the intestine, or the regulatory mechanisms in aquatic species.

The interest of this editor since 1960 in chloride transport and secretion in the frog skin, the ocular epithelia, the opercular epithelium of fishes, and more recently in its hormonal control has prompted the organization of this volume. The authors have contributed with grace and a high standard of knowledge and have updated their material to the time of publication of the volume. To all of them, to Miss Sally Wrigley of my office, who helped in the preparation of this book, and to the staff of Academic Press I want to extend my appreciation for this effort.

José A. Zadunaisky
New York, February 1982

THE STUDY OF THE ANION TRANSPORT PROTEIN ('BAND 3 PROTEIN')

IN THE RED CELL MEMBRANE BY MEANS OF TRITIATED

4,4'-DIISOTHIOCYANO-DIHYDROSTILBENE-2,2-DISULFONIC ACID (^{3}H$_2$DIDS)

H. Passow

H. Fasold[1]

M. L. Jennings[2]

S. Lepke

Max-Planck-Institut für Biophysik and Biochemisches Institut
der Universität, Frankfurt/Main, W-Germany

I. INTRODUCTION

The identification and characterization of transport-related
membrane constituents are greatly facilitated by specific inhibitors that are capable of covalent binding, and that can be obtained in a radioactively labeled form. Such inhibitors become

[1]Biochemisches Institut der Universität, Frankfurt/Main,
West Germany.
[2]Present address: Dept. of Physiology and Biophysics, The
University of Iowa, Iowa City, Iowa

1

indispensable tools when the transporting unit does not possess
enzymatic activity that, like the ATPase-activity of the transport
proteins for Ca^{2+} or alkali ions, can be measured after dissolution
of the membrane and during the subsequent purification procedure.

In the search for useful inhibitors of anion transport across
the red blood cell membrane, it was observed that irreversible inhibition could be obtained by proteolytic enzymes (l) and a
variety of agents (2) [l-fluoro-2,4-dinitrobenzene (N_2ph-F) (3, 4),
trinitrobenzene sulfonate (5), 5-methoxy nitrotropone (3) and
maleic anhydride (6)] that possess as a common feature the capacity
of combinine covalently with amino groups. This indicated that
anion transport was mediated by a protein and that certain amino
groups in this protein played an important role in the maintenance
of its function. Similar work by Knauf and Rothstein (7) added to
the list of inhibitors a fluorescent isothiocyanate derivative
called SITS (4-acetamido-4' isothiocyano stilbene-2,2'-disulfonate,
Fig. lb) that had first been introduced into membrane research by
Maddy (8). This compound acted at much lower concentrations than
all other inhibitors that had been previously studied, and thus it
appeared to possess an unusually high site-specificity. Unfortunately, the fluorescence intensity of this compound proved to be
too low for easy detection on SDS polyacrylamide gel electropherograms of the dissolved red cell membrane. This suggested the use
of radioactively labeled derivatives of SITS for the identification
of the protein to which SITS binding and inhibition of anion transport could be related.

SITS itself is difficult to synthesize in a radioactively
labeled form. For this reason, in the laboratories of Guidotti
and Rothstein, attempts were made to use isothiocyanate derivatives
that resembled SITS in its chemical structure and inhibitory power,
but that could be more easily obtained in a radioactively labeled
form. In our laboratory, we tried to determine the SITS binding
protein by so-called "protection experiments" that did not require

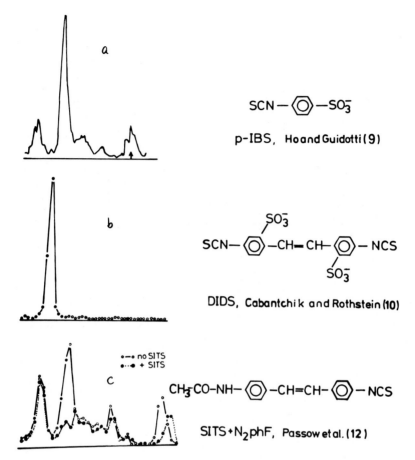

Fig. 1. Labeling patterns of red cell membrane proteins after treatment with the inhibitors indicated in the figure. Band 3 is the major peak in the uppermost tracings and the second big peak in the lower tracing. For description see text. The peak at the front of Fig. 1c represents the lipids.

radioactively labeled SITS but the joint use of unlabeled SITS and commercially available ^{14}C 1-fluoro-2,4-dinitrobenzene (^{14}C-N$_2$ph-F).

Ho and Guidotti (9) applied p-isothiocyano benzene sulfonate (^{35}S-p-IBS), which represents some portion of the SITS molecule. This compound did not fulfill expectations since inhibition of anion transport was accompanied by the labeling of many membrane proteins (Fig. 1a). This was largely due to the fact that, in contrast to

SITS, this compound penetrates and thus reacts with the amino
groups of most of the proteins at the inner membrane surface.
This was not the case with the SITS derivative 4,4-diisothio-
cyano stilbene-2,2'-disulfonate, (DIDS) that was synthesized
by Cabantchik and Rothstein (10). Only when added to permeable
red cell membrane does this compound react like p-IBS with
inward facing membrane proteins. When added to the intact cells,
it labels predominantly [but not exclusively (11)] one membrane protein, the so-called band 3 protein (Fig. 1b). This observation, together with our protection experiments, constituted
the first piece of evidence for an involvement of band 3 protein
in anion transport. In our protection experiments (12, 13), we
showed that the penetrating anion transport inhibitor ^{14}C-N_2ph-
F (1-fluoro-2,4-dinitrobenzene) reacted with any membrane protein
that could be detected on SDS polyacrylamide gel electropherograms
of the red blood cell membrane (Fig. 1c, "no SITS"). If, however,
the dinitrophenylation was carried out after pretreatment of the
red cells with nonradioactive SITS, the labeling pattern remained
unaltered, except that the labeling of the band 3 protein was
drastically reduced, indicating that the common binding sites for
the two inhibitors of anion transport are located on that band
(Fig. 1c, "+ SITS"). Using such protection experiments (12, 13)
or a slightly modified version that involves the use of $^3H_2$DIDS
in addition to N_2ph-F (1-fluoro-2,4-dinitrobenzene) (14, 15) it was
possible to show that all out of a fairly large number of randomly
selected inhibitors of anion transport that do not covalently bind
to the red cell membrane also modify the protein in band 3 (see
ref. 16 and the appendix to this paper). Since the publication of
the described work, SITS and H_2DIDS have found an increasing use
in the study of transport processes and membrane proteins in
erythrocytes and other cells and tissues (e.g., 17 - 23). The
present article is written for readers who wish to acquaint
themselves with the experience that has been derived from the application of radioactively labeled H_2DIDS in the study of anion

transport in red cells. The emphasis will be on practical aspects
of the use of this compound and on the principles of the evaluation
and interpretation of the measurements in the biological system.
Space limitations require that we focus primarily on results from
our laboratory. For detailed information on the work from other
laboratories, notably those in Toronto and Jerusalem, refer to recent reviews of Cabantchik *et al.* (24) and Knauf (25).

II. SYNTHESIS, PROPERTIES, STORAGE AND DETERMINATION OF SPECIFIC
     ACTIVITY OF $^3H_2DIDS$

Figure 2 shows the formulas of DIDS (4,4-diisothiocyano stil-
bene-2,2'-disulfonate, H_2DIDS (dihydro-DIDS), and a few related
compounds that are used for the synthesis of the isothiocyanates.
Depending on the orientation of the isothiocyanate residues with
respect to each other, H_2DIDS may exist in three isomeric forms:
cis, α-trans, and β-trans (11). For the analogous stilbene derivative DIDS there exist in addition cis-trans and rotational isomers

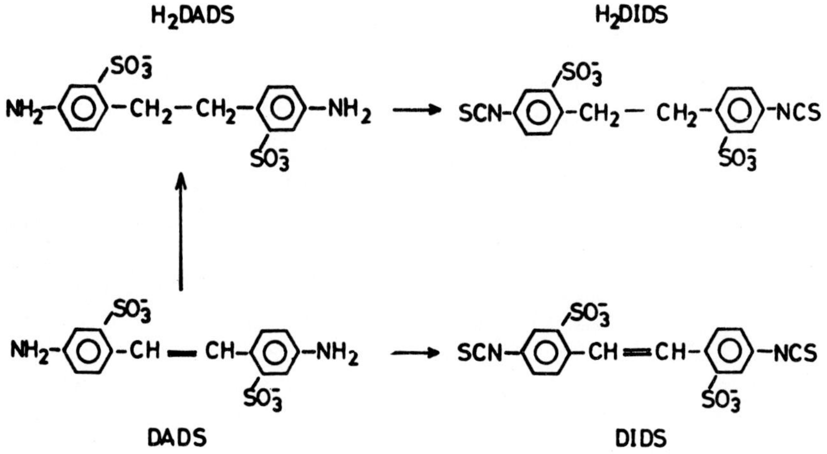

Fig. 2. *DIDS and related compounds.*

at the double bond. The equilibrium between the cis-trans isomers is disturbed by illumination which leads to a change of absorbance and fluorescence.

H_2DIDS is most conveniently synthesized from commercially available 4,4'-diamino stilbene-2,2'-disulfonate (DADS) by hydrogenation of the double bond and subsequent conversion of the amino groups into NCS groups by treatment with thiophosgene. Unlike DIDS, the resulting product is no longer a stilbene but a 1,2-diphenylethane derivative. Since the physiological actions of H_2DIDS are quite similar to those of the originally used DIDS (11, 26), it became customary to use the abbreviation H_2DIDS instead of a chemically more correct one, and thus to emphasize the structural similarities between the two inhibitors rather than the differences.

The commercially available DADS is rather impure and needs purification by silica column chromatography (27). The purified material is tritiated by exposure to tritium gas in the presence of palladium charcoal as a catalyst at pH 6.0. It is essential that tritiation is not only followed by the customary exchange of labile ^{3}H with protons but also by exhaustive hydrogenation until the last traces of any remaining DADS are transformed into dihydro-DADS (H_2DADS). This ensures that the subsequent conversion by thiophosgen of the amino groups into isothiocyanate groups leads to the formation of $^3H_2$DIDS that is free from contaminating DIDS. Since in the red blood cell membrane the reactivity of DIDS exceeds that of $^3H_2$DIDS, such contamination would lead to an underestimation of the binding capacity for $^3H_2$DIDS (for discussion of a pertinent example and details of synthesis, see ref. 11). The synthesis of the unlabeled H_2DIDS follows the same procedure except that the exhaustive hydrogenation is not preceded by tritiation.

$^3H_2$DADS can be stored at -20°C for considerable lengths of time. However, slow radiolysis requires occasional purification by preparative paper electrophoresis. H_2DIDS is susceptible to decomposition or polymer formation. In our laboratory, $^3H_2$DIDS is freshly synthesized from the stored $^3H_2$DADS at least every 4 - 8 weeks. It is dissolved in water, subdivided into small portions, quickly frozen and kept at -16°C until use. There is no detectable change of reactivity for up to about 8 weeks of storage. However, repeated freezings and thawings enhance polymer formation and must be avoided.

The purity of the $^3H_2$DIDS can be checked by measuring IR and UV spectra (Figs. 3 and 4, respectively) and by high voltage paper electrophoresis. The UV spectrum provides a convenient and reasonably sensitive indicator for decomposition of the product. The first signs of decomposition consist of changes of the relative heights of the peaks at 275 and 285 nm. Paper electrophoresis in 1 M pyridine-glacial acetic acid-H_2O, pH 6.5 in combination with autoradiography allow one to detect the formation of polymers, which remain close to the start.

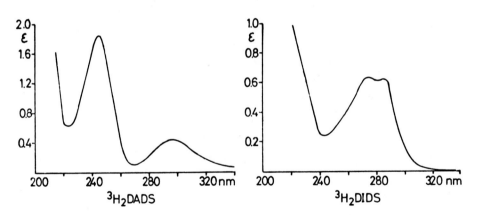

Fig. 3. Absorption spectra of $^3H_2$DADS and $^3H_2$DIDS. The molar extinction coefficient at the maxima for $^3H_2$DIDS are somewhat variable in different preparations. At 275 nm and 285 nm we find 2.64 × 10⁴ (s.d. = ±0.30 × 10⁴) and 2.62 × 10⁴ (s.d. = ±0.28 × 10⁴), respectively. The spectra for $^3H_2$DADS are more reproducible. The molar extinction coefficients at 245 nm and 295 nm are 1.85 × 10⁴ (s.d. = ±0.09) and 0.418 × 10⁴ (s.d. ±0.02 × 10⁴), respectively. The concentrations of $^3H_2$ DADS and $^3H_2$DIDS used for the measurements of the above spectra amounted to 1 × 10⁻⁴M and 2 × 10⁻⁵M, respectively.

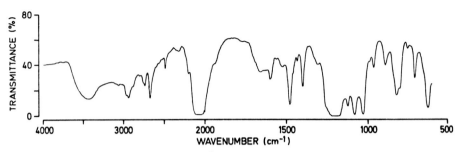

Fig. 4.   IR spectrum of $^{3}H_{2}DIDS$.

For the determination of the specific activity, it is recommended to employ a combination of at least two of the methods described below:

(1) Measure the radioactivity of a $^{3}H_{2}$DIDS sample of known weight.

(2) Calculate the specific activity of $^{3}H_{2}$DADS or the final $^{3}H_{2}$DIDS from measurements of radioactivity and UV absorption at several wavelengths at and near the maxima. The observed agreement between the estimates for $^{3}H_{2}$DIDS and $^{3}H_{2}$DADS indicates that the acidification associated with the thiophosgen treatment does not lead to an exchange of tritium against protons from the medium.

(3) Observe the reaction of the $^{3}H_{2}$DIDS with ^{14}C-glycine of known specific activity or with a fluorescent xanthene derivative (see ref. 11) and determine the ratio between the ^{3}H and either the ^{14}C of the glycine or the extinction of the fluorescent dye. Since ^{14}C-glycine can be obtained at high specific activity, it is easy to prepare, by dilution with nonradioactive glycine, a sample of fairly accurately known specific activity. The thiourea bond formation between $^{3}H_{2}$DIDS and the xanthene derivative takes place with no change of the molar extinction coefficient for the fluorescence emission. On paper electropherograms, the products of the reaction between $^{3}H_{2}$DIDS and the amino acid or the dye yield two bands corresponding to the coupling of one or two of the isothiocyanate groups of H_{2}DIDS with the dye or the amino acid. When re-

peated purifications yield identical ratios for $^3\text{H}/^{14}\text{C}$ or $^3\text{H}/\text{fluor-}$
escence, the product is considered to be pure and the specific ac-
tivity can be calculated from these ratios.

(4) Two other techniques are based on the comparison of the
reaction of the radioactively labeled preparation and a spectro-
scopically pure, unlabeled preparation of $H_2\text{DIDS}$ with the red cells.
Either the effects on anion transport or the capacity for binding
to the red cell membrane are compared. We employ the latter method
as a matter of routine for each newly synthesized batch of $^3H_2\text{DIDS}$.
For this purpose labeled and unlabeled $H_2\text{DIDS}$ are mixed at different molar ratios and the red cells are exposed to these mixtures
for 1 hr at 37°C. After removal of reversibly bound $^3H_2\text{DIDS}$
by washing with albumin (34), hemolysis in a saponin-containing
medium, and after repeated washings, the resulting white ghosts
are dissolved in soluene (Packard) and the radioactivity is
determined. From the specific activity that had been estimated
by one or several of the methods (1) - (3), the number of bound
molecules per cell are calculated. If this number comes out to
be independent of the molar ratio in the reaction mixture, we
feel reassured that the purities of the tritiated and the
independently prepared unlabeled compounds are the same (see
Fig. 5).

III. APPLICATIONS TO THE STUDY OF ANION TRANSPORT ACROSS THE RED
BLOOD CELL MEMBRANE

A. *Relationship between H_2DIDS Binding and Inhibition of Transport*

Mixing of red cells with media that contain an excess of
$H_2\text{DIDS}$ leads, probably in a fraction of a second, to complete inhibition of anion transport (28). This effect is associated with
an equally fast reversible binding that, together with the effect
on transport, slowly becomes irreversible. The high rate at which

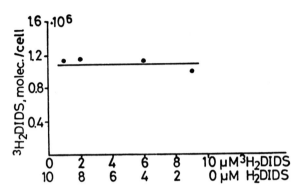

*Fig. 5. H₂DIDS binding per cell, as calculated after expo-
sure of the red cells to mixtures of ³H₂DIDS and H₂DIDS at constant
sum ³H₂DIDS plus H₂DIDS for 1 hr at 37°C, hematocrit (hct) 10%.
When there is no discrimination by the cell surface between the two
preparations, and if the specific activity of ³H₂DIDS is correctly
estimated, a straight line parallel to the abcissa should be obtained.*

reversible binding goes to completion makes H_2DIDS or DIDS highly
suited for an inhibitor stop technique for the measurement of rapid
anion exchange (28). If the red cells are exposed to H_2DIDS at 0°C
and subjected to washes with media containing 0.5% serum albumin at
that temperature, virtually all of the bound H_2DIDS is removed and
anion transport is restored. At more elevated temperatures, H_2DIDS
binding and inhibition of transport become gradually irreversible,
following an approximately exponential time course with an apparent
activation enthalpy of about 23 kcal/mol (11). After long periods
of exposure to a saturating concentration of H_2DIDS, when anion
transport is effectively blocked and irreversible H_2DIDS binding is
virtually complete, there is some bound H_2DIDS (about 15 - 20%)
that can no longer be removed from the membrane by albumin washes
but that comes off when the cells are hemolyzed. Polyacrylamide
gel electropherograms of the supernatant of the hemolysate revealed
no binding to specific cell constituents. Apparently, this fraction of H_2DIDS moves into niches near the outer membrane surface

without undergoing covalent attachment.

      Covalent binding takes place predominantly but not exclusively
at the protein in band 3. With increasing H_2DIDS concentration in
the medium, binding to band 3 reaches a clearly defined plateau
while the binding to other sites continues to increase (Fig. 6).
Among the latter sites only two are distinct enough for easy detection on polyacrylamide gels. One site is located in the region of
bands 4.1 - 4.5, at a molecular weight range of about 60 - 65,000
daltons and represents, perhaps, a degradation product of band 3
that is formed by membrane associated proteinases or by proteinases
released from leucocytes (see ref. 29). Another site is occasionally observed near the front and is associated with the lipids.
Binding to the lipids may be quite pronounced after trypsination
of the surface of intact cells (30) and in untreated, resealed
ghosts (14), suggesting that there is a tendency for reorientation
of the predominantly inward-facing amino lipids phosphatidyl-

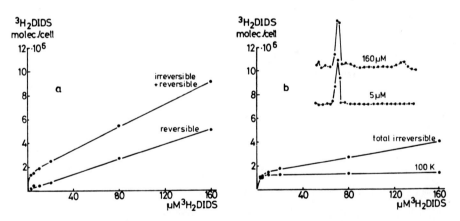

Fig. 6. H_2DIDS binding as a function of H_2DIDS concentration in the medium. 10% hct. 37°C, pH 7.4, incubation time 90
min. "Irreversible + reversible" = binding deduced from decrease
of $^3H_2DIDS$ in the supernatant. "Reversible" = after deduction
of "total irreversible" binding. The latter refers to covalently
bound $^3H_2$ DIDS on the isolated membrane. "100k" = binding to
band 3. Inset in (b) $^3H_2DIDS$ distribution on SDS polyacrylamide
gels after exposure to 5 μM and 160 μM $^3H_2DIDS$. The major peaks
refer to band 3.

TABLE I. Some estimates of the numbers of H_2DIDS binding sites, per cell as determined by means of $^3H_2DIDS$ or related inhibitors

Inhibitors	Molecules/cell Bound at Complete Inhibition	Authors
$[^3H]$-DIDS	$3.0 \cdot 10^5$	Cabantchik and Rothstein (30)
$[^{14}C]$-N_2ph-F/SITS	0.8-$1.2 \cdot 10^6$	Passow et al. (12) and
" /I_2DIDS	1.2-$1.3 \cdot 10^6$	Zaki et al. (13)
" /DAS	$1.0 \cdot 10^6$	
$[^{35}S]$-p-IBS	$3.0 \cdot 10^5$	Ho and Guidotti (9)
DIDS	$1.2 \cdot 10^6$	Halestrap (32)
$^3H_2DIDS$	$1.2 \cdot 10^6$	Lepke et al. (11)
DIDS	$1.0 \cdot 10^6$	
$^3H_2DIDS$	$1.1 \cdot 10^6$	Ship et al. (26)
^{14}C-DIDS	$1.1 \cdot 10^6$	

ethanolamine and phosphatidylserine.

The covalent H_2DIDS binding to band 3 is linearly related to the inhibition of anion transport (11, 26). Inhibition is complete when band 3 is saturated with the agent. The number of sites occupied at complete inhibition is about $1 - 1.2 \times 10^6$/cell (see Table I) and close to the only existing, rather crude estimate of the band 3 molecules per cell (about 1.0×10^6/cell, ref. 31). A linear relationship is also obtained when the binding of DIDS is measured (11, 26). In our hands, DIDS binding is somewhat more specific for band 3 than the binding of H_2DIDS. The numbers of molecules bound to band 3 at complete inhibition are, however, not significantly different for DIDS and H_2DIDS.

The determinations of the numbers of H_2DIDS binding sites on the band 3 protein allow one to calculate turnover numbers. For Cl transport at 25°C, one obtains 2.75×10^4 sec^{-1}. This value is much smaller than that for ion transport through gramicidin pores, but close to the estimate for valinomycin-mediated K^+-transport across lipid bilayers (11).

The linear relationship between transport and binding, and the similarity of the number of H₂DIDS binding sites with the number of band 3 molecules indicate that the anion transport protein migrates in band 3. Further work has ruled out the possibility that inhibition is related to the modification of a small number of transporting units that are hidden among the bulk of the band 3 molecules, and there is good evidence for the assumption that all H_2DIDS molecules that are covalently bound to band 3 are equally involved in the inhibition of anion transport (14).

B. H_2DIDS Binding to Band 3

The findings presented above recall briefly why H_2DIDS could play such a prominent role in the identification of the anion transport protein. They also suggest that a more detailed study of H_2DIDS binding to the band 3 protein should provide information about the molecular mechanism of the transport process. The remaining part of the paper is devoted to a brief review of the recent state of knowledge about H_2DIDS binding to the band 3 protein.

So far we have discriminated between rapid noncovalent binding of H_2DIDS and subsequent slow transition to covalent binding. This represents an oversimplification since the H_2DIDS molecule carries two isothiocyanate groups, each of which is capable of covalent bond formation. Recent work has shown that H_2DIDS is, in fact, capable of crosslinking two different segments of the peptide chain of the band 3 protein (33). Thus, a more complete schema of the reaction of H_2DIDS with the protein involves at least three steps: noncovalent binding, covalent bond formation of one isothiocyanate group with one of the two different sites on the peptide chain, and finally, covalent bond formation of the other isothiocyanate group with the remaining site on the band 3 molecule. Below we shall consider these steps separately.

Noncovalent binding. H_2DIDS can, and actually does, covalently react with any available amino group (e.g., in leaky red cell ghosts, it reacts with all accessible membrane proteins, see

ref. 10). However, the reactivity, at least with respect to band
3, is modulated by the affinity for noncovalent binding of the
H_2DIDS to the region of the protein molecule in which the covalent
bond formation will take place. The structural parameters on
which the affinity depends have been inferred from studies with a
large variety of reversibly-acting inhibitors with different charge,
hydrophobicity, and electron configuration (23a). Such inhibitors
are capable of interacting with the H_2DIDS binding site on band 3
as has been shown by "protection experiments" of the type illustrated in Fig. 7 and described in the appendix (16). It was found that
the negative charge, the capacity to accept electrons, and the hydrophobicity are the essential features that convey to the H_2DIDS
molecule its rather unique capacity for reversible binding to the
inhibitory site on band 3 (34).

Covalent binding. The reactions of the two isothiocyanate
groups can be studied in red cells in which the band 3 protein is
split by external chymotrypsin into two fragments of 60,000 and
35,000 daltons. This splitting produces no change of anion transport (35). If chymotrypsinized red cells are exposed to $^3H_2DIDS$,
most of the bound $^3H_2DIDS$ first reacts with the 60K fragment and
then crosslinks with the 35K fragment to form a product that migrates on SDS polyacrylamide gels in the same location as the
original band 3 protein from which the chymotryptic fragments had
been derived (33). The remainder of the bound $^3H_2DIDS$ reacts
first with the 35K fragment and then crosslinks with the 60K fragment (see Fig. 8). The time course of covalent bond formation is
depicted in Fig. 9. The drawn lines have been calculated by
means of equations that have been derived from the reaction schema
in Fig. 8. They pass reasonably well through the data points, indicating that this schema describes the essence of the behavior
of the system quite adequately. At pH 7.9, the rate at which the
H_2DIDS molecule reacts with the binding site on the 60K fragment
is about 3 - 4 times higher than the rate at which it reacts with
the site on the 35K fragment. After covalent bond formation of

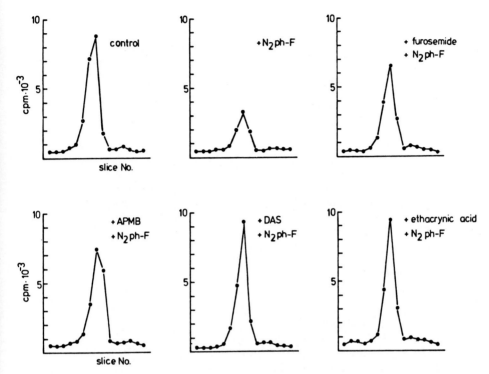

Fig. 7. H_2DIDS distribution profiles on SDS polyacrylamide gel electropherograms of membranes from untreated ("control") and dinitrophenylated red cells that had been exposed to a saturating concentration of $^3H_2DIDS$. Dinitrophenylation (30 min, 37°C, pH 7.4, 0.15 mM N_2pH-F) was carried out prior to the $^3H_2DIDS$ treatment (60 min, 37°C, pH 7.4, 10 µM $^3H_2DIDS$) in the absence ("N_2pH-F") or presence (e.g. "+ furosemide + N_2pH-F") of a reversibly acting inhibitor. The concentration of the latter were 2.0 mM (furosemide, APMB, DAS) or 3.0 mM (ethacrynic acid). The major peak represents $^3H_2DIDS$ on band 3. For explanation see appendix. According to ref. (16).

one of the isothiocyanate groups, the rate of reaction of the other is drastically decreased and at that particular pH, the reaction rate with the binding site on the 60K fragment is close to the reaction rate with the site on the 35K fragment.

Closer inspection of Fig. 9 reveals deviations between the calculated curves and the data points. They are particularly conspicuous for the curve that represents the restoration of band 3

H. Passow *et al.*

Kinetics of crosslinking

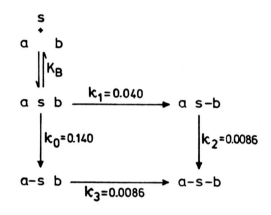

for $s \gg K_B$

$$a\text{-}s\ b = \frac{k_0\ \overline{ASB}}{k_0 + k_1 - k_3}\ (e^{-k_3 t} - e^{-(k_0 + k_1)t})$$

$$a\ s\text{-}b = \frac{k_1\ \overline{ASB}}{k_0 + k_1 - k_2}\ (e^{-k_2 t} - e^{-(k_0 + k_1)t})$$

$$a\text{-}s\text{-}b = \text{const.} + \text{sum of 3 exponentials}.$$

Fig. 8. Reaction schema for crosslinking of sites a and b by s. The numerical values of the rates constants K_0, K_1, K_2, K_3 were derived from the experiment in Fig. 9. $\overline{ASB}$ indicates the sum of the forms a-sb, as-b, a-s-b. The dashes (-) indicate covalent bond formation with a, b, and a as well as b, respectively.

by the establishment of the crosslinks. This observation is interesting since it indicates that the reaction schema in Fig. 8
neglects significant details of the crosslinking reaction. We assume that the H_2DIDS binding site may exist in different conformations in which, perhaps, the distances between sites a and b are
different. The rates of covalent bond formation with the isothiocyanate groups of the reversibly bound H_2DIDS molecules would then

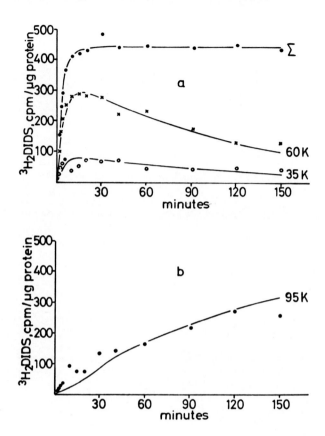

Fig. 9a,b.   Time course of crosslinking by $^3H_2DIDS$ of chymo-tryptic 60K and 35K fragments of band 3.   The curves marked 60K and 35K show disappearance of the corresponding fragments, the curve marked 95K the appearance of the "restored" 95K molecule. The drawn lines were calculated by means of the equations in Fig. 8 using the constants indicated there. Σ represents the sum of co-valently bound $^3H_2DIDS$ on the three fragments. $^3H_2DIDS$ concentra-tion in the medium 10 μM, hct. 10%, 37°C, pH 7.9.   Cells pre-treated with 1 mg/ml chymotrypsin, 1 hr, 37°C.

be different for the different conformers. Space does not permit the discussion of the implications of this assumption in detail, and thus it must suffice to state that in terms of this model most of the observed discrepancies seem to be explainable (36).

Isothiocyanate groups react quite specifically with amino

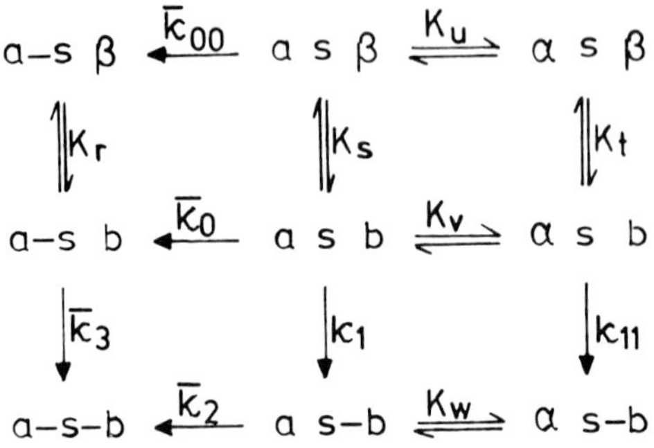

$$a{-}s\ \beta \xleftarrow{\ \overline{k}_{00}\ } a\ s\ \beta \overset{K_u}{\rightleftharpoons} \alpha\ s\ \beta$$

$$\Big\Updownarrow K_r \qquad\qquad \Big\Updownarrow K_s \qquad\qquad \Big\Updownarrow K_t$$

$$a{-}s\ b \xleftarrow{\ \overline{k}_{0}\ } a\ s\ b \overset{K_v}{\rightleftharpoons} \alpha\ s\ b$$

$$\Big\downarrow \overline{k}_3 \qquad\qquad \Big\downarrow k_1 \qquad\qquad \Big\downarrow k_{11}$$

$$a{-}s{-}b \xleftarrow{\ \overline{k}_{2}\ } a\ s{-}b \overset{K_w}{\rightleftharpoons} \alpha\ s{-}b$$

Hydrogen ion equilibria and crosslinking reactions at site a b

$$a + H^+ = \alpha \qquad\qquad a\ b + s \rightleftharpoons a\ s\ b$$

$$b + H^+ = \beta \qquad\qquad \alpha\ b + s \rightleftharpoons \alpha\ s\ b$$

$$a\ \beta + s \rightleftharpoons a\ s\ \beta$$

$$\alpha\ \beta + s \rightleftharpoons \alpha\ s\ \beta$$

Fig. 10. Hydrogen ion equilibria at H_2DIDS sites a and b. The $\overline{k}$'s represent pH independent rate constants for the irreversible reactions of s with a and b, the K's indicate the mass law constants for the hydrogen ion equilibria. The pH dependent constant κ_3 in Fig. 8 is a function of the pH independent rate constant $\overline{k}_3$ defined by the above reaction scheme. $\kappa_3 = \overline{k}_3(1+H^+/K_r)$.

groups. (For an exception: see ref. 37.) The reaction can take place, however, only with the deprotonated forms of these groups. This implies that the rate of reaction should be pH-dependent. In principle, at least, it should be possible to calculate apparent pK values from the variations of the reaction rate with pH. Unfortunately, the reaction rates for the covalent bond formation of the two NCS groups of H_2DIDS depend on the pK values for several different hydrogen ion equilibria that are difficult to separate (Fig. 10). However, it is possible to simplify the situation by confining the measurements to the reaction of such H_2DIDS molecules that are covalently attached through one NCS group to the 60K fragment with the binding site on the 35K fragment, corresponding to the transition a-s b to a-s-b (see Fig. 8). Under these

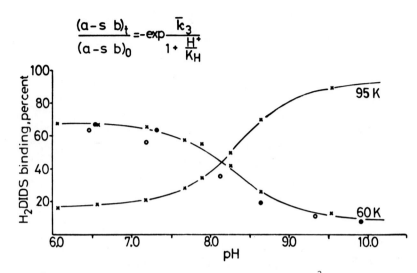

$$\frac{(a-s\ b)_t}{(a-s\ b)_0} = -\exp\frac{\overline{k}_3}{1+\frac{H^+}{K_H}}$$

Fig. 11.   pH dependence of crosslinking by $^3H_2DIDS$ of the chymotryptic 60K fragment with the 35k fragment. 60K: $^3H_2DIDS$ on 60K fragment.   95K: $^3H_2DIDS$ on "reconstituted" band 3. $^3H_2DIDS$ had been covalently attached to chymotrypsin-pretreated cells for 10 min at pH 7.2. This leads predominantly, to a labeling of the 60K fragment of band 3. After removal of excess $^3H_2DIDS$ by albumine washes, the pH's were adjusted to the values indicated on the abscissa and the crosslinking region was allowed to proceed at these pH's for 60 min at 37°C. Under these experimental conditions only the transition a s-b → a-s-b is observed.

conditions, the reaction rate only depends on the pK value of the binding site on the 35K fragment. Figure 11 shows the result of such an experiment. Using the equation indicated in that figure, we arrive at an apparent pK of about 9.9. This value is, of course, not necessarily identical with the apparent pK of the unperturbed system since the thiourea bond formation at the adjacent 60K fragment and the introduction of the two sulfonyl groups of H_2DIDS into the immediate vicinity of the NH_2 group on the 35K fragment reduce the positive charge of that vicinity by three units. This is likely to affect the pK of that latter group. Nevertheless, it is interesting to note that the calculated pK

value is higher than the pK of the amino group on the 60K segment
of 8.5 - 8.6 that had been determined from measurements of the pH
dependence of dinitrophenylation of that group. The numerical
values of the two pK values are sufficiently different from
the value for normally dissociating amino groups (pK = 10.5,
see ref. 38) to suggest that the amino groups that are involved in thiourea bond formation with H_2DIDS are located in the
close proximity of additional positively-charged groups. This
conclusion is supported by the observation that obliteration by
dinitrophenylation of the thiourea bond-forming amino group on
the 60K fragment has little effect on the strong noncovalent H_2DIDS
binding to band 3 (14) which, as was mentioned above, seems to require positive charges at the binding site.

In conclusion, it should be pointed out that the interpretation of our data suggests that, within a diameter of about 20 Å
(the length of one H_2DIDS molecule), there are located at least
five basic amino acid residues. Two of them are the lysine residues involved in covalent bond formation with the isothiocyanate
groups of H_2DIDS, two more are engaged in the noncovalent binding
of H_2DIDS and at least one additional one is located near lys.
b and accounts for the low pK value mentioned above. This cluster
of positively-charged groups should be able to accumulate the substrate anions by electrostatic forces in front of the rate-limiting
barrier as has been discussed previously within the context of the
fixed charge hypothesis (39).

C.   *The Use of $^3H_2DIDS$ in Sidedness Studies*

For an understanding of the question how a transport protein
translocates a substrate across the membrane, it is essential to
find out whether or not certain groups of the protein molecule are
capable of alternating between the outer and inner membrane surface. The nonpenetrating $^3H_2DIDS$ has been used in the search for
an answer to this question.

In our experiments, we first made use of the H_2DIDS analogue

4,4'-diacetamido stilbene-2,2'-disulfonate (DAS) which inhibits
anion transport by combination with the H_2DIDS binding site on band
3 (Fig. 7, ref. 14). This compound was found to act only if applied to the outer membrane surface (13, 40). This indicates that
it is incapable of recruiting H_2DIDS binding sites to the inner
membrane surface and suggests that, unlike a mobile transfer site,
the H_2DIDS binding site cannot cross the membrane. On the other
hand, the reversibly-acting inhibitor 2-(4'-aminophenyl)-6-methyl-
benzenethiazol-3'-7-disulfonate (APMB), which also acts by combining with the H_2DIDS binding site (Fig. 7, ref. 15) inhibits
anion transport at either surface (13). It was found that intracellular APMB is capable of rendering the exofacial H_2DIDS binding
site inaccessible for H_2DIDS and DAS (41). Since it is unlikely
that APMB could recruit DAS binding sites to the inner membrane
surface while DAS itself cannot, it also seems unlikely that the
APMB binding site at the inner membrane surface is identical with
the common binding site for H_2DIDS, DAS, and APMB in the outer
surface. We suspect, therefore, that the demonstrated movements
of the H_2DIDS binding site do not represent a translocation all
the way across the permeability barrier but the transition from
an exposed into a buried state (42).

Recent observations indicate that exofacial H_2DIDS is capable
of rendering inaccessible the endofacial binding site for the inhibitor NAP taurine (N-(4-azido-2-nitrophenyl)-2-aminoethyl sulfonate) (43). The results obtained with APMB and NAP taurine
would suggest that the allosteric interactions between the H_2DIDS
binding sites at the outer membrane surface with the binding sites
for the other inhibitors at the inner membrane surface can be
transmitted across the membrane in either direction.

D. The H_2DIDS Binding Region of the Band 3 Molecule (Fig. 12)

The crosslinking between the chymotryptic 60K and the 35K
fragment demonstrates that two different segments of the band 3
protein participate in the formation of the region in which H_2DIDS

H. Passow *et al.*

The H₂DIDS binding site

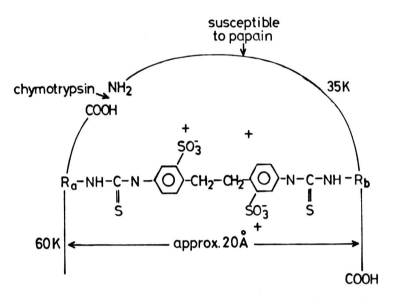

Fig. 12. Schematic representation of the H₂DIDS binding region of the band 3 molecule.

binding takes place (33). After establishment of the crosslink the distance between the binding sites on the two segments is equal to the length of a H_2DIDS molecule, i.e., about 20 Å. The crosslink is established between two amino groups in an environment that is exposed to the outer membrane surface, carries a strong positive charge, and yet has partially hydrophobic qualities. The region may exist in an exposed and a buried state whereby the latter can be induced by a reversible modification of the protein molecule at the inner membrane surface. Conversely, a modification of the H_2DIDS binding region at the outer membrane surface induces an allosteric effect on a transport-related site at the inner membrane surface. An alternative hypothesis would consist of the assumption that the H_2DIDS binding "site" is capable of crossing the permeability barrier and, after removal from the outer membrane surface, becomes exposed to the inner surface (44). The evidence from our laboratory tends to favor the first hypothesis.

Dinitrophenylation of the amino group on the 60K fragment leads to an inhibition of anion transport and prevents the covalent bond formation between H_2DIDS and lys a. Nevertheless, noncovalent H_2DIDS binding to the adjacent positively charged area is still possible (14). This indicates that the amino group a on the 60K fragment is an inhibitory site and that therefore, the 60K fragment is in some way related to anion transport. However, our observation that papain treatment of the chymotrypsinized red cells leads to inhibition of transport without major modification of the 60K fragment but with a partial digestion of the 35K fragment would suggest that the 60K fragment alone (or a subfragment of 17K daltons) does not suffice to perform anion transport, but that the 35K fragment is also involved (33).

The study of the kinetics of anion transport has shown that there exist at least two functionally different sites: a transfer site and one or several modifier sites. The transfer site is supposed to bind the substrate anion and to translocate the bound anion across a rate-limiting barrier. Modifier sites influence binding to the transfer site and/or the translocation of that site across the rate-limiting barrier (2, 45). On the basis of competition studies with reversibly-binding H_2DIDS and the substrate anion Cl^-, it has been suggested that the H_2DIDS binding site is a transfer site (44). Such a view certainly constitutes an oversimplification of a more complex situation. Competition is confined to noncovalently binding H_2DIDS. Thus, competition takes place for those charged groups in the H_2DIDS binding region on the band 3 protein that are not involved in covalent bond formation. Moreover, we found (42) that the rate of dinitrophenylation of the amino group a on the 60K fragment and the consequent inhibition of transport, can be accelerated by suitable substrate anions, whereby the half saturation concentration for Cl^- is about equal to the K_m value for chloride transport. This suggests that amino group a is not directly involved in the binding of the anion to be transported although apparently allosterically linked to a site that participates in the transport of the substrate anion. Further evidence that the H_2DIDS site and the substrate anion binding site are not identical comes from papain proteolyses experiments (Jennings, unpublished). Papain does not change substrate (SO_4) affinity significantly, but strongly (tenfold) reduces the affinity of 4,4'-dinitro stilbene-2,2'-disulfonate (DNDS), a reversibly acting stilbene disulfonate (34). This would suggest that even the sites for noncovalent H_2DIDS binding are not involved in complex formation with the substrate anions and the subsequent anion translocation. The role of the amino group b on the 35K fragment is still unclear, but it could possibly be part of the anion binding region that is responsible for substrate inhibition at high substrate concentration. Regardless of their specific functions in the transport process, all positively-charged groups of the H_2DIDS binding region would contribute to an accumulation of the substrate anions at the rate-limiting barrier.

SUMMARY

This paper is intended to provide information about tritiated H_2DIDS that may be of practical usefulness for the study of anion transport in cells and tissues. It reviewed work primarily from this laboratory with the emphasis on synthesis, storage and determination of specific activity and contained examples of applications to red cells. The examples included the quantitative evaluation of crosslinking experiments, the determination of the binding to the H_2DIDS binding site of reversibly and irreversibly binding inhibitors other than H2DIDS, and the application to the study of the sidedness of inhibitor actions. The paper was concluded by a brief summary of the current ideas about the nature of the H_2DIDS binding region on the protein in band 3 of the red cell membrane.

APPENDIX

The Use of H$_2$DIDS for the Determination of the Binding of Other Inhibitors to Band 3

The modification of H$_2$DIDS binding sites on the protein in band 3 by other covalently binding inhibitors can be easily measured by "back-titration" with H$_2$DIDS of the unmodified sites. For this purpose, the red cells are first exposed for a defined length of time to the chosen inhibitor, e.g., N$_2$ph-F. The excess of the inhibitor is then removed by washing and the cells are exposed to an excess of ^{3}H$_2$DIDS for a sufficient time period to allow the labeling of all unmodified ("protected") band 3 molecules. The difference between ^{3}H$_2$DIDS binding to untreated and to pretreated red cells indicates the number of band 3 molecules that have been modified at the H$_2$DIDS binding site by the chosen inhibitor (11, 15). With some modification, the method can also be applied to the determination of binding of reversibly-acting inhibitors to band 3 (15, 42). Since H$_2$DIDS has an exceptionally high affinity for the binding site on band 3, these inhibitors will usually not be successful in competing with H$_2$DIDS for the site at which reversible binding takes place. Hence they will exert little effect on the rate of covalent ^{3}H$_2$DIDS binding which depends on the number of binding sites on band 3 that are occupied in the time average by reversibly-bound ^{3}H$_2$DIDS. However, the reaction between N$_2$ph-F and the H$_2$DIDS binding site on band 3 is not preceded by measurable noncovalent binding. Hence it is possible to expose the red cells to N$_2$ph-F in the presence of a reversibly-acting inhibitor of one's choice for a certain length of time. Subsequently, the inhibitor and the unreacted N$_2$ph-F are removed. The cells are then exposed to an excess of ^{3}H$_2$DIDS and the H$_2$DIDS binding capacity of band 3 is determined (Fig. 7). This method was successfully employed to demonstrate the modification of band 3 by a large number of reversibly-binding inhibitors

of anion transport, among others APMB, DAS, (4,4'-diacetamido
stilbene-2,2'-disulfonate) persantin, dinitrophenol, and the important diuretics furosemide, piretanide, and ethacrynic acid
(Fig. 7) (14 - 16, 42).

REFERENCES

1. H. Passow. Effects of pronase on passive ion permeability of
    the human red blood cell. *J. Memb. Biol. 6*: 233-258, 1971.
2. S. Gerhardt, A. Schöppe-Fredenburg, and K. F. Schnell. In-
    hibition of sulfate self-exchange in human red blood cells.
    In "Erythrocytes, Thrombocytes, Leucocytes" (E. Gerlach, K.
    Moser, E. Deutsch, and W. Wilmans, eds.), pp. 87-95. Thieme,
    Stuttgart, 1973.
3. H. Passow and K. F. Schnell. Chemical modifiers of passive
    ion permeability of the erythrocyte membrane. *Experientia
    25*: 460-468, 1969.
4. J. Poensgen and H. Passow. Action of 1-fluoro-2,4-dinitro-
    benzene on passive ion permeability of the human red blood
    cell. *J. Memb. Biol. 6*: 210-232, 1971.
5. L. Zaki, C. Gitler, and H. Passow. The effect of trinitro-
    benzene sulfonate on passive ion permeability of the human
    red blood cell. *Abstr. XXV Intern. Congr. Physiol. Sci. 86*:
    471, 1975.
6. A. L. Obaid, A. F. Rega, and P. Garrahan. The effects of
    maleic anhydride on the ionic permeability of red cells.
    J. Memb. Biol. 9: 385-401, 1972.
7. P. A. Knauf and A. Rothstein. A chemical modification of
    membranes. I. Effects of sulfhydryl and amino reactive
    reagents on anion and cation permeability of the human red
    blood cell. *J. Gen. Physiol. 58*: 190-210, 1971.
8. H. Maddy. A fluorescent label for the outer components of

the plasma membrane. *Biochem. Biophys. Acta 88*: 390-399, 1964.

9. M. K. Ho and G. A. Guidotti. A membrane protein from human erythrocytes involved in anion exchange. *J. Biol. Chem. 250*: 675-685, 1975.

10. Z. I. Cabantchik and A. Rothstein. Membrane proteins related to anion permeability of human red blood cells. I. Localization of disulfonic stilbene binding sites in proteins involved in permeation. *J. Memb. Biol. 15*: 207-226, 1974.

11. S. Lepke, H. Fasold, M. Pring, and H. Passow. A study of the relationship between inhibition of anion exchange and binding to the red blood cell membrane of 4,4'-diisothiocyano-stilbene-2,2'-disulfonic acid (DIDS) and its dihydro derivative (H_2DIDS). *J. Memb. Biol. 29*: 147-177, 1976.

12. H. Passow, H. Fasold, L. Zaki, B. Schuhmann, and S. Lepke. Membrane proteins and anion exchange in human erythrocytes. *In* "Biomembranes: Structure and Function"; Proceedings of the 9th FEBS Meeting, Budapest 1974, Vol. 35 (G. Gardos and I. Szasz, eds.), 1975: 197-214.

13. L. Zaki, H. Fasold, B. Schuhmann, and H. Passow. Chemical modification of membrane proteins in relation to anion exchange in human red blood cells. *J. Cell. Physiol. 86*: 471-494, 1975.

14. H. Passow. The binding of 1-fluoro-2,4 dinitrobenzene and of certain stilbene-2,2'-disulfonic acids to anion permeability-controlling sites on the protein in band 3 of the red blood cell membrane. *In* "Cell Membrane Receptors for Drugs and Hormones: A Multidisciplinary Approach, (R. W. Straub and L. Bolis, eds.). New York, Raven Press, 1978: 203-218.

15. H. Passow, M. Pring, B. Legrum-Schuhmann, and L. Zaki. The action of 2-(4'-amino-phenyl)-6-methylbenzene-thiazol-3,7'-disulfonic acid (APMB) on anion transport and the protein in band 3 of the red blood cell membrane. *In* "Biochemistry of

Membrane Transport" Proceedings in Life Sciences (G. Semenza and E. Carafoli, eds.), FEBS-Symposium. Berlin: Springer, 1977: 306-315.

16.   L. Zaki, W. Ruffing, E. M. Gärtner, H. Fasold, R. Motais, and H. Passow. Band 3 as site of action of reversibly binding inhibitors of anion transport across the red cell membrane. *Abstr. 11th Meeting of the Federation of European Biochemical Societies,* Nr. A4-17-671, 1977.

17.   K. Heintze, P. Olles, K. U. Petersen, and J. R. Wood. Effects of a disulphonic stilbene on fluid and elektrolyte transport in guinea-pig isolated gall-bladder. *J. Physiol. 384*: 152P-153P.

18.   Petzinger, E., Grundmann, E., Veil, L. B., Frimmer, M., and Fasold, H. Inhibitory effects of DIDS in response of isolated hepatocytes to phalloidin. *Nauyn-Schmiedeberg's Arch. Pharmacol. 304:* 303-307, 1978.

19.   B. Hellman, A. Lermark, J. Sehli, M. Söderberg, and I. B. Talgedal. The pancreatic β-cell recognition of insulin secretagogues. VII. Binding and permeation of chloromercuribenzene-p-sulphonic acid in the plasma membrane of pancreatic β-cells. *Arch. Biochem. Biophys. 158*: 435-441, 1973.

20.   K. J. Ullrich, G. Capasso, G. Rumrich, F. Papavassiliou, and S. Klöss. Specificity and sodium dependence of the active sugar transport in the proximal convolution of the rat kidney. *Pflügers Archiv 368*: 245-252, 1977.

21.   G. Ehrenspeck and W. A. Brodsky. Effects of 4-acetamido-4'-isothiocyano-2,2'-disulfonic stilbene on ion transport in turtle bladders. *Biochem. Biophys. Acta 419*: 555-558, 1976.

22.   F. J. Koschier, W. B. Kinter, H. H. Church, M. F. Stokols, J. M. Goldinger, and S. K. Hong. Autoradiographic analysis of the renal slice accumulation of H_2DIDS (4,4'-diisothiocyano-2,2'-dihydrostilbene disulfonate). *Fed. Proc. 38*: 1046, Abstract No. 4330, 1979.

23.   C. Levinson, R. J. Corocoran, and E. H. Edwards. Interaction of tritium-labeled H_2DIDS (4,4'-diisothiocyano-1,2-diphenyl-

ethane-2,2'-disulfonic acid) with the Ehrlich mouse ascites tumor cell. *J. Memb. Biol. 45*: 61-81, 1979.

23a. R. Motais and J. L. Cousin. A structure activity study of some drugs acting as reversible inhibitors of chloride permeability in red cell membranes: influence of ring substituents. *In* "Cell Membrane Receptor for Drugs and Hormones: A Multidisciplinary Approach," (R. W. Straub and L. Bolis, eds.). New York, Raven, 1978: 219-225.

24. Z. I. Cabantchik, P. A. Knauf, and A. Rothstein. The anion transport system of the red blood cell. The role of membrane protein evaluated by the use of "probes." *Biochem. Biophys. Acta 515*: 239-302, 1978.

25. P. A. Knauf. Erythrocyte anion exchange and the band 3 protein: Transport kinetics and molecular structure. *In* "Current Topics in Membranes and Transport," (F. Bronner and A. Kleinzeller, eds.), *12*, 249-363, 1979.

26. S. Ship, Y. Shami, W. Breuer, and A. Rothstein. Synthesis of tritiated 4,4'-diisocyano-2,2'-stilbene-disulfonic acid (^{3}H DIDS) and its covalent reaction with sites related to anion transport in human red blood cells. *J. Memb. Biol. 33*: 311-323, 1977.

27. A. Kotaki, M. Naoi, and K. A. Yagi. A diaminostilbene dye as a hydrophobic probe for proteins. *Biochem. Biophys. Acta 229*: 547-556, 1971.

28. Ch. Ku, M. L. Jennings, and H. Passow. Comparison of the inhibitory potency of reversibly acting inhibitors of anion transport on chloride and sulfate movements across the human red cell membrane. *Biochem. Biophys. Acta 553*: 132-141, 1979.

29. G. Tarone, N. Hamasaki, M. Fukuda, and V. T. Marchesi. Proteolytic digestion of human erythrocyte band 3 by membrane associated protease activity. *J. Memb. Biol. 48*: 1-12, 1979.

30. Z. I. Cabantchik and A. Rothstein. Membrane proteins related

to anion permeability of human red blood cells. II. Effects
of proteolytic enzymes on disulfonic stilbene sites of surface proteins. *J. Memb. Biol. 15*: 227-248, 1974.

31. Th. L. Steck. Organization of proteins in the human red blood
cell membrane. *J. Cell. Biol. 62*: 1-19, 1974.

32. A. P. Halestrap. Transport of pyruvate and lactate in human
erythrocytes. Evidence for the involvement of the chloride
carrier and a chloride-independent carrier. *Biochem. J. 156*:
193-207, 1976.

33. M. L. Jennings and H. Passow. Anion transport across the
erythrocytes membrane, *in situ* proteolysis of band 3 protein,
and cross-linking of proteolytic fragments by 4,4'-diisothio-
cyano-dihydrostilbene-2,2'-disulfonate. *Biochem. Biophys.
Acta 554*: 498-519, 1979.

34. M. Barzilay, S. Ship, and Z. I. Cabantchik. Anion transport
in red blood cells. *J. Memb. Biol. 2*: 255-281, 1979.

35. H. Passow, H. Fasold, S. Lepke, M. Pring, and B. Schuhmann.
Chemical and enzymatic modification of membrane proteins and
anion transport in human red blood cells. *In* "Membrane
Toxicity," (M. W. Miller and A. E. Shamoo, eds.). Plenum
Press, 1977: 353-379.

36. L. Kampmann, S. Lepke, and H. Passow. The kinetics of intra-
molecular crosslinking of the band 3 protein by 4,4'-diiso-
thiocyano dihydrostilbene-2,2'-disulfonic acid (H_2DIDS).
"Protides in Biological Fluids," in press.

37. P. D. Wood, R. M. Epand, and M. A. Moscavello. Localisation
of the basic protein and lipophilin in the myelin membrane
with a nonpenetrating reagent. *Biochem. Biophys. Acta 467*:
120-129, 1977.

38. C. Tanford. The interpretation of hydrogen ion titration
curves of proteins. *In* "Advances in Protein Chemistry,"
Vol. 17, (C. B. Anfinsen, M. L. Anson, and J. T. Edsall, eds.).
new York: Academic, 1962: 69-165.

39. S. Lepke and H. Passow. The permeability of the human red

blood cell to sulfate ions. *J. Memb. Biol. 6*: 158-182, 1971.

40. J. H. Kaplan, K. Scorah, H. Fasold, and H. Passow. Sideness of the inhibitory action of disulfonic acids on chloride equilibrium exchange and net transport across the human erythrocyte membrane. *FEBS Letters 62*: 182-185, 1976.

41. H. Passow and L. Zaki. Studies on the molecular mechanism of anion transport across the red blood cell membrane. *In* "Molecular Specialization and Symmetry in Membrane Function," (A. K. Solomon and M. Karnovsky, eds.). Cambridge, Mass. and London, England. Harvard University Press, 1978: 229-250.

42. H. Passow, H. Fasold, E. M. Gärtner, B. Legrum, W. Ruffing, and L. Zaki. Anion transport across the red cell membrane and the conformation of the protein in band 3. *Proc. New York Academy of Sciences 341*: 361-383, 1980.

43. S. Grinstein, L. McCulloch, and A. Rothstein. Transmembrane effects of irreversible inhibitors of anion transport in red blood cells. Evidence for mobile transport sites. *J. Gen. Physiol. 73*: 493-514, 1979.

44. Y. Shami, A. Rothstein, and P. A. Knauf. Identification of the Cl-transport site of human red blood cells by kinetic analysis of the inhibitory effect of a chemical probe. *Biochem. Biophys. Acta 508*: 357-363, 1978.

45. M. Dalmark. Chloride and water distribution in human red cells. *J. Physiol. 250*: 65-84, 1975.

ARGUMENTS IN SUPPORT OF A SINGLE TRANSPORT SITE ON
EACH ANION TRANSPORTER IN HUMAN RED CELLS

R. B. Gunn

O. Fröhlich

Department of Pharmacological
and Physiological Sciences
University of Chicago
Chicago, Illinois

I. INTRODUCTION

The anion exchange system of erythrocytes mediates the
physiologically crucial exchange of chloride and bicarbonate anions
across the plasma membrane of these cells as they flow through the
capillaries of the lung and peripheral tissues. This Hamburger
shift was first described by Nasse in 1878 (1). The carbonic
anhydrases A and B in the erythrocyte cytoplasm rapidly convert
CO_2 and H_2O into protons (H^+) and bicarbonate (HCO_3^-) ions. This
interconversion produces a second form of CO_2 for transport besides
the CO_2 physically dissolved in the blood. Additionally, the
erythrocyte membrane contains the anion transport system which
exchanges the intracellular bicarbonate with the plasma chloride
and, thereby, further increases the CO_2 carrying capacity of the
venous blood by using the plasma bicarbonate space as well as the

CHLORIDE TRANSPORT
IN BIOLOGICAL MEMBRANES
33

intraerythrocyte compartment. In the laboratory this anion exchange system can also mediate the self-exchange of small inorganic
anions with their isotopes as well as heteroexchange between chloride, bicarbonate, nitrate, bromide, iodide, sulfate, phosphate,
dithionite, and other anions. This general anion exchange mechanism involves amino groups on the band 3 protein that makes up
about half of the intrinsic membrane protein of human erythrocytes.
There are $0.8 - 1.2 \times 10^6$ copies of this 95,000 dalton protein per
red cell depending on the individual donor. The peptide backbone
of this protein crosses the phospholipid core of the membrane perhaps several times (2) and, thereby, it is believed, forms the
transport mechanism. Because chloride-bicarbonate exchange is the
major mediated mechanism of erythrocyte membranes, and because
erythrocytes are so easily obtained, scientists in several laboratories are currently investigating its molecular basis and
mechanism.

In this report we review the data which demonstrate the
asymmetric kinetics of the anion exchange mechanism. We also discuss the data used to distinguish a single reciprocating site model
from a double-site sequential model.

Anion transport across the erythrocyte membrane has several
interesting features. The most important feature is that isotopic
exchange of chloride is $10^3 - 10^4$ times the net transport of chloride even when electroneutrality of the overall transport does not
constrain the net chloride flux, i.e., even when the membrane is
more permeable to potassium than to chloride (3). This obligatory
one-for-one exchange flux has a high temperature coefficient, shows
saturation kinetics and competition among the anions listed above,
and can be inhibited by diuretics, local anesthetics, and amino
reactive reagents.

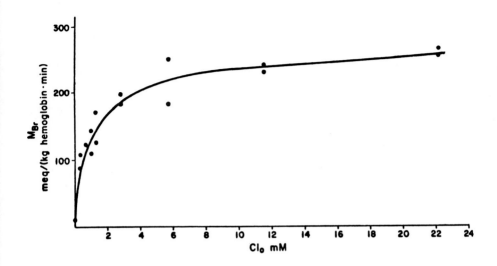

FIG. 1.   The stimulation of bromide efflux from human erythrocytes by increasing external chloride concentrations.   Chloride was
substituted for isotonic-isoionic citrate-sucrose solutions to vary
the external chloride concentration.   Each point is the flux calculated from the product of the measured initial rate coefficient of
^{82}Br efflux (1% hematocrit) and the measured initial bromide content of the cells.   The maximum flux calculated by making the best
fit of all the data to a rectangular hyperbola was 324 ± 15 mEq/(kg
hemoglobin/min), the $K_{Cl\ 1/2-out} = 0.8 \pm 0.1$ mM.   Experiment P-23, pH
7.8, 0^0C, 0.90 kg hemoglobin/kg dry cell solids.   Reprinted with
permission J. Gen. Physiol. (4).

II.   DATA AND INTERPRETATION

Figure 1 contains data that support these points.   In this

experiment we measured, by making an isotonic-isoionic substitution

with a citrate-sucrose solution, the bromide efflux from cells

containing bromide as the sole permeant anion into a series of

solutions with different concentrations of chloride.   Citrate at

pH 7.8 appears to be an untransported, indifferent anion.   At zero

external chloride, bromide leaves the red cells at a measurable

rate on this scale but this flux includes exchange for traces of
external bicarbonate and net flux accompanying cations which are
more permeant at high membrane potentials (4,5,6). If valinomycin,
the potassium ionophore, has been added to increase the tracer exchange of potassium to 10 times the chloride flux, the bromide flux
measured in this way is not detectably increased. Thus, there is
an obligatory exchange of internal anion with another external
anion and efflux occurs only (expecting 0.01% of maximum) if the
chloride or other permeant anion is present on the outside. This
chloride-stimulated bromide efflux is saturated when external chloride is about 20 mM. The analysis of this heteroexchange of
bromide for chloride is complex and not easily understood without
first considering the exchange of chloride isotopes where only one
species of anion is transported in both directions.

The concentration dependence of the equilibrium self-exchange
flux of chloride is shown in Fig. 2. The cells were pretreated
with nystatin to alter their salt content after the method of Cass
and Dalmark (7), and the equilibrium self-exchange of chloride
isotope was measured at 0^0C, pH 7.4. It is important to keep in
mind that the chloride concentrations are equal on the two sides
of the membrane in this experiment and that both concentrations
are varied together along the abscissa. The activation of the flux
by both internal and external chloride together involves at least
two processes. At low concentrations, the flux is activated by
chloride and tends to saturate at intermediate concentrations.
This is most likely the result of chloride binding to a limited
number of transport sites to form the complexes necessary for
transport. At high chloride concentrations the flux declines.
Dalmark has attributed this to a second site or modifier site (8).
His data suggest that chloride and other halides interact with a
second site with a low affinity and noncompetitively inhibit the
operation of the transport unit. An alternative suggestion is
that the decline in self-exchange flux at high anion concentrations reflects the interaction of two anion transport sites.

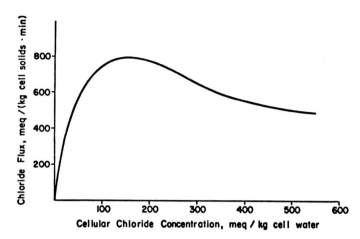

FIG. 2. Chloride self-exchange flux as a function of cellular chloride concentration. Equal external and internal chloride concentrations increase together along the abscissa. Cell cation and anion content were altered by pretreatment with nystatin in a solution of the desired chloride concentration. The flux is norma- lized to a constant number of cells having one kg of dry solids. The self-inhibition evident above physiological concentrations may be due to a complexation of chloride at a site separate from the transport locus which is responsible for the activation of the flux at low concentrations. Reprinted with permission Ann. N.Y. Acad. Sci. (39).

We will discuss these two possibilities later and now focus on the interactions between chloride and the transport site.

Our objective was to experimentally dissect the self-exchange flux into the component partial reactions for influx and efflux, and to determine by kinetic analysis whether the behavior of the molecular mechanism for anion transport was of the ping-pong or sequential (simultaneous) type as defined by Cleland (9).

In order to reduce the interference of the modifier site on the magnitude of the observed fluxes, we measured chloride and bromide fluxes at only low concentrations where the reactions with the transport site dominate the changes in the observed fluxes.

Consider the reaction scheme shown in Fig. 3. This is a

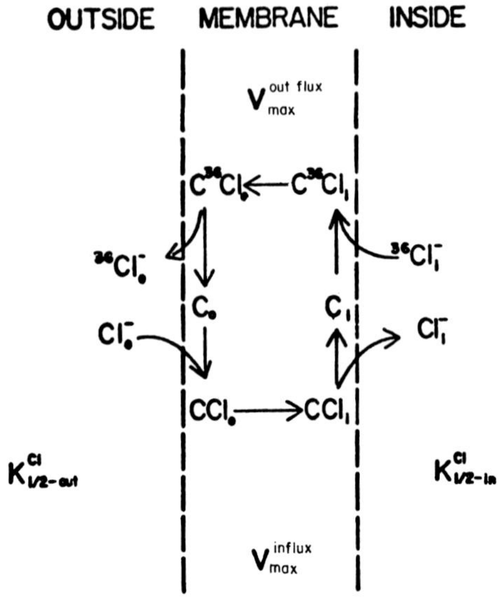

OUTSIDE   MEMBRANE   INSIDE

FIG. 3.   A symbolic scheme of possible partial reactions for the exchange of inside $^{36}Cl^-$ with outside Cl^-. C_o and C_i represent the transport site on the transport mechanism which can complex with external chloride ions (Cl_o^-) and internal chloride ions (Cl_i^-) respectively.   Since the exchange is one-for-one, no interconversions of the uncomplexed sites themselves are allowed and no arrow connects them.   Reprinted with permission Ann. N.Y. Acad. Sci. (39).

kinetic scheme and not a molecular picture. It is a symbolic representation of the kinds of reaction steps which must occur when two chloride anions are exchanged across the membrane. The conformation in which extracellular chloride reacts with the transporter before it can be transported is denoted C_o. In this conformation, the transport site, or the first transport site if there is a series of them, is exposed to the outside solution. We say it has access to the outside chloride ions which are undergoing random thermal motion in their aqueous environment. In a like manner, C_i is symbolic for the conformation of the transporter in which the transport site has access to the internal chloride ions. The question which interests us, is, how are these two classes of sites

related? Clearly the chloride is transported and in some way
leaves the transport site on C_o in a manner which allows it to
enter the inside solution. Assuming microscopic reversibility, the
site on C_o must at some time have access to the internal solution
when the Cl desorbs into that solution. If the site originally on
C_o has at least temporary access to the inside solution, then it
becomes the transport site on the C_i conformation by definition.
This transformation of C_o with a chloride attached is indicated
by an arrow to the right at the bottom of Fig. 3. This process
which quite properly is the translocation reaction, most likely is
a conformational change within the transport apparatus. The movement may not span more than a few Ångstroms and most probably does
not involve a movement as large as that experienced by a valinomycin when it acts as a potassium ionophore in the membrane. From
this perspective we may now consider the self-exchange of chloride-
36 from the inside of the cell on the right hand side for Fig. 3
with the unlabeled chloride on the outside of the cell on the left
hand side of Fig. 3.

The chloride-36 at the upper right complexes to the inward-
facing transport site when the transporter conformation is C_i. The
resulting complex, $C^{36}Cl_i$, in some unknown way undergoes a conformational change which results in the complex facing the outside
solution. This complex, $C^{36}Cl_o$, then dissociates the ^{36}Cl to the
outside solution. In a like manner, external facing sites on
transporters in the C_o conformation react with external chloride
ions to form complexes which undergo a conformational change reaction in the opposite direction and the CCl_i complex can dissociate
Cl into the cell interior.

Since the exchange is nearly always obligatory and one anion
must be transported into the cell for each that is transported out
of the cell, almost no short circuiting can be permitted in the
form of C_o or C_i undergoing conformational reactions without an
anion complexed to them. For this reason no arrow connects C_o and
C_i in the center of the scheme. While these statements are valid

for the range of exchange fluxes for halides, it may not be true that C_o and C_i never interconvert. If one permits this transport to occur once for every 10^4 anion exchanges, one can model the anion conductance as well as the exchange (5,10,11,12).

There are four experimentally observable parameters in the scheme shown in Fig. 3: $K_{1/2-in}$ and $K_{1/2-out}$, and V_{max-in} and $V_{max-out}$. The $K_{1/2-out}$ equals the concentration of outside anion which stimulates the exchange to half its maximum while the inside concentration is held constant. The $K_{1/2-in}$ equals the analogous inside anion concentration. The other two observables, $V_{max-out}$, and V_{max-in}, are the maximum fluxes when stimulated by increasing the outside or inside concentration, respectively. In this highly coupled system the influx always equals the outflux, but it is not necessary that V_{max-in} equals $V_{max-out}$. For example, at a fixed, nonsaturating, internal chloride concentration, $V_{max-out}$ will reflect the maximum transport inward, but the outward flux rate limits the cycle. If the concentrations on the two sides are reversed, the V_{max-in} (obtained if necessary by further raising the internal concentration) may be larger if the fixed concentration on the outside is saturating. As a digression, we have found that V_{max-in}^{max} and $V_{max-out}^{max}$ for bromide are about equal ($\pm$ 30%) by measuring the exchange of bromide with chloride in both directions at saturating concentrations of chloride [as calculated from Fig. 1 this value is 324 $\pm$ mEq/(kg hemoglobin · min) at pH 7.8 and 0^0C].

III. PING-PONG *versus* SEQUENTIAL KINETICS

As the scheme for chloride-36 and chloride exchange is written in Fig. 3, the mechanism will have ping-pong kinetics. However, there is an alternative mechanism that cannot be excluded by previously published data. This is a sequential mechanism which may be described in terms of Fig. 3 if a further constraint is placed upon the system. This constraint is that the two transport reactions indicated by the horizontal arrows are physically coupled. That is, two anions are simultaneously transported in opposite directions on a transport mechanism which has the features of both conformations C_o and C_i at the same time. We now wish to review the kinetic differences between these two mechanisms and some evidence we have published elsewhere (4) that speaks to the issue of whether this is a ping-pong or a sequential mechanism.

The ping-pong reaction is shown at the top of Table I with individual rate constants indicated for the steps we have previously outlined. The countertransported anions, chloride-35, on the outside, and chloride-36, on the inside of the cell, can combine with C_o and C_i, the two conformations of the transport site. A central characteristic of a ping-pong mechanism is that only *binary* complexes are formed between the transporter and anion. At any time there is never more than one anion complexed to each transport machinery.

At the bottom of Table I is the sequential reaction. The transporter, C, in a sequential mechanism has *two* loci for binding anions. On the left it reacts with $^{36}Cl_i$ from the inside and Cl_o from the outside to form the *ternary* complex $^{36}Cl_iCCl_o$ which undergoes the conformational transport reaction with rate constant k_9. On the right, the ^{36}Cl and Cl are unloaded and the empty transporter, C, which is regenerated is the same as the original transporter on the far left of the scheme.

The ping-pong scheme at the top is like a simple carrier but more realistically is a double-gated channel or a lock-carrier (13,14) similar to the type proposed by Patlak (15). The sequential scheme, on the other hand, is like a revolving door which turns only if one person from inside and one person from outside have entered into the door. And then they are both simultaneously switched to the opposite sides in a single transport step.

These two mechanisms might be experimentally distinguished if one measures the dependence of $K_{1/2-out}$ and $V_{max-out}$ on the internal anion concentration.

TABLE I. Ping-pong and sequential reaction schemes. C is the transport site for chloride-35 and chloride-36. In the ping-pong mechanism each C has only one site and can either complex a Cl^- or $^{36}Cl^-$ but not both simultaneously and these complexes can have access to either the inside (i) or the outside (o) but not both simultaneously. In the sequential mechanism each C has two transport sites which both must be complexed with either Cl^- or $^{36}Cl^-$, one from each side of the membrane, before translocation can take place with the rate constant k_9 or k_{-9} in the two directions across the membrane. Neither mechanism implies a classical mobile carrier; while both imply that transport sites can reciprocate their access to ions for complexation reactions with anions in the two adjacent solutions.

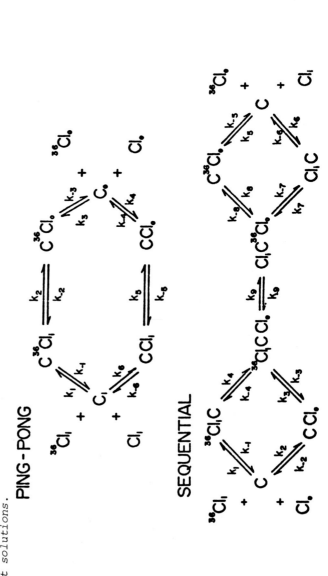

PING-PONG

SEQUENTIAL

TABLE II. *Flux equations for ping-pong and sequential mechanisms. The flux, V, per total number of transporters, C_t, is a complex function of constants, As and Bs, and the concentration of chloride outside, Cl_o, and inside, Cl_i. In both mechanisms, if Cl on one side is held constant, the flux is a hyperbolic function of the anion concentration on the opposite side and obeys Michaelis-Menten kinetics.* *

Ping-Pong

$$\frac{V}{C_t} = \frac{(A_1 Cl_i) Cl_o}{A_2 Cl_i + (A_3 + A_4 Cl_i) Cl_o}$$

Sequential

$$\frac{V}{C_t} = \frac{(B_1 Cl_i) Cl_o}{(B_2 + B_3 Cl_i) + (B_4 + B_5 Cl_i) Cl_o}$$

The coefficients A_1 through A_4 and B_1 through B_5 as expressed by the individual rate constants are: (See Table I for the measuring of the rate constants in the respective models.).

Ping-pong Model:

$$A_1 = k_1 k_{-1} k_2 k_{-2} k_3 k_{-3}$$
$$A_2 = k_1 k_2 k_{-3} (k_{-1} k_{-2} + k_{-1} k_{-3} + k_2 k_{-3})$$
$$A_3 = k_{-1} k_{-2} k_3 (k_{-1} k_{-2} + k_{-1} k_{-3} + k_2 k_{-3})$$
$$A_4 = k_1 k_3 [k_2 k_{-3} (k_{-1} + k_2 + k_{-2}) + k_{-1} k_{-2} (k_{-3} + k_2 + k_{-2})]$$

Sequential Model:

$$B_1 = k_{-1} k_{-2} k_9 (k_{-3} + k_{-4}) (k_1 k_{-2} k_4 + k_{-1} k_2 k_3)$$
$$B_2 = k_{-1}^2 k_{-2}^2 (k_{-3} + k_{-4}) (k_{-3} + k_{-4} + k_9 + k_{-9})$$
$$B_3 = k_{-1} k_{-2} \{ k_{-1} k_3 [k_{-4} k_9 + (k_{-4} + k_{-9}) (k_{-3} + k_{-4})]$$
$$\quad + k_1 k_{-2} (k_{-3} + k_{-4}) (k_{-3} + k_{-4} + k_9 + k_{-9}) \}$$
$$B_4 = k_{-1} k_{-2} \{ k_{-2} k_4 [k_{-3} k_9 + (k_{-3} + k_{-9}) (k_{-3} + k_{-4})]$$
$$\quad + k_{-1} k_2 (k_{-3} + k_{-4}) (k_{-3} + k_{-4} + k_9 + k_{-9}) \}$$
$$B_5 = k_{-1} k_{-2} \{ k_1 k_4 [k_{-2} k_{-9} + (k_{-2} + k_{-3}) (k_{-3} + k_{-4} + k_9)]$$
$$\quad + k_2 k_3 [k_{-1} k_{-9} + (k_{-1} + k_{-4}) (k_{-3} + k_{-4} + k_{-9})] \}$$
$$\quad + k_{-9} [k_{-2} k_{-4} (k_1 k_{-2} k_4 + k_{-1} k_2 k_3 + k_{-1} k_3 k_4)$$
$$\quad + k_{-1} k_{-3} (k_1 k_{-2} k4 + k_{-1} k_2 k_3 + k_{-2} k_3 k_4)]$$

The flux equations for the two mechanisms are shown in Table II. The equations for the sequential mechanism have been simplified by neglecting terms which have squares of concentrations because, experimentally, we have confined ourselves to low anion concentrations for another reason: to avoid the effects of anions on the modifier sites.

V is the flux, C_t the total number of transporters so that the ratio of V over C_t is the turnover number. The A's and B's are constants. They are complex combinations of the individual k's for the steps shown in Table I (see footnote to Table II). When the flux is measured at fixed internal chloride, Cl_i, as a function of external chloride, Cl_o, then the flux obeys a hyperbolic, Michaelis-Menten equation in both schemes. In each case, the numerator on the right-hand side is a constant times Cl_o and the denominator is a constant plus another constant times Cl_o. This functional relationship was demonstrated for external chloride stimulation of bromide efflux in Figure 1.

The $K_{1/2\text{-out}}$ (the concentration of external chloride to give half-maximal flux) is, of course, the ratio of the first and second constants in the denominators as shown explicitly in Table III.

In the ping-pong mechanism note that $K_{1/2\text{-out}}$ is itself a hyperbolic function of the internal or trans-anion concentration. In contrast, for a sequential mechanism this is not necessarily so. Only if B_2 is very small will $K_{1/2\text{-out}}$ be a Michaelis-Menton function of internal chloride Cl_i. It is not widely appreciated that the observed $K_{1/2}$ values for mediated transport processes should be functions of transconcentrations. The clear functional relationships derived here are, in part, due to the tight coupling of the two unidirectional fluxes, but the interdependence of $K_{1/2}$ values measured on one side of the membrane on the concentration of the transport species on the other side is quite general and often worth investigating.

The maximum efflux in both the ping-pong and sequential

TABLE III. *The dependence of outside parameters of chloride flux on internal chloride concentration. $K_{1/2-out}$ and $V_{max-out}$ in a ping-pong mechanism and $V_{max-out}$ in a sequential mechanism are hyperbolic functions of internal chloride concentration. $K_{1/2-out}$ in a sequential mechanism is a concave, monotone increasing function and may be linear or hyperbolic in limiting cases. The ratio, $V_{max-out}/K_{1/2-out}$, is either a constant or a hyperbolic function of Cl_i, depending on the mechanism.*

	$K_{1/2-out}$	$V_{max-out}$	$V_{max-out}/K_{1/2-out}$
Ping-pong	$\dfrac{A_2 Cl_i}{A_3 + A_4 Cl_i}$	$\dfrac{C_t A_1 Cl_i}{A_3 + A_4 Cl_i}$	$\dfrac{C_t A_1}{A_2}$
Sequential	$\dfrac{B_2 + B_3 Cl_i}{B_4 + B_5 Cl_i}$	$\dfrac{C_t B_1 Cl_i}{B_4 + B_5 Cl_i}$	$\dfrac{C_t B_1 Cl_i}{B_2 + B_3 Cl_i}$

mechanisms are hyperbolic functions of internal chloride concentration and so the dependence of $V_{max-out}$ on Cl_i cannot be used to distinguish the two models.

Perhaps the best measure to distinguish these two systems is the ratio of $V_{max-out}$ to the $K_{1/2-out}$ shown in the third column of Table III. For the ping-pong mechanism this ratio is constant. It is independent of the internal chloride concentration, while the same ratio in the sequential mechanism has a hyperbolic dependence on internal chloride. Note again that if B_2 is very small, the ratio becomes a constant and the two mechanisms may not be distinguishable.

The experimental results indicate that at pH 7.8 at 0°C the mechanism has ping-pong kinetics. Each flux was measured as the efflux of chloride-36 from resealed ghosts (4) into an isotonic-isoionic solution of sucrose-citrate and chloride at pH 7.8 at 0°C. These effluxes were measured at a fixed internal chloride concentration into eight to ten different solutions in duplicate. The activation of the chloride efflux by external chloride was hyperbolic. The $K_{1/2-out}$ and $V_{max-out}$ were estimated together with their standard deviations using the statistical method of

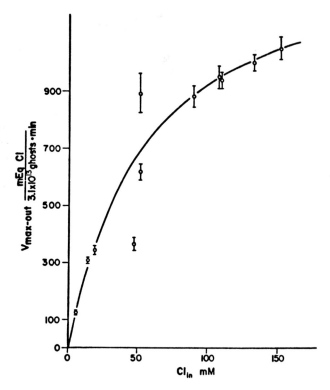

FIG. 4. $V_{max-out}$ as a function of internal chloride concentration. Each point is the result of the statistical analysis of 16 - 20 separate flux measurements at different external chloride concentrations. The best estimate of $V_{max-out}$ plus and minus the standard error of that estimate are given; thus, over two-thirds of the best estimates should fall within the bars if the scatter in the data is random. Except for two series of experiments, these estimates of $V_{max-out}$ fall along a hyperbolic curve as expected from both kinetic schemes. The maximum $V_{max-out}$ was calculated to be 1450 ± 170 mEq Cl/(3.1×10^{13} ghosts/min) and the half maximum $V_{max-out}$ was achieved at 57 ± 13 mM internal chloride.

Wilkinson (16). A similar series of 16 - 20 fluxes were measured at 11 different internal chloride concentrations in resealed human red cell ghosts prepared by a modification of the methods of Bodemann and Passow (17) and Schwoch and Passow (18).

The $V_{max-out}$ is shown in Fig. 4 as a function of the internal chloride concentration. Aside from a pair of experiments the $V_{max-out}$ does follow a hyperbolic curve drawn as a solid line.

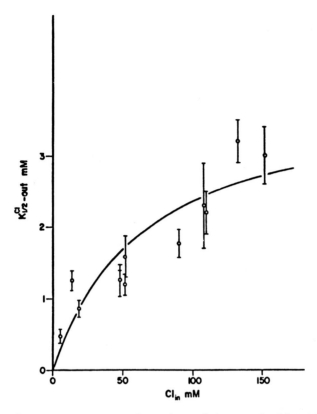

FIG. 5. $K_{1/2\text{-}out}$ as a function of internal chloride concentration. The estimates and their standard errors of $K_{1/2\text{-}out}$ from the same 11 series of experiments shown in Fig. 4 also fall along a hyperbolic curve. The scatter is somewhat greater but it should be kept in mind that this is a tertiary graph of parameters derived from secondary graphs of flux as a function of external chloride which are themselves parameters derived from measurements of the time course of radioactive tracer loss from intact erythrocytes or resealed erythrocyte ghosts.

This result is expected for both ping-pong and sequential mechanisms.

In Figure 5, $K_{1/2\text{-}out}$ values are graphed as a function of internal chloride concentration. The solid line is the best hyperbolic curve which would agree with a ping-pong model. The data do not fit as well to this curve as one might hope nor do they fit well to the best monotone concave curve which would intercept the

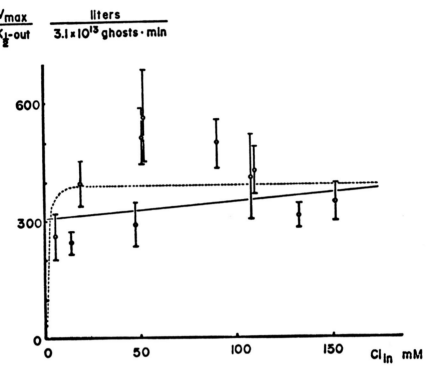

FIG. 6.   The ratios of V_{max} and $K_{1/2-out}$ values from Figs.
4 and 5 with their standard errors as a function of internal
chloride concentration.   The best fit straight line is the solid
line with slope equal to 0.47 ± 0.51 that is not significantly
different from zero.   This agrees with the prediction of a ping-
pong mechanism.   The dotted line is the best fit hyperbolic function with a half-maximum value at 0.2 mM internal chloride.   As
explained in the text, this hyperbolic function is expected from
a sequential mechanism but one which infrequently desorbs the ion
at one of the two transport sites and thus degenerates into a
ping-pong mechanism transporting only one ion at a time.   The
mechanism appears to behave as a ping-pong mechanism though an
underlying sequential machinery cannot be excluded by these data
at pH 7.8 and 0°C.   Redrawn from Fig. 11 in Reference 4.   We
thank Dr. J. M. Salhany for pointing out the correct slope given
above.

ordinate at $K_{1/2-out}$ ≈ 0.4 mM. This intercept would be an estimate
of B_2/B_4 in the sequential model.

      Figure 6 shows the ratio of the V_{max} and $K_{1/2}$ values from

Figs. 4 and 5 as a function of internal chloride. The slope of the best least squares straight line for this data is not significantly different from zero. The ping-pong model predicts a horizontal straight line. Alternatively, the dotted curve line through the data is that for a sequential model. The Cl_{in} concentration at which the curve has reached half of its maximum value should equal B_2/B_3 as can be seen from inspection of the last column in Table III. We must be chary in deciding between these two models. It may be that the mechanism is ping-pong and the lower values of $V_{max-out}/K_{1/2-out}$ at lower Cl_{in} are fortuitous. It may be that the mechanism is sequential with a very low B_2 value so that we have only collected data at internal chloride concentrations at which the curve is nearly saturated. But what does it mean for a sequential mechanism to have a small value of B_2? One possible way for B_2 to be small is for the off-reaction described by either k_{-1} or k_{-2} to be very slow relative to the other, as can be seen in the footnote to Table II. This negligible off-rate would allow only one site to be fully functional in mediating exchange during any given transport cycle. The anion at the other site would rarely leave the mechanism but act only as "nontransported" co-factor. The transporter then would have one remaining site which would carry an anion in a ping-pong fashion.

This example demonstrates the limits of our analysis. We cannot exclude the possibility that the single site we infer from the data is always the second site of a sequential two-site model. Nor, as we shall see, can we exclude the possibility that the single site is only the first site of a sequential two-site model.

The steady-state kinetics of anion exchange can never un-equivocally prove that the mechanism operates by only forming com-plexes with one anion at a time, but they might prove that a two-site simultaneous (sequential) model is required. The reasons lie in the fundamental principles of these experiments. To prove that a ping-pong mechanism with a single alternating site operates the erythrocyte anion exchange one must show that a ternary transport

complex between the transporter and two transportable anions never
occurs. The data reviewed here (4) and that by Knauf (14) show
that over a broad range of experimental conditions no ternary complex was evident. During our measurements the fraction of the total transporter in any particular conformation was constant. We
failed to find any deviations in the flux from that predicted by
a model with only one transported anion bound at a time to the
transporter. And we can conclude that the fraction of transporters
with two transported anions must be small or zero. However this
does not mean that it is zero. More importantly, even if the ternary complex is rare and fleeting and never involves more than
0.1% of the transporter's conformations, it may be the only transporting conformation. The size of the pool of a particular transport configuration gives no information about the flux through the
pool unless the size is zero. Thus while a ternary complex could
be a small fraction of the transporters, it may have a very high
turnover rate and thereby mediate all of anion transport, and still
escape detection by steady state transport kinetics. From this argument it is evident that transient kinetic measurements with
rapidly transported anions are needed. Recent work in other laboratories supports a ping-pong type transport mechanism.

Firstly, there are data consistent with the ping-pong mechanism, such as the apparent recruitment of transport sites to one
side of the membrane. These are chloride-sulfate heteroexchange
experiments (19), membrane-labeling experiments (20, 21) and experiments describing the trans-effect of anions on the inhibitory
potency of extracellular inhibitors (22). But these data are not
inconsistent with a sequential mechanism. Nor were they designed
with an experimental eye toward emphasizing kinetic differences
between the two mechanisms.

Secondly, there are data that are very difficult to explain
except with a ping-pong scheme. Jennings (private communication)
has shown that chloride efflux from red cells in exchange for the
much more slowly countertransported phosphate anion is saturated,

and therefore independent of the intracellular chloride concentration Cl_i above Cl_i = 1 mM, although chloride self exchange exhibits
a $K_{1/2-in}^{max}$ = 55-60 mM (4); and we have shown for Br-Cl and I-Cl (23
and unpubl. obs.) that the maximum rate of heteroexchange of these
anions can be predicted from the maximum self-exchange rates of the
two anions separately, if a ping-pong mechanism is assumed. No
such prediction can be made in a sequential scheme because the
translocation rate of the ternary complex with the anions X and Y
has no a *priori* relation to the rate of the ternary complex with
two X's and the complex with two Y's. But in a ping-pong scheme
the reaction cycle of heteroexchange is simply the sum of two half
cycles contained in the self-exchange reaction cycles of X and Y.
While these predictions could be fortuitous, the increasing number
of correct predictions using ping-pong kinetics make a sequential
scheme less likely. We have been able to exclude one particular
sequential mechanism, namely an ordered sequential reaction in
which the inside transport site must be loaded with an anion before the outside site is loaded. This model is incompatible with
the observed competitive kinetics between external chloride and
external dinitrostilbene disulfonate (DNDS) (24-26).

Thirdly, Salhany *et al.* (27) have recently claimed that the
anion transport mechanism is much more complex and requires a
sequential kinetic scheme to describe the data. They argued that
the phenomenon of self-inhibition is an important feature which is
due to a strong interaction between two anion transport sites. In
the simple ping-pong scheme, self-inhibition is ignored because it
is assumed that it is due to an additional low-affinity site, the
so-called modifier site (8), whose effect on transport is probably
negligible at the low concentrations which suffice to saturate
transport (4). Salhany and his co-workers find additional evidence for their view in their dithionite influx experiments that
show deviations from Michaelis-Menten behavior consistent with a
negatively cooperative behavior. They proposed a "partially con-
certed dimeric model" (28) in which two single-site transporter

molecules can interact with each other. This model is based on
the observation that the transport protein, the band 3 protein of
the red cell membrane (29), occurs in the form of dimers (30) or
higher polymer (31). The model therefore describes the kinetics
of a sequential-type scheme because it involves a ternary complex
of the transporter (dimer) with two anions. This more complicated
kinetic scheme is able to give a "better" fit to data than the
ping-pong scheme and it does so because it has additional adjust-
able parameters. It should be noted, however, that Salhany's
measurements followed dithionite transport and may be difficult to
interpret because of the dissociation of dithionite into the monovalent anion SO_2^- (27, 32), albeit with the equilibrium favoring the
dimeric, divalent form. The monomeric form of dithionite, in
analogy with other monovalent anions such as chloride, bicarbonate,
and nitrate, may be transported at a much faster rate than the divalent dithionite $S_2O_4^{2-}$, which is a very slowly transported anion
like sulfate. Even if only the dimeric form were transported, the
observed fluxes are of the same order of magnitude as the anion
conductance. Consequently, net dithionite flux into red cells may
reflect both heteroexchange with internal chloride and net conductance. Dithionite may, in addition, require the cotransport of
protons as does sulfate which complicates the reaction scheme
further for dithionite (33). The introduction of the conductance
pathway by permitting C_o and C_i to interconvert in the ping-pong
scheme (Table I) will give a set of equations which are similar to
those of a sequential mechanism. Therefore we are unconvinced by
these data that a ternary complex of two transported anions and
the transporter is the transport complex, but do agree that at low
rates of transport where the loaded exchange cycle does not dominate the measured flux, kinetics similar to a sequential scheme
may be observed. Salhany's spectrophotometric methods are much
more sensitive than our tracer flux measurements and therefore may
pick up subtleties in the reaction scheme that we miss.

IV. ANION EXCHANGE IS ASYMMETRIC

These data clearly demonstrate the asymmetry of the anion exchange system. While previous studies (34-37) using inhibitors have strongly suggested that there was an asymmetry in the transport sites facing the two surfaces of the membrane, those experiments could not prove asymmetry because the access of the inhibitors to the transport sites might have been asymmetric and not the transport system itself. In this study the probe for the transport site is a transported anion, and thus the apparent asymmetry is a property of the transport system. In Fig. 5, the values of the $K_{1/2\text{-out}}$ range from 0.4 mM to a maximum value of 3 mM. In contrast, the $K_{1/2\text{-in}}$ is 15-100 fold greater. The maximum value of $K_{1/2\text{-in}}$ corresponds to the concentration of internal chloride at which $K_{1/2\text{-out}}$ is half-maximal. This is 55 mM in Fig. 5. It can also be shown that the maximum $K_{1/2\text{-in}}$ as given in Fig. 4 corresponds to the internal concentration causing half-maximal $V_{max\text{-out}}$, which is again about 55 mM. The asymmetry is, therefore, a factor of about 15. If the translocation reactions are rate-limiting compared to the surface complexations (as yet an unproven assumption) then the factor of 15 is the ratio of the apparent affinities of the transport site when it faces the two solutions. The affinity on the outside would be 15 times that on the inside. The asymmetry of apparent affinities is in the direction expected from the greater negative internal surface charge due to phosphatidylserine in the inner leaflet of the lipid bilayer (38). Whether this corresponds is fortuitous remains to be established. Even without assuming the relative slowness of the transport step, the transport mechanism is very asymmetric and has very different $K_{1/2}$ values on the two sides.

In conclusion, erythrocyte anion exchange is a facilitated transport process. Anion exchange has very asymmetric $K_{1/2}$ values for chloride on the internal and external surfaces. The dependence of $K_{1/2}$ and V_{max} on transmembrane anion concentrations in-

dicates that the molecular mechanism behaves like a ping-pong
mechanism but this could be due to a rather special case of an
underlying sequential mechanism.

SUMMARY

    The molecular mechanism of rapid anion transport in human
red cells which mediates the Hamburger Shift could involve a single
transport locus alternately forming complexes with one anion at a
time from the two solutions on either side of the membrane (ping-
pong mechanism); or, alternatively, could involve two transport
sites facing opposite sides, both of which must be loaded to form
a ternary complex before simultaneous transport in opposite directions (sequential mechanism). The flux of chloride via these two
mechanisms may behave differently when it is measured as a function
of chloride concentration on the cis and trans sides. We have
measured initial chloride fluxes at several fixed internal chloride concentrations at pH 7.8 and 0°C. The membrane asymmetry
for chloride and bromide transport is striking. The concentration
of anion which will half saturate the flux, $K_{1/2}$, is 15 times
greater on the inside than on the outside for both anions separately. The dependence of $K_{1/2-out}$ and $V_{max-out}$, the maximum transport
measured by increasing external chloride concentration are hyperbolic functions of internal chloride concentration. The ratio
$V_{max-out}/K_{1/2-out}$ as a function of internal chloride is that predicted by a ping-pong mechanism or a degenerate sequential
mechanism. This degenerate sequential mechanism could transport a
single anion while the second anion usually remains attached to the
carrier. The kinetics we observed are essentially those of a ping-
pong mechanism with a single site alternatively having access to
the two sides of the red cell membrane at which only one anion is
complexed, transported, and unloaded to the opposite side. However
these data do not rule out a sequential (simultaneous) mechanism.
The reasons for this have been discussed.

ACKNOWLEDGMENT

This work has been supported in part by research grants HL-20365 and GM-28893. Dr. Gunn is the recipient of Research Career Development Award 5K04HL00208 from the USPHS.

REFERENCES

1. H. Nasse. Untersuchungen über den Austritt und Eintritt von Stoffen (Transsudation and Diffusion) durch die Wand der Haargefässe. *Pflüger's Arch. ges. Physiol.* *16*:604, 1878.
2. M. J. A. Tanner, D. G. Williams, and R. E. Jenkins. Structure of the erythrocyte anion transport protein. *Ann. N.Y. Acad. Sci.* *341*:455-464, 1980.
3. M. J. Hunter. A quantitative estimate of the non-exchange-restricted chloride permeability of the human red cell. *J. Physiol.* *218*:P49-P56, 1971.
4. R. B. Gunn and O. Fröhlich. Asymmetry in the mechanism for anion exchange in human red cell membranes: Evidence for reciprocating sites which react with one transported anion at a time. *J. Gen. Physiol.* *74*:351-374, 1979.
5. P. L. LaCelle and A. Rothstein. The passive permeability of the red blood cell to cations. *J. Gen. Physiol.* *50*:171-188, 1966.
6. J. A. Donlon and A. Rothstein. The cation permeability or erythrocytes in low ionic strength media of various tonicities. *J. Membr. Biol.* *1*:37-52, 1969.
7. A. Cass and M. Dalmark. Equilibrium dialysis of ions in nystatin-treated red cells. *Nature* *244*:47, 1973.
8. M. Dalmark. Effects of halides and bicarbonate on chloride transport in human red cells. *J. Gen. Physiol.* *67*: 223-234, 1976.
9. W. W. Cleland. The kinetics of enzyme-catalysed reactions

with two or more substrates or products. I. Nomenclature and
rate equations. *Biochim. Biophys. Acta* 67:104-137, 1963.

10. O. Fröhlich. Chloride net transport of the human erythrocyte
at low extracellular chloride concentrations. *Biophys. J.*
33:2a, 1981.

11. O. Fröhlich, M. Milanick and R. B. Gunn. Anion transport in
red cells. Kinetic modelling and transport parameters. In
preparation.

12. B. Vestergaard-Bogind and U. V. Lassen. Membrane potential
of Amphiuma red cells: Hyperpolarizing effect of phloretin.
In "Comparative Biochemistry and Physiology of Transport."
(L. Bolis, K. Block, S. E. Luria, and F. Lynen, eds.), 1974,
North Holland, Amsterdam, p. 346.

13. R. B. Gunn. Transport of anions across red cell membranes.
In "Transport Across Single Biological Membranes. Vol. II
of Membrane Transport in Biology." (G. Giebisch, D. C.
Tosteson, and H. H. Ussing, eds.), 1979, Springer Verlag,
Berlin, pp. 59/80.

14. P. A. Knauf. Erythrocyte anion exchange and the band 3 protein: transport kinetics and molecular structure. *Curr.*
Top. Membrane Transp. 12:249-363, 1979.

15. C. S. Patlak. Contributions to the theory of active transport. II. The gate type non-carrier mechanism and generalizations concerning tracer and flow efficiency, and measurement
of energy expenditure. *Bull. Math. Biophys.* 19:209-235,
1957.

16. G. N. Wilkinson. Statistical estimations of enzyme kinetics.
Biochem. J. 80: 324-332, 1961.

17. H. Bodemann and H. Passow. Factors controlling the resealing
of the membrane of human erythrocyte ghosts after hypotonic
hemolysis. *J. Membr. Biol.* 8:1-26, 1972.

18. G. Schwoch and H. Passow. Preparation and properties of human erythrocyte ghosts. *Mol. Cell. Biochem.* 2:197-218,
1973.

19. M. L. Jennings. Apparent "recruitment" of SO_4 transport sites by the chloride gradient across the human erythrocyte membrane. *In* "Membrane Transport in Erythrocytes, Alfred Benzon Symposium 14." (U. V. Lassen, H. H. Ussing, and J. O. Wieth, eds.), 1980, Munksgaard, Copenhagen, pp. 450-463.

20. S. Grinstein, L. McCullough, and A. Rothstein. Transmembrane effects of irreversible inhibitors of anion transport in red blood cells. Evidence for mobile transport sites. *J. Gen. Physiol.* *73*:493-514, 1979.

21. H. Passow, L. Kampmann, H. Fasold, M. Jennings, and S. Lepke. Mediation of anion transport across the red blood cell membrane by means of conformational changes of the band 3 protein. *In* "Membrane Transport in Erythrocytes, Alfred Benzon Symposium 14." (U. V. Lassen, H. H. Ussing, and J. O. Wieth, eds.), 1980, Munksgaard, Copenhagen, pp. 345-367.

22. P. A. Knauf, T. Tashis, S. Grinstein, and W. Furuya. Spontaneous and induced asymmetry of the human erythrocyte anion exchange system as detected by chemical probes. *In* "Membrane Transport in Erythrocytes, Alfred Benzon Symposium 14." (U. V. Lassen, H. H. Ussing, and J. O. Wieth, eds.), 1980, Munksgaard, Copenhagen, pp. 389-403.

23. R. B. Gunn, O. Fröhlich, and M. Milanick. Chloride-bicarbonate exchange in human red cells: Predictions and calculations of a simple ping-pong model. *J. Supramol. Struct.* Suppl. 4, Abstr. 162, 1980.

24. O. Fröhlich and R. B. Gunn. Chloride transport kinetics of the human red blood cell studied with a reversible stilbene inhibitor. *Fed. Proc.* *39*:1714, 1980.

25. O. Fröhlich and R. B. Gunn. Binding of cis and trans isomers of 4,4'-dinitrostilbene-2,2'-disulfonate (DNDS) to the erythrocyte anion transporter. *In* "Adv. Physiol. Sci. Vol. 6. Genetics, Structure and Function of Blood Cells," (S. R. Hollán, G. Gárdos, and B. Sarkadi, eds.), Pergamon, 1981, pp. 275-280.

58   R. B. Gunn and O. Fröhlich

26.   O. Fröhlich. The external anion binding site of the human
      erythrocyte anion transporter: DNDS binding and competition
      with chloride. *J. Membrane Biol.*, in press.
27.   J. M. Salhany and J. C. Swanson. Kinetics of passive anion
      transport across the human erythrocyte membrane. *Biochemistry*
      17:3354-3362, 1978.
28.   J. M. Salhany, E. D. Gaines, and R. Sullivan. Steady state
      and transient state kinetics of erythrocyte anion exchange.
      Evidence for cooperativity in substrate and inhibitor binding
      suggesting site-site interactions within the band 3 protein
      dimer. *Biophys. J. 33*:3a, 1981.
29.   G. Fairbanks, T. L. Steck, and D. F. H. Wallach. Electro-
      phoretic analysis of the major polypeptides of the human
      erythrocyte membrane. *Biochemistry 10*:2606-2717, 1971.
30.   T. L. Steck. Cross-linking of the major proteins of the iso-
      lated erythrocyte membrane. *J. Mol. Biol. 66*:295-305, 1972.
31.   R. S. Weinstein, J. Khodadad, and T. L. Steck. The band 3
      protein intramembrane particle of the human red blood cell.
      In "Membrane Transport in Erythrocytes, Alfred Benzon Sym-
      posium 14." (U. V. Lassen, H. H. Ussing, and J. O. Wieth,
      eds.), 1980, Munksgaard, Copenhagen, pp. 35-46.
32.   D. O. Lambeth and G. Palmer. The kinetics and mechanisms of
      reduction of electron transfer proteins and other compounds
      of biological interest by dithionite. *J. Biol. Chem. 248*:6095-
      6103, 1973.
33.   M. L. Jennings. Proton fluxes associated with erythrocyte
      membrane anion exchange. *J. Membr. Biol. 28*:187-205, 1976.
34.   P. A. Knauf, S. Ship, W. Breuer, L. McCullough, and A. Roth-
      stein. Asymmetry of the red cell anion exchange system.
      Different mechanisms of reversible inhibition by N-(4-azido-
      2-nitrophenyl)-2-aminoethylsulfonate (NAP-taurine) at the in-
      side and outside of the membrane. *J. Gen. Physiol. 72*:631-
      649, 1978.
35.   S. Lepke and H. Passow. Asymmetric inhibition by phlorizin

of sulfate movements across the red blood cell membrane. *Biochim. Biophys. Acta 298*:529-533, 1973.

36. K. F. Schnell, S. Gerhardt, S. Lepke, and H. Passow. Asymmetric inhibition by phlorizin of halide movements across the red blood cell membrane. *Biochim. Biophys. Acta 318*:474-477, 1973.

37. J. H. Kaplan, K. Scorah, H. Fasold, and H. Passow. Sidedness of the inhibiting action of disulfonic acids on chloride equilibrium exchange and net transport across the human erythrocyte membranes. *FEBS Lett. 62*:182-185, 1976.

38. R. F. A. Zwaal, B. Roelofsen, D. Comfurius, and L. L. M. van Deenen. Organization of phospholipids in human red cell membranes as described by the action of various purified phospholipases. *Biochim. Biophys. Acta 406*:83-96, 1975.

39. R. B. Gunn and O. Fröhlich. The kinetics of the titratable carrier for anion exchange in erythrocytes. *Ann. N.Y. Acad. Sci. 341*:384-393, 1980.

FUNCTIONAL SITES OF THE RED CELL ANION EXCHANGE PROTEIN:

USE OF BIMODAL CHEMICAL PROBES

Philip A. Knauf

Department of Radiation Biology and Biophysics
University of Rochester School of Medicine and Dentistry
Rochester, New York

Sergio Grinstein

Research Institute
The Hospital for Sick Children
Toronto, Canada

For over 100 years, it has been known that the human red
blood cell possesses a very rapid anion exchange system (1).
Physiologically, this system permits the exchange of chloride for
bicarbonate ions across the red cell membrane, thereby increasing
the CO_2-carrying capacity of the blood (2). Despite its extreme
rapidity (the half-time for chloride exchange at $38°C$ (3) is only
50 msec), this system involves no energy input and simply permits
anions to equilibrate mutually in accordance with their concentration gradients.

Although it is conspicuous in the membrane, the anion exchange system is difficult to identify once the membrane is dis-

61

rupted, since it is not associated with any enzyme activity and
its affinity for substrate anions is low (4). This problem has
been circumvented by use of chemical probes, inhibitors which interact with the system selectively and with high affinity. Radioactively labeled probes which react covalently with the system
have been used by Cabantchik and Rothstein and others to identify
the transport system with band 3, a 93,000 dalton intrinsic membrane glycoprotein (5 - 9). Independent experiments at the same
time by Passow and co-workers (10, 11), using a different technique, also provided evidence for the involvement of band 3.
Since band 3 comprises over 25% of the total membrane protein of
the red cell (12), the membrane appears to be highly specialized
for anion transport. The transport system itself seems to be
highly specialized, in that it catalyzes the one-for-one exchange
of anions very rapidly, but permits very little (less than one
part in 10,000 of the exchange) net flow of anions (13 - 15).
These observations have led to the concept that some portion of
band 3 may serve as a "carrier" which is capable of binding anions
and translocating them across the membrane (16).

This paper will focus on three questions relating to the
mechanism and molecular nature of this transport system as revealed by the use of selected chemical probes: (1) What is the
functional nature of the sites which are labeled by these chemical
probes? (2) What parts of the band 3 protein are labeled? and
(3) Can these probes detect a conformational change in the band 3
protein which is associated with the transport of anions?

I. ACTION OF BIMODAL PROBES AS REVERSIBLE INHIBITORS:
    IDENTIFICATION OF INHIBITORY SITES

Two types of probes which have been most useful are DIDS
(4,4'-diisothiocyano-stilbene-2,2' disulfonate) [or its dihydro
derivative, H_2DIDS (4,4'-diisothiocyano-1,2-diphenylethane-2,2'-

disulfonate)] and the photoreactive agent NAP-taurine [*N*-(4-azido-
2-nitrophenyl)-2-amino ethyl sulfonate][1] (Fig. 1). All of these
probes are "bimodal." They can be used under certain conditions
as reversible inhibitors of the anion exchange system. Under other
conditions they react covalently with their binding sites and can,
therefore, be used as labels for those sites. Because of this
feature, it is possible to determine the mechanism of reversible
inhibition by classical kinetic techniques, and then to use the
probes to label the inhibitory sites (17, 18). Under certain conditions the probes cross the membrane very slowly (5, 20), so that
their actions on the two sides of the membrane may be studied independently.

The erythrocyte anion exchange system possesses at least two
sites that can bind anions (Fig. 2). Anions (A) bound to the substrate site (C) can be translocated across the membrane by a spon-

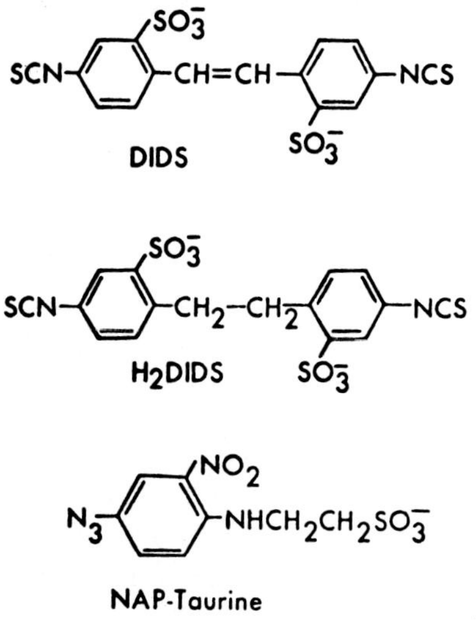

FIG. 1.   Structures of some bimodal chemical probes.

[1]*Abbreviations: DIDS: 4,4'-diisothiocyano-stilbene-2,2'disulfonate; H₂DIDS: 4,4'-diisothiocyano-1,2-diphenylethane-2-2'-
disulfonate; NAP-taurine: N-(4-azido-2-nitrophenyl)-2-amino ethyl
sulfonate; SDS: sodium dodecyl sulfate.*

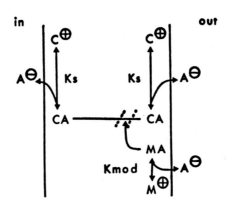

FIG. 2. *Carrier model for the red cell anion transport system. Anions (A) bind to the substrate site (C) to form a complex (CA) which can be transported across the membrane. Binding of anions to the modifier site (M) interferes with the transport of the carrier-anion complex. For chloride, the dissociation constant for the substrate site (K_S) is 65 mM (22), considerably less than the corresponding value (K_{mod}) for the modifier site [335 mM, (4) - (4)]. The model is shown here as a symmetrical carrier. Recent evidence (18, 34, 48, 55) suggests that the apparent K_S may be different at the two sides of the membrane. This asymmetry, however, does not affect the inhibition kinetics. (Reproduced from reference 18, with permission.)*

taneous conformational change of the transport "carrier" (21, 49, 59). In addition to this transport site, which is half-saturated by chloride at 65 mM (22), there is a second site with a lower affinity for chloride [half-saturated at 335 mM chloride (4)].
When anions are bound to this modifier site (M), anion exchange is inhibited. In principle, the chemical probes DIDS and NAP-taurine could inhibit transport by blocking the substrate site, binding to the modifier site, or acting noncompetitively at some other site. These possibilities can be most readily distinguished by examining the effects of the substrate chloride on the inhibition of chloride exchange by these probes.

The capacity of the probe to react covalently with the site introduces considerable complexity into such studies of the competition between substrate and inhibitor. Fortunately, in the case of DIDS, it has been shown that the covalent reaction is not necessary for inhibition, since other disulfonic stilbenes which have no reactive groups are effective inhibitors of anion exchange (5, 6, 11, 17, 23). Even the reactive probes DIDS or H_2DIDS appear to inhibit by initially combining reversibly with the site, after which the covalent reaction occurs (5, 8, 9, 24).

Thus, if conditions are chosen so that the covalent reaction is slow, the probes can be used as purely reversible inhibitors, and their competition with substrate can be readily studied. In the case of H_2DIDS, the covalent reaction takes place very slowly at low temperature (8, 9). Because of this, it was possible to use H_2DIDS as a reversible inhibitor of chloride exchange at $0°C$, and to study the effects of chloride concentration on the inhibitory potency of H_2DIDS (25). Cells were equilibrated with different KCl concentrations in the presence of nystatin, so that in all cases the internal and external chloride concentrations were nearly equal (26). In this way, possible effects of membrane potential or chloride gradients across the membrane unrelated to direct competition between substrate and inhibitor were avoided. After the nystatin was washed away, cells were loaded with ^{36}Cl and the equilibrium exchange flux of isotope was measured in the presence of various H_2DIDS concentrations.

Under these conditions the inhibitory effects of H_2DIDS could be rapidly reversed when the free H_2DIDS was complexed by albumin (25). The effects of chloride concentration on H_2DIDS inhibition were assessed by using the Hunter-Downs (27) plot (Fig. 3), which has the advantage that a straight line is obtained over the entire range of substrate concentrations, even though there are two distinct anion binding sites, the substrate and modifier sites (17, 18, 25). In the Hunter-Downs plot, $I(1-i)/i$ is plotted on the y axis, where I is the inhibitor concentration and i is the fractional inhibition. If the inhibitor binds reversibly to a single inhibitory site [as H_2DIDS appears to do (25)], $I(1-i)/i$ is equivalent to the apparent K_i, that is, to the concentration of inhi-

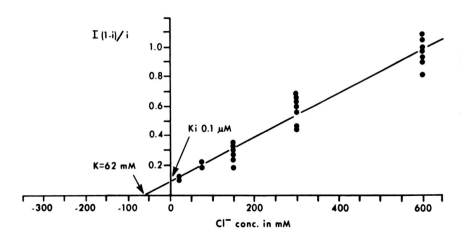

FIG. 3. *Hunter-Downs plot for H_2DIDS inhibition of chloride exchange at $0°C$. The concentration of H_2DIDS required for 50% inhibition (apparent K_i) in μM, calculated as $I(1-i)/i$, where I is the inhibitor (H_2DIDS) concentration and i is the fractional inhibition, is plotted on the y axis as a function of the external chloride concentration in mM. Conditions were chosen such that the internal and external chloride concentrations are nearly equal. The x intercept gives the chloride concentration required to half-saturate the site at which H_2DIDS binds. The y intercept is the concentration of H_2DIDS required to 50% inhibit the chloride flux as the chloride concentration approaches zero. Data from ref. (25)*

bitor required to produce 50% inhibition. From the data in Fig. 3, it is apparent that as the chloride concentration is raised, more H_2DIDS is required to inhibit chloride exchange by 50%. Thus, there is some competition between H_2DIDS and chloride.

Quantitative information concerning the nature of this interaction is provided by the x intercept of the Hunter-Downs plot (Fig. 3). The line cuts the axis to the left of zero at a point corresponding to the chloride affinity of the inhibitor binding site (17, 18, 25, 28). In the case of H_2DIDS, the inhibitory site is half-saturated with chloride at 62 mM, corresponding closely with the chloride concentration which half-saturates the transport site, about 65 mM (22).[2] The x intercept is significantly different ($p < 0.001$) from the value predicted for action at the modifier site, in which case an x intercept of -335 mM (4, 18) would have been expected.

On the basis of these data, H_2DIDS seems to act at the substrate site of the anion exchange system. The y intercept of the Hunter-Downs plot (Fig. 3), which indicates the concentration of inhibitor required to half-saturate the transport system when the chloride concentration approaches 0 (17, 18, 25, 28), is only 0.1 μM. H_2DIDS is, therefore, one of the most potent known inhibitors of the anion exchange system. DIDS, which competes with H_2DIDS (8, 9) and which probably binds to the same site, has a similarly high affinity (29).

Although external DIDS or H_2DIDS binds to the substrate site with high affinity, internal (cytoplasmic) DIDS or H_2DIDS has no effect on transport whatsoever (30). These results are similar to those obtained earlier by Passow and co-workers (11, 31) who used less potent stilbene derivatives. The data strongly suggest that the transport site is asymmetric in terms of its accessibility to the disulfonic stilbene inhibitors, such that these inhibitors can reach the site from the extracellular, but not from the cytoplasmic side of the membrane.

The photoreactive compound NAP-taurine is a bimodal probe par excellence. NAP-taurine is not reactive with proteins or lipids in the dark, but upon exposure to intense light it is converted to a highly reactive aryl nitrene (32). Thus, NAP-taurine can be used in the dark as a classical reversible inhibitor and then, after ex-

---

[2] These conclusions are not affected by differences in the affinities for chloride at the two sides of the membrane. All ping-pong models in which the half-saturation concentrations for chloride at the two sides of the membrane differ also predict that for an inhibitor which acts at the substrate site the x-intercept on the Hunter-Downs plot will equal the negative of the substrate site chloride dissociation constant (measured with $Cl_i = Cl_o$) (18, 50).

posure to light, the membrane components at or near its binding
site can be labeled and identified (33).

Unlike DIDS or H_2DIDS, NAP-taurine at 37°C appears to cross
the membrane, probably by utilizing the inorganic anion exchange
system (37). At 0°C, however, it crosses the membrane very slowly
(20); therefore its effects at the inside and outside of the membrane can be studied independently.

From its behavior as a transported substrate, NAP-taurine
would be expected to inhibit anion exchange by binding to the substrate site of the transport system. When NAP-taurine is present
at the cytoplasmic side of the membrane, the effects of chloride
concentration on inhibition bear out this prediction (Table I).
Increasing the chloride concentration causes a pronounced increase
in the apparent K_i. The inhibitory site is half-saturated with
chloride at only 36 mM (28). Because of the scatter in the data,
this is not significantly different ($p > 0.2$) from the chloride concentration required to half-saturate the substrate site (65 mM). It
is, however, significantly different ($p > 0.001$) from the value expected for action at the modifier site. Cytoplasmic NAP-taurine
is not a very effective inhibitor of chloride exchange, with a K_i
at 0 chloride concentration of 730 μM. The affinity of cytoplasmic
NAP-taurine as a substrate for the transport system is also rather

TABLE I. *Side-Dependent Effects of H_2DIDS and NAP-Taurine on*
Chloride Exchange

	NAP-taurine		H_2DIDS	
	Outside	Inside	Outside	Inside
K_i^a (0 mM Cl⁻)	20 μM	730 μM	0.1 μM	No effect
K_i^b (140 mM Cl⁻)	37 mM	3600 μM	0.3 μM	-
K_{Cl}	165 mM	36 mM	62 mM	-
Probable site	Modifier	Substrate	Substrate	No effect

[a] K_i *is the concentration of inhibitor required to 50% inhibit*
chloride exchange.
[b] K_{Cl} *is the concentration of chloride required to half-saturate*
the inhibitory site. K_i at 0 mM Cl⁻ was determined by extrapolation
from Hunter-Downs plots such as that shown in Fig. 3.

low, which is consistent with the concept that cytoplasmic NAP-taurine inhibits by acting at the substrate site (28).

Extracellular NAP-taurine acts in quite a different fashion (28). In the first place, it is a far more potent inhibitor, with a K_i at 0 chloride concentration of only 20 μM (Table I), making it over 35 times as potent an inhibitor as cytoplasmic NAP-taurine. In the second place, it seems to act at a site with a much lower chloride affinity, which is significantly different ($p < 0.001$) from that expected for competitive inhibition at the substrate site. Although more complex models for the action of external NAP-taurine, involving additional sites besides the substrate and modifier, cannot be rigorously excluded, the data are consistent with the concept that external NAP-taurine inhibits by binding to the modifier site (28).

Inasmuch as NAP-taurine seems to be transported by the anion exchange system, it is probable that extracellular NAP-taurine also binds to the substrate site with a low affinity similar to that seen for cytoplasmic NAP-taurine. The inhibitory effect of external NAP-taurine, however, is dominated by its much higher affinity binding to the modifier site. Like the substrate site, the modifier site seems to be asymmetric in terms of its accessibility to inhibitors, since it is affected strongly by external, but not by internal NAP-taurine. The modifier site is also asymmetric in that it is affected by external, but not internal Cl (34).

II. LABELING OF THE BAND-3 PROTEIN

If conditions are chosen so that the probes can react covalently, and if the probes are radioactively labeled, they can be used to label the membrane components which are located near the functional sites of the transport system. For example, external NAP-taurine can be used as a label for the modifier site after photoactivation, while external H_2DIDS can be used to label the substrate site.

While it is most probable that the labeling occurs near these functional sites, the size of the probes (35) and the possibility
that allosteric effects may be responsible for some of their apparently competitive interactions with substrate (36), makes it
possible that the labeled sites are actually located at some distance from the functional entities.

    After reaction with intact cells, both probes are largely
localized in the 93,000 dalton integral membrane protein known as
band 3 (6, 8, 9, 11, 16, 32, 37, 38). From work in several laboratories, this protein is known to traverse the membrane at least
once (17, 18, 39, 40, 41), and probably several times (19, 42).
At the extracellular side of the membrane (Fig. 4), there is a site
of cleavage (site 1) by chymotrypsin (40, 41, 43) or pronase (43,
44), which generates a 38,000 dalton^3 C-terminal fragment (C)

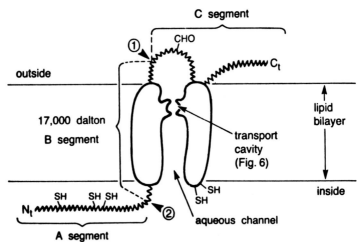

    FIG. 4. *Proposed arrangement of a single band 3 monomer in
the membrane. Sites of proteolytic cleavage at the outside and inside of the membrane are labeled 1 and 2 respectively. N_t and C_t
are the N and C-termini of the polypeptide chain; CHO is the site
of glycosylation; SH indicates reactive sulfhydryl groups. Band 3
structure taken from references 19, 42, 45, 46, 62, and 68; transport mechanism from references 16, 17, 18, and 55.*

    ^{3}*Molecular weights assigned to the various band 3 segments
differ among different laboratories. To avoid confusion molecular
weights will be taken as those determined by Steck (62).*

containing all of the carbohydrate of band 3 (45) and a 55,000
dalton nonglycosylated segment (A+B). Both of these fragments are
tightly associated with the membrane (41). Cleavage with chymotrypsin or trypsin at the inner surface (site 2) of the membrane
(41, 46) removes a water soluble (41) N-terminal (40, 45) ∿40,000
dalton segment (A). Treatment with chymotrypsin at both surfaces
(38, 41) results in formation of a 17,000 dalton segment (B) which,
since it is the product of proteolytic cleavage at both sides of
the membrane, must cross the membrane (18, 41). Recent evidence
(19, 42) indicates that the 38,000 dalton C-segment also traverses
the membrane.

When intact cells are labeled with NAP-taurine in the presence
of intense light, band 3 is heavily labeled, as shown in Fig. 5.
This labeling is decreased by increasing chloride concentration
(38) and, after pronase treatment of intact cells, becomes primarily localized in the A + B segment of band 3 (16, 37, 38). Labeling
of this segment is correlated with inhibition of transport, such
that the number of molecules per cell associated with 100% inhibition is about 9×10^5. This is similar to the number of band 3
monomers (12), which suggests that there is one NAP-taurine binding
site (and therefore one modifier site) per band 3 monomer.

If ghosts from cells labeled with extracellular NAP-taurine
are stripped of peripheral proteins by dilute alkali treatment,
and treated with chymotrypsin to cleave band 3 at sites 1 and 2
(Fig. 4), almost all of the label is found in the 17,000 dalton
transmembrane B-segment (Fig. 5B). This portion of band 3 would,
therefore, appear to contain part or all of the anion binding
structure which is recognized kinetically as the modifier site.

When intact cells are reacted with H_2DIDS, the label is again
primarily found in the 17,000 dalton B-segment (46), which suggests that this part of band 3 makes up at least part of the substrate site. This concept is supported by the finding that
vesicles from which most other proteins, including the A segment
of band 3, have been removed are nearly as effective in transport-

ing sulfate as are untreated ghosts (46). Since these vesicles may contain some of the 38,000 dalton C-segment of band 3, it is possible that this portion of band 3 also forms part of the substrate site. This is particularly likely because H_2DIDS is capable of reacting with the 38,000 dalton C-segment of band 3 (47) and crosslinking it to the 17,000 dalton B-segment.

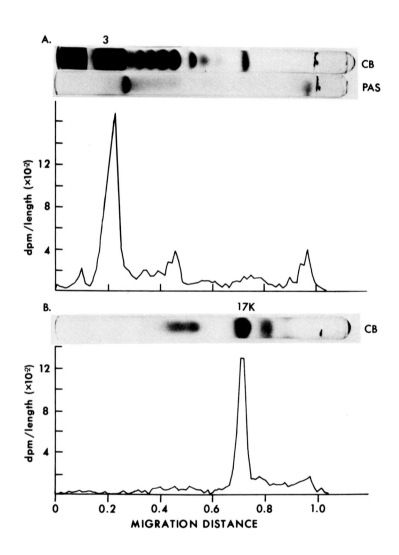

III. MECHANISM OF TRANSPORT

In addition to identifying functional sites, chemical probes may be used to investigate the mechanism of anion exchange. Two types of models have been suggested to account for the very tightly coupled one-for-one exchange of anions manifested by the red cell anion exchange system (17, 18, 48). According to the simultaneous model, anions must be bound to separate sites at opposite sides of the membrane before they are simultaneously exchanged. This model provides an appealingly direct explanation for the tight coupling of anion influx and efflux. According to the ping-pong model, there is only one substrate (anion) binding site, which is alternately exposed to the inside or outside. To account for the tight coupling of inward and outward flows of anions with this model, it is assumed that the unloaded carrier (without an anion bound to it) crosses the membrane very slowly, if at all (15, 17, 18, 21). Thus, once the transport site has carried an anion from outside to

FIG. 5. Distribution of external NAP-taurine labeled membrane proteins on 7.5% polyacrylamide gels. Intact red cells were labelled with 300 μM ^{35}S-NAP-taurine and were then washed with buffer containing 0.5% albumin to remove unreacted NAP-taurine. Ghosts were prepared from the cells, and some of the ghosts were extracted with dilute alkali and treated with chymotrypsin as described in reference 38. Samples of the ghost suspensions were solubilized in SDS and were electrophoresed on 7.5% polyacrylamide gels according to the method of Fairbanks et al. (12). To each gel was applied an amount of ghost suspension corresponding to about 3.7×10^8 cells, assuming complete recovery of the ghosts at all stages of treatment. Gels were slightly overloaded so that minor bands could be seen. Photographs show gels stained with either Coomassie brilliant blue (CB) (for protein) or Periodic acid-Schiff reagents (PAS) (for sialoglycoprotein). On the graphs are plotted the number of disintegrations per minute of ^{35}S divided by l/l_o, where l is the length of the gel slice and l_o is the distance from the top of the gel to the Pyronin-Y marker dye, against the distance from the top of the gel relative to that of the marker dye. A. Ghosts from intact labeled cells; B. Chymotrypsin treated, alkali-stripped ghosts (Adapted from reference 38, with permission.).

inside, and has released the anion, it cannot return to the outside unless it has an anion bound to it. The system thus retains
a memory for the direction of the last transport event in terms
of the orientation (inside or outside-facing) of the transport
site.

These models can readily be tested by using chemical probes,
particularly those such as internal NAP-taurine and external
H_2DIDS or DIDS, which bind to the transport site. According to
the ping-pong model, after reaction with extracellular DIDS, most
of the substrate sites will be tied up by DIDS at the external
membrane surface, and, therefore, fewer sites will be available
for reaction with internal NAP-taurine. Such a transmembrane
effect is not predicted by the simultaneous model with noninter-
acting sites.

Experimentally, Grinstein et al. (30) reacted intact cells
with DIDS, washed away the DIDS, made ghosts, prepared inside-out
vesicles, and then observed the effect of the DIDS-pretreatment on
the amount of NAP-taurine labeling at the cytoplasmic surface. As
is shown in Table II, DIDS pretreatment caused a substantial reduction in internal NAP-taurine binding, as predicted by the ping-
pong model. DIDS primarily affected the binding of NAP-taurine to

TABLE II. *Transmembrane Effects of DIDS Binding to the Outside on
NAP-Taurine Binding to the Inside Surface of the Membrane[a]*

Irradiation time (min)	Sites per cell × 10^{-5}		
	Control	DIDS	Difference
1	3.16	0.81	2.35
5	7.87	3.13	4.74
1	2.07	0.58	1.49
3	4.50	1.57	2.93
5	7.29	1.77	5.52

[a]*Intact cells were treated with 5 µM DIDS. Inside-out ghost
vesicles from these cells or from control cells were then exposed
to light for the indicated times in the presence of 230 µM NAP-
taurine.*

band 3, but effects on other proteins were also seen, possibly related to perturbation of the membrane structure by DIDS.

A potentially less disruptive method for producing a trans-membrane effect on the availability of the substrate site to a probe involves the use of the substrate chloride itself. For the ping-pong model, the ratio of outside-facing to inside-facing unloaded carriers (E_o/E_i) will be related to the Donnan ratio (Cl_i/Cl_o).[4] For the particular case of a carrier with equal affinities for chloride at the two sides of the membrane, E_o/E_i will equal Cl_i/Cl_o (17, 18, 49). If the chloride concentration inside the cell is reduced, a larger proportion of the carriers will face toward the inside. The effectiveness of an inhibitor such as H_2DIDS, which combines with the substrate site when it is facing outward, should therefore be reduced.

The internal chloride concentration can be reduced by resealing ghosts with various mixtures of chloride and citrate, an anion which interacts weakly with the transport system (34). When this is done, it is found that the concentration of H_2DIDS required to reversibly inhibit chloride exchange by 50% is increased when the internal chloride concentration is decreased (50, 51), in qualitative agreement with the ping-pong model. Recent studies demonstrate that when the chloride gradient is in the opposite direction $(Cl_i > Cl_o)$ the inhibitory potency of H_2DIDS is increased, providing further support for this model (51, 52, 53).

By using a similar method, but with external NAP-taurine as a probe, it is possible to see whether or not the accessibility of the modifier site to NAP-taurine is affected by the conformation of the substrate site. In fact, changes in the chloride gradient did have an effect on the ability of extracellular NAP-taurine to inhibit anion exchange (16, 28, 50, 52, 53). The results suggest

---

[4]*This is a consequence of the fact that the unidirectional influx and efflux of chloride are equal at equilibrium, and is completely unrelated to the associated membrane potential changes (17, 18).*

that the modifier site is accessible to NAP-taurine only when the substrate site is in the outward-facing conformation. Similar studies with a noncompetitive inhibitor, niflumic acid, demonstrate that even this inhibitor, which does not bind to either the substrate or modifier sites (69), can only interact with its binding site when the transport site is in the outward-facing conformation (54). Thus, there seem to be strong allosteric interactions between the transport site and the binding sites for various chemical probes.

These studies with probes lend considerable support to the ping-pong model, which is also favored by other kinetic evidence (17, 18, 48, 55). In addition, they demonstrate that the transport site can exist in different conformations, which can be distinguished by the probes. This concept is supported by Passow et al.'s recent demonstration (36, 56) that two different conformational states of the transport protein can be distinguished on the basis of the reactivity of a certain lysine residue with the covalent chemical modifier, fluorodinitrobenzene.

For a ping-pong transport model with two different conformations of the transport system, there would seem to be no a priori reason why the chloride dissociation constants for the transport site at the two sides of the membrane, K_{in} and K_{o}, should be equal. Similarly, the rate constant for the conformational change from inside-facing to outside-facing, designated k, need not be equal to the rate constant for the reverse conformational change, k'. With either sort of asymmetry, or a combination of the two, the ratio of E_{o}/E_{i} will not be equal to 1, even when $Cl_{i} = Cl_{o}$. The ratio of E_{o}/E_{i} with equal Cl at the two sides of the membrane can be defined as an asymmetry factor, A, which is given by the following equation (50, 52):

$$A = \left(\frac{E_{o}}{E_{i}}\right)_{Cl_{i}=Cl_{o}} = \frac{k \, K_{o}}{k' K_{in}} = \frac{K_{1/2}^{o,max}}{K_{1/2}^{i,max}}$$

As the equation indicates, A can be determined by measuring the apparent half-saturation concentration for chloride when it is varied at only one side of the membrane, keeping chloride high and constant at the other side. The ratio of the apparent half-saturation constant for the outside of the membrane to that for the inside gives a value for A (50).

Such measurements are available in the literature, but they give contradictory values for A. Data by Schnell et $al.$ (34) suggest an A value of about 2.5, whereas Gunn and Fröhlich (48) obtained results indicating an A value of about 0.06. From a quantitative analysis of the effects of chloride gradients on the inhibitory potencies of chemical probes such as H_2DIDS and NAP-taurine, it has been possible to obtain an independent estimate of A (52). The data strongly indicate that A < 1, thereby supporting the observations of Gunn and Fröhlich (48) which suggest that more than 15 times as many unloaded transport sites face the cytoplasm as face the outside medium. Recent experiments with niflumic acid (Knauf and Mann, unpublished observations) indicate that the loaded transport sites exhibit a similar asymmetry, which provides evidence that the asymmetry is primarily due to a difference in the rate constants, k and k', rather than to a difference in the chloride dissociation constants. Thus, the system seems to be intrinsically asymmetric, with as yet unknown structural features favoring the inside-facing over the outside-facing conformation. The protonated form of the transport site, which carries sulfate, appears to have similar characteristics, with more of the sites in the inside-facing form (57).

IV. MODEL FOR THE TRANSPORT SYSTEM

The chemical probe data, together with other information concerning the kinetics of anion transport and the structure of the band 3 protein, can be incorporated into a tentative model for the

transport system, as shown in Figs. 4 and 6. The substrate site
probably contains a positively charged guanidino group of arginine
(Fig. 6A). The characteristics of anion binding to a guanidino
group would fit the observed anion selectivity of the system (4,
18), and the high pK_a of arginine would explain the insensitivity
of the transport system to pH changes in the range from 7 to 11
(58). There also seems to be a second group which becomes charged
as the pH is decreased below 7. This may be an amino group with a
very low pK_a (due to the adjacent arginine). When it is charged,
it converts the monovalent anion exchange system to a form that
can transport divalent anions, such as sulfate (59). In addition

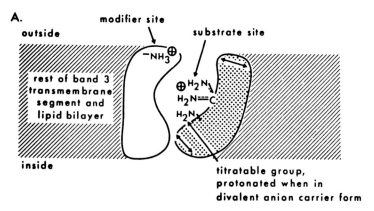

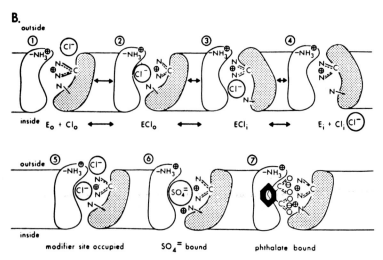

to the transport site, there may be an additional positively
charged group, perhaps an amino or guanidino group, near the outside of the transport cavity, which is the modifier site. This
positive charge could normally act to increase the local anion
concentration near the substrate site, but binding of anions at
high concentrations to this low affinity site may block the operation of the transport site.

These sites are assumed to be adjacent to a hydrophilic transport cavity, through which anions pass during transport. This cavity may be formed at the interface of two domains of the membrane-
associated portion of band 3, probably involving parts of the
17,000 dalton B-segment and 38,000 dalton C-segment. These domains

FIG. 6. A. Model for the transport site. The substrate
site probably contains a guanidino group of arginine, while the
titratable group which converts the monovalent anion carrier to
the divalent carrier is likely to be an amino group. The modifier
site is shown as a charged amino group, but could as well be a
guanidino group. The transport cavity is depicted as being formed
by the juxtaposition of two domains of a single band 3 monomer,
consisting of portions of the 17,000 dalton B-segment and the
38,000 dalton C-segment. For simplicity, the reorientation of the
domains which causes a change in the accessibility of the substrate
site is assumed to involve movement of only one of the domains
(shaded), but in fact both might move slightly. The possible location of the transport site in the band 3 protein is shown in
Fig. 4. B. Operation of the transport site. The site in the
outward facing conformation (1) can bind chloride (2) and then can
undergo a reorientation to the inward-facing form (3), after which
the chloride is released (4), and the cycle is repeated in the opposite direction. It is assumed that binding of the anion is necessary for the conformational change to occur. If the modifier
site is occupied by chloride (5), the conformational change cannot
take place, so transport is inhibited. The rate of the conformational reorientation is critically dependent on the nature of the
ion bound to the transport site, and is most rapid with chloride
or bicarbonate present. Sulfate can bind to the system (6) if the
amino group is titrated to form $-NH_3^+$, but sulfate does not fit the
system as well as chloride and so is transported more slowly.
The hydrophilic portion of amphiphilic substrates, such as
phthalate (7), can interact with the transport cavity, while the
hydrophobic part passes through the adjacent hydrophobic region
(in this case, above the plane of the page). (From reference 18,
with permission.)

probably consist of alpha-helical segments (60) of the same band
3 monomer, with a primary structure where all of the hydrophilic
residues are located on one side of each helix. Such a structure
would be consistent with the relatively high content of hydrophilic amino acid residues in these portions of the band 3 molecule (35, 61). Although band 3 exists in the membrane as a dimer
(17, 18), most of the evidence so far suggests that individual
monomers function independently in transport (18, 62).

From thermodynamic considerations, as well as from direct
evidence that band 3 is asymmetrically arranged in the membrane
with different segments facing inside and outside, it is unlikely
that the transport event involves rotation or flip-flop of band 3,
nor any other major reorientation which would involve the dis-
turbance of many noncovalent interactions. It would seem more
likely that the system functions as a lock-carrier (59, 63, 64,
65), in which the transport cavity remains stationary, while the
position of the diffusion barrier changes, as shown in Fig. 6B,
panels 1-4. According to this model, a subtle reorientation of
protein domains forming the transport cavity changes the transport
site from inside-facing to outside-facing or vice versa. This
change is spontaneous, but can only occur when an anion is present
in the cavity. Such a model would be completely consistent with
the carrier kinetic model of transport (59, 64).

When a chloride ion is bound to the modifier site (as in
panel 5 of Fig. 6B), the system is locked in one conformation and
transport is blocked. NAP-taurine, since it binds to the modifier
site only when the transport site is in the outward-facing form,
must lock the system in the outward-facing conformation. When the
transport site faces inward (as in panels 3 or 4) the modifier site
is blocked and cannot interact with NAP-taurine.

As shown in panel 6 of Fig. 6B, when the second group in the
transport cavity is charged, the system can transport divalent
anions such as sulfate. The system is also capable of transporting organic anions (66), even ones with large hydrophobic groups

such as phthalate (panel 7). This suggests that there is a hydrophobic region adjacent to the transport cavity through which these hydrophobic moieties can pass. Although large hydrophobic groups are permitted, the size of the polar portion of transported organic anions is severely limited. The intercharge spacing for divalent anions cannot exceed about 4 Å (18, 66). From this evidence, the dimensions of the transport cavity would appear to be small compared to the thickness of the bilayer. Anions, therefore, probably gain access to the actual transport cavity by diffusing through aqueous channels (Fig. 4) formed by the 17,000 dalton B-segment of band 3 (and probably also the 38,000 dalton C-segment). Free diffusion across the channel is, of course, blocked by the barrier adjacent to the transport cavity.

Although this model incorporates much of the available information concerning the transport system, the details of the transport mechanism are still largely speculative. When further information becomes available concerning the molecular nature of the transport and modifier sites, it should be possible to fill in some of the details and to evaluate the overall soundness of the model. As studies of other transport systems progress, it should also be possible to determine to what extent similar models may apply to other anion transport systems, as well as to other facilitated diffusion processes.

A final comment might be made on the possible teleological reason why the erythrocyte requires such a complex, tightly coupled system for anion exchange. There would seem to be no advantage to the red cell in having an anion exchange system which does not carry current, since there is no evidence that membrane potential regulation plays an important role in red cell function, as it does, for example, in muscle and nerve. The answer seems to lie in the requirement for a very high anion-cation selectivity. While anion exchange must take place in less than a second, it is important that cation fluxes be low, since the red cell regulates its volume by a system of cation pumps and leaks. If the leak

fluxes were higher, much more metabolic energy would have to be
expended to enable the pump to maintain a steady state. An
aqueous pore lined with positive fixed charges could only achieve
the high anion-cation selectivity of the red cell membrane if the
effective fixed charge concentration were over 100 M (67). In contrast the lock-carrier model can achieve a very high selectivity
without such an enormous positive charge concentration, not only
because of the anion selectivity of the binding site, but also because the conformational reorientation can occur only when an anion
is bound. The fact that only a small conformational change is required for transport permits a very large turnover number and thus
a very high anion flux. Such a mechanism would permit the red cell
to exchange anions with a half-time of only 50 msec (3), while
maintaining the permeability for potassium over a million times
smaller than the chloride permeability (as measured by isotope
exchange) (18).

V. SUMMARY

The red blood cell-anion exchange system contains at least
two anion binding sites. Anions bound to the substrate (transport)
site can be translocated across the membrane. When anions are
bound to a lower affinity site, the modifier site, transport is
inhibited. Under certain conditions, the chemical probes H_2DIDS
and NAP-taurine act as reversible inhibitors of anion (chloride)
exchange. From the effects of chloride on their inhibitory potency, it appears that extracellular H_2DIDS and cytoplasmic NAP-
taurine bind to the substrate site, while extracellular NAP-
taurine acts at the modifier site. When conditions are altered
so that the probes can react covalently with their binding sites,
both extracellular NAP-taurine and extracellular H_2DIDS selectively label a 17,000 dalton transmembrane segment of the 93,000 dalton protein known as band 3. This segment, therefore, appears to

form at least part of both the modifier and substrate sites.
These data and others suggest a model in which adjacent protein
domains in the transmembrane segments of band 3 form a transport
site. Its orientation (to the inside or outside) can be altered
by a spontaneous conformational change which occurs only when an
anion is bound to the site. The conformation of the substrate
site affects the accessibility of the modifier site. Anions gain
access to these sites through a hydrophilic channel formed by the
transmembrane band 3 segments.

ACKNOWLEDGMENTS

The authors wish to express their appreciation to the Medical
Research Council (Canada) and to the National Institutes of Health
for financial support. This paper has been assigned University of
Rochester Report No. UR-3490-2025.

REFERENCES

1. H. Nasse. Untersuchungen über den Austritt und Eintritt von
    Stoffen (Transsudation and Diffusion) durch die Wand der
    Haargefässe. *Arch. Gesamte Physiol. Menschen Tiere 16*:604-
    634, 1878.
2. R. B. Gunn (1979). Transport of anions across red cell mem-
    branes. *In* "Membrane Transport in Biology," (G. Giebisch
    et al., eds.), Vol. 2, pp. 59-79. Springer-Verlag, New York.
3. J. Brahm. Temperature-dependent changes of chloride trans-
    port kinetics in human red cells. *J. Gen. Physiol. 70*:283-
    306, 1977.
4. M. Dalmark. Effects of halides and bicarbonate on chloride
    transport in human red blood cells. *J. Gen. Physiol. 67*:223-

234, 1976.

5. Z. I. Cabantchik and A. Rothstein. The nature of the membrane sites controlling anion permeability of human red blood cells as determined by studies with disulfonic stilbene derivatives. *J. Membr. Biol. 10*:311-330, 1972.

6. Z. I. Cabantchik and A. Rothstein. Membrane proteins related to anion permeability of human red blood cells. I. Localization of disulfonic stilbene binding sites in proteins involved in permeation. *J. Membr. Biol. 15*:207-226, 1974.

7. M. Ho and G. Guidotti. A membrane protein from human erythrocytes involved in anion exchange. *J. Biol. Chem. 250*:675-683, 1975.

8. S. Lepke, H. Fasold, M. Pring, and H. Passow. A study of the relationship between inhibition of anion exchange and binding to the red blood cell membrane of 4,4'-diisothiocyanostilbene-2,2'-disulfonic acid (DIDS) and of its dihydro derivative (H_2DIDS). *J. Membr. Biol. 29*:147-177, 1976.

9. S. Ship, Y. Shami, W. Breuer, and A. Rothstein. Synthesis of tritiated 4,4'-diisothiocyano-2,2'-stilbene disulfonic acid (($^3H_2$)DIDS) and its covalent reaction with sites related to anion transport in red blood cells. *J. Membr. Biol. 33*:311-324, 1977.

10. H. Passow, H. Fasold, L. Zaki, B. Schuhmann, and S. Lepke (1975). Membrane proteins and anion exchange in human erythrocytes. *In* "Biomembranes: Structure and Function," (G. Gardos and I. Szasz, eds.). North-Holland Publications, Amsterdam, pp. 197-214.

11. L. Zaki, H. Fasold, B. Schuhmann, and H. Passow. Chemical modification of membrane proteins in relation to inhibition of anion exchange in human red blood cells. *J. Cell. Physiol. 86*:417-494, 1975.

12. G. Fairbanks, T. L. Steck, and D. F. H. Wallach. Electrophoretic analysis of the major polypeptides of the human erythrocyte membrane. *Biochemistry 10*:2606-2617, 1971.

13. M. J. Hunter. A quantitative estimate of non-exchange-restricted chloride permeability of the human red cell. *J. Physiol. (London)* 218:49P-50P, 1971.

14. M. J. Hunter. Human erythrocyte anion permeabilities measured under conditions of net charge transfer. *J. Physiol. (London)* 268:35-49, 1977.

15. P. A. Knauf, G. F. Fuhrmann, S. Rothstein, and A. Rothstein. The relationship between anion exchange and net anion flow across the human red blood cell membrane. *J. Gen. Physiol.* 69:363-386, 1977.

16. A. Rothstein, Z. Cabantchik, and P. Knauf. Mechanisms of anion transport in red blood cells: Role of membrane proteins. *Fed. Proc.* 35:3-10, 1976.

17. Z. I. Cabantchik, P. Knauf, and A. Rothstein. The anion transport system of the red blood cell: The role of membrane protein evaluated by the use of "probes." *Biochim. Biophys. Acta (Biomembr. Rev.)* 515:239-302, 1978.

18. P. A. Knauf. Erythrocyte anion exchange and the band 3 protein: Transport kinetics and molecular structure. *Curr. Top. Membr. Transp.* 12:249-363, 1979.

19. A. Rao. Disposition of the band 3 polypeptide in the human erythrocyte membrane. The reactive sulfhydryl groups. *J. Biol. Chem.* 254:3503-3511, 1979.

20. J. V. Staros, F. M. Richards, and B. E. Haley. Photochemical labeling of the cytoplasmic surface of the membranes of intact human erythrocytes. *J. Biol. Chem.* 250:8174-8178, 1975.

21. R. Gunn, M. Dalmark, D. Tosteson, and J. Wieth. Characteristics of chloride transport in human red blood cells. *J. Gen. Physiol.* 61:185-206, 1973.

22. P. C. Brazy and R. B. Gunn. Furosemide inhibition of chloride transport in human red blood cells. *J. Gen. Physiol.* 68:583-599, 1976.

23. M. Barzilay, S. Ship, and Z. I. Cabantchik. Anion transport in red blood cells. I. Chemical properties of anion recogni-

tion sites as revealed by structure-activity relationships of aromatic sulfonic acids. *Membr. Biochem.* 2:227–254, 1979.

24. E. T. Rakitzis, P. J. Gilligan, and J. F. Hoffman. Kinetic analysis of the inhibition of sulfate transport in human red blood cells by isothiocyanates. *J. Membr. Biol.* 41:101–115, 1978.

25. Y. Shami, A. Rothstein, and P. A. Knauf. Identification of the Cl⁻ transport site of human red blood cells by a kinetic analysis of the inhibitory effects of a chemical probe. *Biochim. Biophys. Acta* 508:357–363, 1978.

26. M. Dalmark. Chloride and water distribution in human red cells. *J. Physiol. (London)* 250:65–84, 1975.

27. J. L. Webb. *Enzyme and Metabolic Inhibitors,* Volume 1. New York: Academic Press, 1963.

28. P. A. Knauf, S. Ship, W. Breuer, L. McCulloch, and A. Rothstein. Asymmetry of the red cell anion exchange system: Different mechanisms of reversible inhibition by N(4-azido-2-nitrophenyl)-2-aminoethylsulfonate (NAP-taurine) at the inside and outside of the membrane. *J. Gen. Physiol.* 72: 607–630, 1978.

29. J. Funder, D. C. Tosteson, and J. O. Wieth. Effects of bicarbonate on lithium transport in human red cells. *J. Gen. Physiol.* 71:721–746, 1978.

30. S. Grinstein, L. McCulloch, and A. Rothstein. Transmembrane effects of irreversible inhibitors of anion transport in red blood cells: Evidence for mobile transport sites. *J. Gen. Physiol.* 73:493–514, 1979.

31. J. H. Kaplan, K. Scorah, H. Fasold, and H. Passow. Sidedness of the inhibitory action of disulfonic acids on chloride equilibrium exchange and net transport across the human erythrocyte membrane. *FEBS Lett.* 62:182–185, 1976.

32. J. V. Staros and F. M. Richards. Photochemical labeling of the surface proteins of human erythrocytes. *Biochemistry 13*: 2720–2726, 1974.

33. P. A. Knauf and A. Rothstein. Use of NAP-taurine as a photo-
     affinity probe for the human erythrocyte anion exchange sys-
     tem. *Ann. N.Y. Acad. Sci. 346*:212-230, 1980.

34. K. F. Schnell, E. Besl, and A. Manz. Asymmetry of the
     chloride transport system in human erythrocyte ghosts.
     Pflügers Arch. 375:87-95, 1978.

35. A. Rothstein, M. Ramjeesingh, S. Grinstein, and P. A. Knauf.
     Protein structure in relation to anion transport in red cells.
     Ann. N.Y. Acad. Sci. 341:433-442, 1980.

36. H. Passow, L. Kampmann, H. Fasold, M. Jennings, and S. Lepke.
     (1980). Mediation of anion transport across the red blood
     cell membrane by means of conformational changes of the band
     3 protein. *In* "Alfred Benzon Symposium No. 14," (H. H.
     Ussing *et al.*, eds.). Copenhagen, Munksgaard, pp. 345-367.

37. Z. Cabantchik, P. Knauf, T. Ostwald, H. Markus, L. Davidson,
     W. Breuer, and A. Rothstein. The interaction of an anionic
     photoreactive probe with the anion transport system of the
     human red blood cell. *Biochim. Biophys. Acta 455*:526-537,
     1976.

38. P. A. Knauf, W. Breuer, L. McCulloch, and A. Rothstein.
     N-(4-Azido-2-nitrophenyl)-2-aminoethylsulfonate (NAP-taurine)
     as a photoaffinity probe for identifying membrane components
     containing the modifier site of the human red blood cell
     anion exchange system. *J. Gen. Physiol. 72*:631-649, 1978.

39. M. S. Bretscher. A major protein which spans the human
     erythrocyte membrane. *J. Mol. Biol. 59*:351-357, 1971.

40. L. K. Drickamer. Fragmentation of the 95,000 dalton trans-
     membrane polypeptide in human erythrocyte membranes. Arrange-
     ment of the fragments in the lipid bilayer. *J. Biol. Chem.
     251*:5115-5123, 1976.

41. T. L. Steck, B. Ramos, and E. Strapazon. Proteolytic dissec-
     tion of band 3, the predominant transmembrane polypeptide of
     the human erythrocyte membrane. *Biochemistry 15*:1154-1161,
     1976.

42. D. G. Williams, R. E. Jenkins, and M. J. A. Tanner. Structure of the anion-transport protein of the human erythrocyte membrane. Further studies on the fragments produced by proteolytic digestion. *Biochem. J. 181*:477-493, 1979.

43. Z. I. Cabantchik and A. Rothstein. Membrane proteins related to anion permeability of human red blood cells. II. Effects of proteolytic enzymes on disulfonic stilbene sites of surface proteins. *J. Membr. Biol. 15*:227-248, 1974.

44. W. W. Bender, H. Garan, and H. C. Berg. Proteins of the human erythrocyte membrane as modified by pronase. *J. Mol. Biol. 58*:783-797, 1971.

45. K. Drickamer. Orientation of the band 3 polypeptide from human erythrocyte membranes. Identification of NH_2-terminal sequence and site of carbohydrate attachment. *J. Biol. Chem. 253*:7242-7248, 1978.

46. S. Grinstein, S. Ship, and A. Rothstein. Anion transport in relation to proteolytic dissection of band 3 protein. *Biochim. Biophys. Acta 507*:294-304, 1978.

47. M. L. Jennings and H. Passow. Anion transport across the red cell membrane, *in situ* proteolysis of band 3 protein, and crosslinking of proteolytic fragments by 4,4'-diisothiocyanodihydrostilbene-2,2'-disulfonate (H_2DIDS). *Biochim. Biophys. Acta 554*:498-519, 1979.

48. R. B. Gunn and O. Fröhlich. Asymmetry in the mechanism for anion exchange in human red blood cell membranes: Evidence for reciprocating sites that react with one transported anion at a time. *J. Gen. Physiol. 74*:351-374, 1979.

49. M. Dalmark. Chloride transport in human red cells. *J. Physiol. (London) 250*:39-64, 1975.

50. W. Furuya. Investigations into the anion exchange mechanism of human erythrocytes using chemical probes and chloride gradients. M.Sc. Thesis, University of Toronto, 1980.

51. W. Furuya and P. A. Knauf. Use of a probe to detect conformational changes in the human erythrocyte anion exchange sys-

tem. *Ped. Proc. 38*:1064, 1979.

52. P. A. Knauf, T. Tarshis, S. Grinstein, and W. Furuya (1980). Spontaneous and induced asymmetry of the human erythrocyte anion exchange system as detected by chemical probes. *In* "Alfred Benzon Symposium No. 14," (H. H. Ussing *et al.*, eds.). Munksgaard, Copenhagen, pp. 389-403.

53. P. A. Knauf, W. Furuya, and T. Tarshis. Asymmetry of the human erythrocyte anion exchange system as detected by chemical probes using chloride gradients. *Fed. Proc. 39*:1715, 1980.

54. P. A. Knauf, N. Mann, and F. Law. Niflumic acid senses the conformation of the transport site of the human red cell anion exchange system. *Biophys. J. 33*:49a, 1981.

55. R. B. Gunn and O. Fröhlich (1980). Asymmetry in the mechanism of anion exchange and evidence for reciprocating sites that react with one transported anion at a time. *In* "Alfred Benzon Symposium No. 14," (H. H. Ussing *et al.*, eds.). Munksgaard, Copenhagen, pp. 431-443.

56. H. Passow, H. Fasold, E. M. Gärtner, B. Legrum, W. Ruffing, and L. Zaki. Anion transport across the red blood cell membrane and the conformation of the protein in band 3. *Ann. N.Y. Acad. Sci. 341*:361-383, 1980.

57. M. L. Jennings (1980). Apparent "recruitment" of SO_4 transport sites by the Cl gradient across the human erythrocyte membrane. *In* "Alfred Benzon Symposium No. 14," (H. H. Ussing *et al.*, eds.). Munksgaard, Copenhagen, pp. 450-463.

58. J. Funder and J. Wieth. Chloride transport in human erythrocytes and ghosts: A quantitative comparison. *J. Physiol. (London) 262*:679-698, 1976.

59. R. B. Gunn (1978). Considerations of the titratable carrier model for sulfate transport in human red blood cells. *In* "Membrane Transport Processes," (J. F. Hoffman, ed.), pp. 61-77. Raven, New York.

60. G. Guidotti. The structure of intrinsic membrane proteins.

J. Supramol. Struct. 7:489-497, 1977.

61. A. Rothstein, M. Ramjeesingh, and S. Grinstein (1980). The arrangement of transport and inhibitory sites in band 3 protein. *In* "Alfred Benzon Symposium No. 14," (H. H. Ussing *et al.*, eds.), pp. 329-340. Munksgaard, Copenhagen.

62. T. L. Steck. The band 3 protein of the human red cell membrane; A review. *J. Supramol. Struct.* 8:311-324, 1978.

63. O. Jardetzky. Simple allosteric model for membrane pumps. *Nature (London) 211*:969-970, 1966.

64. C. S. Patlak. Contributions of the theory of active transport: II. The gate type noncarrier mechanism and generalizations concerning tracer flow, efficiency, and measurement of energy expenditure. *Bull. Math. Biophys. 19*:209-235, 1957.

65. S. J. Singer. Thermodynamics, the structure of integral membrane proteins, and transport. *J. Supramol. Struct. 6*: 313-323, 1977.

66. L. Aubert and R. Motais. Molecular features of organic anion permeability in ox red blood cell. *J. Physiol. (London) 246*: 159-179, 1975.

67. A. K. Solomon. Red cell membrane structure and ion transport. *J. Gen. Physiol. 43 (Suppl. 1)*:1-15, 1960.

68. A. Rao and R. A. F. Reithmeier. Reactive sulfhydryl groups of the band 3 polypeptide from human erythrocyte membranes. Location in the primary structure. *J. Biol. Chem. 254*:6144-6150, 1979.

69. J. L. Cousin and R. Motais. Inhibition of anion permeability by amphiphilic compounds in human red cell: Evidence for an interaction of niflumic acid with the band 3 protein. *J. Membr. Biol. 46*:125-153, 1979.

ANION AND PROTON TRANSPORT THROUGH LIPID

BILAYERS AND RED CELL MEMBRANES

John Gutknecht

Anne Walter

Duke University Marine Laboratory,

Beaufort, North Carolina

Coupled transports of cations and anions occur in a variety
of cell and epithelial membranes. Perhaps the best characterized
of these electroneutral, passive ion transport processes occurs
in the red-cell membrane (1). In 1972 Gunn (2) formulated a
"titratable carrier" model to explain the principal features of
anion transport in red cells, e.g., rapid nonconductive anion exchange, pH-dependent fluxes, low membrane conductance, and coupled
transports of anions and protons (3, 4, 5).

The similarities between anion exchange through red cell and
liquid ion exchanger membranes (5, 6) led us to model the red-cell
anion transport system using a lipid bilayer containing a syn-
thetic "titratable carrier," Amberlite LA-2 (7, 8). Amberlite
LA-2 (Rohm and Haas, Philadelphia, Pennsylvania) is a long-chain
secondary amine, n-lauryl(trialkylmethyl) amine. This molecule
is sufficiently hydrophobic so that it can be built into a lipid

bilayer membrane by simply adding it to the membrane forming solution.

We studied the properties of planer bilayers made from a solution of egg lecithin, Amberlite, and *n*-decane (mole ratios =
1:2:82). The membranes were formed on a polyethylene support by
the brush technique of Mueller and Rudin (9). The membrane chamber
was designed so that electrical properties and tracer fluxes could
be simultaneously measured (7). By measuring ionic transference
numbers, total membrane conductance, and one-way ion fluxes, we
were able to estimate both conductive (ionic) and nonconductive
(electroneutral) ion fluxes. Most of the halide fluxes were
measured with $^{82}Br^-$, which is available in much higher specific
activities than $^{36}Cl^-$. Net H^+ (or OH^-) fluxes were measured by
means of a small pH electrode on one side of the membrane (8).

I. IONIC EXCHANGE AND CONDUCTANCE THROUGH LECITHIN-AMBERLITE
   BILAYERS

Figure 1 summarizes the properties of lecithin-Amberlite-
decane bilayers. Details are given elsewhere (7, 8). Figure 1
shows the Amberlite-free base (C) reacting with H^+ and Br^- to form
a charged complex (CH^+) and a neutral complex (CHBr). These two
chemical reactions and three forms of Amberlite provide mechanisms
for at least four types of carrier-mediated ion transport, i.e.,
nonconductive Br^- exchange (path 1), nonconductive HBr net flux
(path 2), conductive Br^- net flux (path 3), and conductive H^+ net
flux (path 4).

The most conspicuous transport property of lecithin-Amberlite
bilayers is a rapid, nonconductive Br^- exchange via path 1. The
Br^- permeability of this electrically silent pathway is about
10^{-4} cm sec^{-1} at pH 6. This Br^- exchange flux is proportional to
$[H^+]$ over the pH range of 7 to 4 and the Br^- flux saturates at
pH < 4. The nonconductive exchange flux is always > 100 times

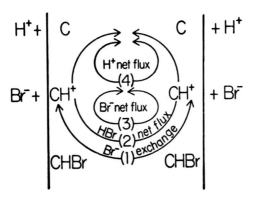

FIG. 1. Model for halide and proton transport through lipid bilayers containing a long-chain secondary amine, Amberlite LA-2, which functions as monovalent, titratable anion carrier. The amine mediates at least four types of transport, i.e., nonconductive halide exchange (path 1), nonconductive halide-proton cotransport (path 2), conductive halide transport (path 3), and conductive proton transport (path 4).

larger than the conductive Br^- flux, which is estimated from the membrane conductance and the transference number for Br^-. The Br^- exchange flux is proportional to $[Br^-]$ over a wide range of $[Br^-]$, and the concentration dependence indicates a 1:1 stoichiometry between Br^- and the transport mechanism (7).

Proton transport occurs by two different pathways, as shown in Fig. 1. First, a rapid, nonconductive flux of HBr is mediated by the neutral complex, CHBr (path 2). By means of a pH electrode, we observed a rapid net H^+ flux which is produced either by a Br^- gradient with symmetrical pH or by a pH gradient with symmetrical Br^-, but not by a pH gradient in halide-free solutions. The H^+ permeability of this electrically silent pathway is $>10^{-3}$ cm sec^{-1} and is partly rate-limited by diffusion through the aqueous unstirred layer (8).

The second H^+ pathway is conductive (path 4). Net H^+ fluxes through this path are too slow to measure by the pH electrode technique, but the pathway can be electrically studied. By measuring H^+ diffusion potentials and membrane conductances in halide-free solutions, we found the conductive H^+ permeability to be about 10^{-5} cm sec^{-1} (8).

We have not demonstrated a conductive Br^- flux via path 3. In fact, the observed Br^- conductance is higher than can be explained by the translocation of CH^+, which is the rate-limiting step in path 3. The current-voltage relations suggest the presence of two different charge carriers in the membrane, i.e., CH^+ and Br^-, and the CH^+ conductance (path 4) can only be observed in Br^--free solutions. The Br^- conductance may be due to an indirect (not carrier-mediated) effect of Amberlite. For example, in lecithin-Amberlite monolayers we observed a large pH-dependent increase in surface potential (i.e., surface charge plus dipole potential), which may increase the rate of Br^- "simple" diffusion through the lipid bilayer (not shown in Fig. 1). Further work is needed to define the mechanism of the Br^- conductive flux, which has a permeability of about 10^{-8} cm sec^{-1} at pH 6. (Although this Br^- permeability is much lower than the H^+ permeability in path 4, the Br^- conductance is higher because $[Br^-] \gg [H^+]$.)

In addition to the halide and proton fluxes, we measured Na^+, SO_4^{2-}, urea, and water fluxes through lecithin-Amberlite bilayers (Table I). The Na^+ and SO_4^{2-} permeabilities are several orders of magnitude smaller than the halide and proton permeabilities and are not affected by pH. Thus, lecithin-Amberlite bilayers are highly selective for monovalent anions and protons. Furthermore, there is no evidence for aqueous channels in lecithin-Amberlite bilayers. For example, the urea permeability (10^{-6} cm sec^{-1}) is similar in lecithin-Amberlite bilayers and pure lecithin bilayers (10) and is not affected by pH. The osmotic water permeability of lecithin-Amberlite bilayers is also similar to that of pure lecithin bilayers ($4-5\times10^{-3}$ cm sec^{-1} at 24°C).

TABLE I. Similarities between Lecithin-Amberlite Bilayers and Mammalian Red Cell Membranes.

Transport property	Lecithin-Amberlite bilayer (pH ∿ 6)[a]	Mammalian red cell membrane (pH ∿ 7)[a]	Reference for red cell data
Halide permeability of the exchange pathway (cm sec^{-1})	10^{-4}	10^{-4}	34
Halide permeability of the conductance pathway (cm sec^{-1})	10^{-8}	10^{-8}	18, 19, 20
Halide/sodium tracer permeability ratio (P_{halide}/P_{Na})	$>10^{4}$	10^{6}	5, 20
Halide/sulfate tracer permeability ratio (P_{halide}/P_{SO_4})	10^{4}	10^{4}	19
Proton/halide conductive permeability ratio (P_H/P_{halide})	10^{3}	10^{4}	19, 20
Total membrane conductance (mho cm^{-2})	10^{-6}	10^{-6}	20
Permselectity of conductance pathway (G_{halide}/G_{Na})	$>10^{1}$	10^{2}	5, 18, 19, 20
Coupled transport of anions and protons?	yes	yes	3, 4
Turnover number for anion carrier (sec^{-1})	$>10^{3}$	10^{4}	21
Urea permeability (cm sec^{-1})	10^{-6}	10^{-4}	35
Osmotic water permeability (cm sec^{-1})	10^{-3}	10^{-2}	12

[a]*The data on lecithin-Amberlite bilayers are from Gutknecht and Walter (7, 8 and unpublished). Because most of the listed parameters are highly pH and temperature dependent, this comparison is based only on orders of magnitude.*

II.   SIMILARITIES BETWEEN LECITHIN-AMBERLITE BILAYERS AND RED-CELL MEMBRANES

Table I shows some similarities between the permeability properties of lecithin-Amberlite bilayers and mammalian red-cell membranes. In both types of membranes the anion exchange fluxes are enormous and are >99% electrically silent (lines 1 and 2). In both membranes the halide/Na$^+$ and halide/SO$_4^{2-}$ tracer permeability ratios are very high (lines 3 and 4). In both membranes the proton/halide conductive permeability ratio is very high (line 5). In both membranes the total membrane conductance is very low (line 6) and the permselectivity of the conductance pathway (G_{halide}/G_{Na}) is lower than the halide/Na$^+$ tracer permeability ratio (lines 3 and 7). Furthermore, both membranes display coupled cotransports of anions and protons under appropriate conditions (line 8). Also, the apparent turnover number of Amberlite in bilayers may be of the same order of magnitude as that of the red-cell anion carrier (line 9). The urea permeability of mammalian red-cell membranes is about 100-fold higher than that of lecithin-Amberlite bilayers (line 10), which probably reflects the presence of a facilitated diffusion pathway in the mammalian red cell. Chicken red cells, which show no facilitated diffusion of urea, have a urea permeability coefficient of 10^{-6} cm sec^{-1} (11), similar to that of lecithin-Amberlite bilayers and pure egg-lecithin bilayers (7, 10). Finally, the osmotic water permeability of lecithin-Amberlite bilayers is somewhat lower than that of mammalian red cells, which apparently have aqueous membrane channels (12).

III.   IONIC EXCHANGE AND CONDUCTANCE THROUGH THE RED-CELL MEMBRANE

Figures 2 and 3 summarize schematically some features of Gunn's titratable carrier-model for anion and proton transport

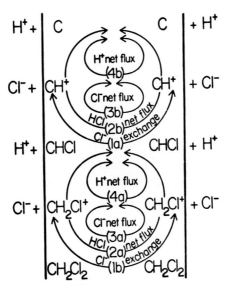

FIG. 2. *Titratable carrier model for halide and proton transport through the red cell membrane. The "divalent" carrier mediates the same four types of transport as shown in Fig. 1, but each transport can occur by two pathways. The numbers 1 - 4 indicate decreasing transport rates, and the letters a and b suggest primary and secondary pathways for each type of transport.*

through the red-cell membrane (2, 3, 4, 13). C indicates the "carrier" and CH^+ and CH^{2+} are the monovalent and divalent forms, that interact with monovalent and divalent anions. Figures 2 and 3 are presented to facilitate comparison between the simple Amberlite model and the more complicated red-cell model. The Amberlite (monovalent carrier) model provides one pathway for each of the four carrier-mediated transport processes. However, the red-cell (divalent carrier) model has two possible pathways for each type of transport, i.e., Cl^- exchange (1a and 1b), HCl net flux (2a and 2b) Cl^- net flux (3a and 3b) and H^+ net flux (4a and 4b). The numbers 1 - 4 indicate decreasing transport rates, and the letters a and b suggest the primary and secondary pathways. The upper pathway is characterized by an apparent $pK_a > 11$ (14) and a $K_{1/2}$ for Cl^- exchange of about 65 mM (15). The lower pathway has an apparent pK_a of 6.2 (14), but the Cl^- concentration dependences of this pathway is not known. An additional titratable site may become protonated at very low pH (1, 13) (not shown in Figs. 2 and 3). Finally, halides may also interact with a "modifier site," which has a considerably lower affinity for halides than the primary transport site (1, 16, 17).

   Nonconductive Anion Exchange (Pathways 1a and 1b). The most conspicuous red-cell transport process is the rapid nonconductive exchange which occurs via pathway 1a. At physiological Cl^- concentrations and neutral pH, the exchange flux through this pathway is about 10^{-8} mol $cm^{-2}sec^{-1}$, yielding a Cl^- permeability coefficient of about 10^{-4} cm sec^{-1} (Table I). The properties of this transport process have been studied and reviewed by numerous authors. Briefly, the process shows enormous electroneutral anion exchange rates, saturation kinetics, high activation energy, high selectivity for monovalent anions, and competitive and noncompetitive inhibition, especially by the specific anion transport inhibitors, SITS (4,4'-diisothiocyano-2,2'-stilbene-disulfonic acid), DIDS (4-acetamido, 4'-isothiocyano-2,2'-stilbene-disulfonic acid), etc. (1, 2, 13).

   According to our interpretation of the model, Cl^- exchange should occur via both pathways 1b (CH_2Cl_2) and 1a (CHCl). The existence of path 1b is strongly suggested by Jennings' (4) recent demonstration of HCl cotransport over the pH range 5.7 to 7.5. However, for unknown reasons, the rate of transport through path 1b is at least 1000 times slower than Cl^- exchange via path 1a (4). Thus, Cl^- exchange via path 1a is inhibited by low pH (2, 14). Chloride exchange is also inhibited by very high $[Cl^-]$, but this may be due primarily to noncompetitive interaction with a "modifier site" (1, 17).

   Nonconductive HCl Transport (Pathways 2a and 2b). Recent work by Jennings (4) strongly suggests the existence of a nonconductive HCl cotransport involving the low-pK_a anion transport site (Fig. 2, path 2a). Jennings equilibrated red-cells to dif-

ferent pH's over the range of 5.7 to 7.5 in CO_2-free solutions.
He then perturbed the equilibrium by changing either external pH
or external $[Cl^-]$, and then measured the rate of return to pH
equilibrium by means of an external pH electrode. The pH dependence
of this equilibration process was compatible with a mechanism involving HCl cotransport, but not Cl^--OH^- exchange. The HCl cotransport was (i) largely inhibited by DIDS and phloretin, (ii) not sensitive (<10%) to membrane voltage, (iii) highly temperature dependent (like Cl^- exchange), and (iv) at least 1000 times slower than
the maximum rate of Cl^- exchange. For technical reasons, Jennings
did not conduct similar experiments at high pH. Thus, the existence
of HCl cotransport via path 2b is hypothetical. The available data
show only that the rate of nonconductive H^+ (or OH^-) transport is
20 times slower at pH 7.5 than at pH 5.8 (4).

   Conductive Cl^- Transport (Pathways 3a and 3b). The "net" or
conductive Cl^- permeability of the red cell is very low, about
10^{-8} cm sec^{-1} (18, 19). However, the conductive Na^+ or K^+ permeability is even lower, about 10^{-10} cm sec^{-1} (18, 19, 20). Thus,
the conductive net flux pathway for Cl^- can only be studied in the
presence of agents such as valinomycin which greatly increase the
cation permeability. The partial inhibition of the Cl^- conductive
flux by DIDS, SITS, and other inhibitors of Cl^- exchange suggests
that both processes involve a common element (1, 19). Furthermore,
the activation energies for both processes are about 16 - 22 kcal/
mol over the range of 15° - 37°C (18, 21).
   The conductive anion flux has been described as a "slippage"
in the electroneutral exchange mechanism (19, 22). According to
the "slippage" hypothesis, the exchange carrier normally moves
back and forth via path 1a. However, the carrier occasionally
crosses in the CH^+ form, which results in a net Cl^- flux (path 3b).
Note that an analogous conductive Cl^- flux is associated with the
low pK site (path 3a). In fact, the twofold increase in the Cl^-
conductance as pH falls from 7.1 to 6.8 (19) is the result we expect for a titratable carrier with a pK of 6.2. Recently, Hunter

(18) noted a quantitative similarity between the abilities of halides to inhibit Cl^- exchange and increase halide conductance $(I^- > Br^- > Cl^-)$. This result, along with the pH dependence of the conductive flux, suggests that path 3a is the primary conductive pathway at neutral pH. However, to clearly distinguish these two pathways, we need additional data on the halide conductive flux as a function of different halide ions and their concentrations over a wide range of pH.

One difference between the properties of the Cl^- exchange and Cl^- conductive pathways is that Cl^- exchange is inhibited by phlorizin and phloretin (23, 24, 25), whereas the Cl^- net conductive flux is either stimulated (24) or not affected (23) by phlorizin. Phloretin and related compounds are known to increase cation and decrease anion conductance by altering the surface (dipole) potential (26). If the rate limiting step in Cl^- conductance is the translocation of a "cation," i.e., CH^+ or CH_2Cl^+ (paths 3a or 3b), then a stimulation of Cl^- conductance by phlorizin is possible. Since Cl^- exchange via path la is normally 10^4 times faster than the Cl^- conductance via path 3a or 3b, an inhibitory effect of phlorizin on Cl^- conductance should not occur until the translocation of CHCl becomes slower than the translocation of CH^+ or CH_2Cl^+.

Since the inhibition of the Cl^- conductive flux by DIDS is incomplete (19), we must consider the possibility that a fraction of the Cl^- conductive flux occurs by "simple" diffusion through membrane lipids or a parallel polar pathway (not shown in Figs. 2 and 3). In this regard, we note that the conductive P_{Cl^-} of 10^{-8} cm sec^{-1} is roughly similar to the values of P_{Cl^-} in many animal cells, as well as some lipid bilayer membranes (7, 19). In red cells only the Cl^- exchange flux, not the Cl^- conductance, is extraordinarily high.

Conductive H^+ Fluxes (Paths 4a and 4b). Although the titratable carrier model predicts the possibility of conductive H^+ fluxes (paths 4a and 4b), these are the slowest and most difficult fluxes to characterize. Jennings (4) noted that a small fraction ($< 10\%$)

of the net H^+ flux was sensitive to the membrane potential.   Knauf
et al. (19) observed a conductive H^+ (or OH^-) flux which was less
than 10% of the conductive Cl^- flux and was also inhibited by
DIDS.   These small fluxes might occur via either paths 4a or 4b,
and the rate-limiting steps would presumably be the translocation
of either CH^+ or CH_2Cl^+.   Since halide conductance appears to be
controlled primarily by the low-pK site, we have designated path
4a as the primary pathway for H^+ conductance.   Paths 4a and 4b
might be distinguished experimentally by the different pH dependencies and different $[Cl^-]$ dependencies of the H^+ conductances.
The experiments will be difficult, however, due to the very low
H^+ conductance and artifacts caused by traces of CO_2, HCO_3^-, and
any other weak acid or weak base which can cross the cell membrane, e.g., lactate and phosphate (4).

   Chloride-Sulfate Exchange.   An elegant test of the titratable
carrier model was performed by Jennings (3).   He equilibrated red
cells with respect to Cl^- and pH, and then placed the cells in a
CO_2-free, high-SO_4^{2-} medium and measured the external pH change
associated with net $Cl^- - SO_4^{2-}$ exchange (countertransport).   The results are shown schematically in Fig. 3.   As predicted by Gunn's
titratable carrier model, one H^+ is taken up for each SO_4^{2-} and Cl^-
exchanged.   Apparently the divalent carrier (CH_2^{2+}) transports SO_4^{2-},
and the monovalent carrier (CH^+) transports Cl^-, and thus an H^+
must be cotransported inwardly with each SO_4^{2-} ion.

   H^+ or OH^- Transport?   The idea that OH^- ions participate in
pH equilibration through the red-cell membrane has been generally
accepted because of the high anion permselectivity of the membrane
(4, 19, 20).   However, the distinction between OH^- and H^+ transport
is difficult because the net effects of the two processes are identical.   Thus, most of the evidence for OH^- transport is equally
compatible with H^+ transport.

   Jennings (4) recently studied the pH dependence of H^+ (or OH^-)
transport through the red-cell membrane, and his results strongly
suggest a mechanism of HCl cotransport rather than $Cl^- - OH^-$ exchange.

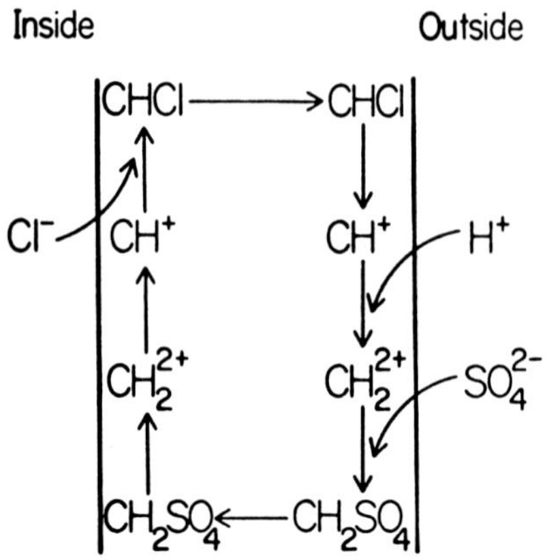

Inside   Outside

$$CHCl \longrightarrow CHCl$$

$$Cl^- \nearrow CH^+ \qquad CH^+ \diagup H^+$$

$$CH_2^{2+} \qquad CH_2^{2+} \diagup SO_4^{2-}$$

$$CH_2SO_4 \longleftarrow CH_2SO_4$$

FIG. 3. *Titratable carrier model for halide-sulfate exchange and sulfate-proton cotransport through the red-cell membrane. For clarity, only the loading steps and net flux directions are shown. [Modified from Jennings (3).]*

However, he studied H^+ transport only over the pH range 5.7 - 7.5, i.e., the low-pK pathway (path 2a). Thus, the question of HCl cotransport versus $Cl^- - OH^-$ exchange through the high-pK pathway remains open.

Knauf *et al.* (19) estimated the conductive OH^- or H^+ permeability to be about 10^{-3} cm sec^{-1} at pH 7, which is $>10^4$ times larger than the conductive P_{Cl^-}. If OH^- is the transported species then the OH^-/Cl^- selectivity ratio must be extraordinarily high (see Wright and Diamond, ref. 27). The possibility of OH^- transport via the high-pK pathway seems even less likely due to the fact that 1 mM OH^- does not inhibit Cl^- exchange (14). On the other hand, a conductive P_{H^+}/P_{Cl^-} ratio of about 10^3 and a conductive P_{H^+}/P_{Na^+} of $>10^4$ are seen in lecithin-Amberlite bilayers (Table I). Thus, a protonated amine can provide the necessary selectivity for this type of transport process.

The only obvious way to distinguish phenomenologically between OH^- and H^+ transport is to measure the pH dependence of the transport process. However, even this simple test can be misleading.

For example, to distinguish between an HCl cotransport (path 2b) and Cl^--OH^- exchange (path 1a), one might measure the pH dependence of the pH equilibration process in a CO_2-free system, i.e., the Jennings (4) experiment. If the pK_a of the anion binding site is ~ 11, then Cl^--OH^- exchange via path 1a should increase with increasing pH over the range of 7 to 9. However, HCl cotransport via path 2b might show a similar dependence on pH, because the rate-limiting step may be the translocation of C, and $[C]$ is also proportional to $[OH^-]$. Thus, HCl cotransport can show a paradoxical (inverse) pH dependence when the pH is much lower than the pK_a of the carrier.

The strongly basic nature and high ionization enthalpy of the high-pK site suggests that the site may be a quaternary ammonium or a guanidino group (14). If the site is a quaternary ammonium, then the initial reactions in the model should be written: $C^+ + Cl^- \rightleftarrows CCl$, and $C^+ + OH^- \rightleftarrows COH$, where C^+ has a formal positive charge. If the high-pK site is a quaternary ammonium, then the question of H^+ versus OH^- transport is answered, because only OH^- participates in the reaction. However, the probability that the high-pK site is a quaternary ammonium seems remote in view of the key role of the Band 3 protein in anion transport. Thus, the primary anion transport site is more likely to be the guanidino group or an amino group with a high pK.

IV. CONCLUSION

The phenomena of anion-proton cotransport and rapid, nonconductive anion exchange both provide strong evidence for a carrier-mediated transport process (as opposed to a simple channel mechanism) in lecithin-Amberlite bilayers and red-cell membranes. The slow apparent turnover numbers for anion transport through lecithin-Amberlite bilayers and red-cell membranes are also consistent with the notion of a carrier-mediated process. However, the carrier mechanisms must differ greatly between these two

types of membranes. Amberlite probably functions as a classical
mobile carrier which moves back and forth through the membrane in
a manner similar to that of valinomycin and various other ionic
carriers (7, 28). The red-cell anion carrier, in contrast, apparently functions as a "lock-carrier" in which only a small part
of the carrier undergoes a conformational change during anion transport (13, 29). The motion accompanying the conformational change
and the ion translocation resulting from this motion are the two
principal notions in the word "carrier."

The key role of the Band 3 protein in red-cell anion-proton
transport is widely recognized and under intensive investigation
(1). However, the analogy with liquid anion exchange membranes has
provided useful insights about the transport process (2, 5). The
existence of a hydrophobic barrier in the transport pathway is suggested by the predominant transport of electroneutral ion pairs
(2, 4, 5) and the failure of small polar molecules such as water
and urea to utilize the pathway (11). The large decrease in activation energy for $Cl^--HCO_3^-$ exchange between low ($0°-10°$) and high
($30°-37°$) temperatures (30) resembles that observed for water
transport through lipid bilayers below and above the lipid phase
transition temperature (31). Other evidence for the role that
lipids may play in red-cell anion exchange is presented by Snow
et al. (32), Cabantchik *et al.* (1), and Obaid and Crandall (33).

In conclusion, titratable carrier models can explain most of
the data on halide and proton transport through lecithin-Amberlite
bilayers and red-cell membranes. The titratable carrier model can
also account for the major features of divalent anion transport in
red cells (1, 13). We have tried to rationalize all the available
red-cell data in terms of the titratable carrier model, utilizing
especially Jennings' recent work on $Cl^--SO_4^{2-}$ exchange and HCl cotransport. Finally, we have suggested some additional tests of the
model which may lead to clearer distinctions among possible transport pathways and, hopefully, to insights about the molecular
events involved in anion and proton transport.

SUMMARY

The "titratable carrier" model explains many features of anion and proton transport through red cell membranes, e.g., high anion selectivity, pH-dependent fluxes, high anion exchange rates, low membrane conductance, and cotransport of anions and protons. All these features are also displayed by lipid bilayers made from egg lecithin and a long-chain secondary amine, n-lauryl(trialkyl-methyl)amine, which behaves as a monovalent, titratable anion carrier. Both types of carriers permit at least four types of passive ion fluxes: nonconductive anion exchange, nonconductive HCl transport, conductive Cl^- transport, and conductive H^+ transport. In the red cell, each type of transport may occur by two different pathways, because the carrier is "divalent" and exists in at least six different forms, e.g., C, CH^+, CHCl, CH_2^{2+}, CH_2Cl^+, and CH_2Cl_2. The existence of CH_2Cl_2 is implied by the net electroneutral transport of HCl over the pH range 5.7 - 7.5. Depending on the driving force, net Cl^- transport occurs by either conductive or nonconductive pathways, whereas Cl^- exchange occurs primarily via CHCl. The postulated high-pK site (C, CH^+), could conceivably be a quaternary amine (COH, C^+) which would provide a mechanism for Cl^--OH^- exchange. However, the limited data suggest that the high-pK site is either a guanidino or amino group (CH^+), which could facilitate either HCl cotransport or Cl^--OH^- exchange, depending on the Cl^-/OH^- selectivity of the carrier and on the relative abilities of the free base (C) and CH^+ to cross the membrane.

ACKNOWLEDGMENTS

For valuable comments on the manuscript we thank Drs. M. L. Jennings and P. A. Knauf. This work was supported by NIH grants HL 12157, ES 02289 and GM 28844.

REFERENCES

1. Z. I. Cabantchik, P. A. Knauf, and A. Rothstein. The anion
    transport system of the red blood cell: The role of membrane
    protein evaluated by the use of "probes." *Biochim. Biophys.*
    Acta 515:239-302, 1978.

2. R. B. Gunn. A titratable carrier model for both mono- and
    divalent anion transport in human red blood cells. *In*
    "Oxygen Affinity of Hemoglobin and Red Cell Acid-Base Status,"
    (M. Rørth and P. Astrup, eds.), pp. 823-827. Munksgaard,
    Copenhagen, 1972.

3. M. L. Jennings. Proton fluxes associated with erythrocyte
    membrane anion exchange. *J. Membr. Biol. 28*:187-205, 1976.

4. M. L. Jennings. Characteristics of CO_2-independent pH equi-
    libration in human red blood cells. *J. Membr. Biol. 40*:365-
    391, 1978.

5. J. O. Wieth. The selective ionic permeability of the red
    cell membrane. *In* "Oxygen Affinity of Hemoglobin and Red
    Cell Acid-Base Status," (M. Rørth and P. Astrup, eds.), pp.
    265-278. Munksgaard, Copenhagen, 1972.

6. G. M. Shean and K. Sollner. Carrier mechanisms in the move-
    ment of ions across porous and liquid ion exchanger membranes.
    Ann. N.Y. Acad. Sci. 137:759-776, 1966.

7. J. Gutknecht, J. S. Graves, and D. C. Tosteson. Electrically
    silent anion transport through lipid bilayer membranes con-
    taining a long-chain secondary amine. *J. Gen. Physiol. 71*:
    269-284, 1978.

8. J. Gutknecht and A. Walter. Coupled transport of protons and
    anions through lipid bilayer membranes containing a long-chain
    secondary amine. *J. Membr. Biol. 47*:59-75, 1979.

9. P. Mueller and D. O. Rudin. Translocators in bimolecular
    lipid membranes: Their role in dissipative and conservative
    bioenergy transductions. *In* "Current Topics in Bioenergetics,"
    (O. R. Sanadi, ed.), Vol. 3, pp. 157-249. Academic Press,

New York, 1969.

10. A. Finkelstein. Water and nonelectrolyte permeability of lipid bilayer membranes. *J. Gen. Physiol.* *68*:127-135, 1976.

11. J. Brahm and J. O. Wieth. Separate pathways for urea and water, and for chloride in chicken erythrocytes. *J. Physiol.* *266*:727-749, 1977.

12. R. I. Sha'afi, G. T. Rich, G. T. Sidel, V. W. Bossert, and A. K. Solomon. The effect of the unstirred layer on human red cell water permeability. *J. Gen. Physiol.* *50*:1377-1399, 1967.

13. R. B. Gunn. Considerations of the titratable carrier model for sulfate transport in human red blood cells. *In* "Membrane Transport Processes," (J. F. Hoffman, ed.), pp. 62-77. Raven Press, New York, 1978.

14. J. Funder and J. O. Wieth. Chloride transport in human erythrocytes and ghosts: A quantitative comparison. *J. Physiol.* *262*:679-698, 1976.

15. P. Brazy and R. B. Gunn. Furosemide inhibition of chloride transport in human red blood cells. *J. Gen. Physiol.* *68*: 583-599, 1976.

16. M. Dalmark. Effects of halides and bicarbonate on chloride transport in human red blood cells. *J. Gen. Physiol.* *67*: 223-234, 1976.

17. P. A. Knauf, S. Ship, W. Breuer, L. McCulloch, and A. Rothstein. Asymmetry of the red cell anion exchange system. *J. Gen. Physiol.* *72*:607-630, 1978.

18. M. J. Hunter. Human erythrocyte anion permeabilities measured under conditions of net charge transfer. *J. Physiol.* *268*:35-49, 1977.

19. P. A. Knauf, G. F. Fuhrmann, S. Rothstein, and A. Rothstein. The relationship between anion exchange and net anion flow across the human red blood cell membrane. *J. Gen. Physiol.* *69*:363-386, 1977.

20. D. C. Tosteson, R. B. Gunn, and J. O. Wieth. Chloride and

hydroxyl ion conductance of sheep red cell membranes. *In* "Erythrocytes, Thrombocytes, Leukocytes," (E. Gerlach, K. Moser, and E. Deutch, eds.), pp. 62-66. Georg Thieme Verlag, Stuttgart, 1973.

21. J. Brahm. Temperature-dependent changes in chloride transport kinetics in human red cells. *J. Gen. Physiol.* *70*:283-749, 1977.

22. U. V. Lassen, L. Pape, and B. Vestergaard-Bogind. Chloride conductance of the *Amphiuma* red cell membrane. *J. Membr. Biol.* *39*:27-48, 1978.

23. S. Gerhardt and K. F. Schnell. The inhibition of the anion net flux across the human erythrocyte membrane. *Pflugers Arch.* *355*:R74-R89, 1975.

24. J. Kaplan and H. Passow. Effects of phlorizin on net chloride movements across the valinomycin treated erythrocyte membrane. *J. Membr. Biol.* *19*:179-194, 1974.

25. J. O. Wieth, M. Dalmark, R. B. Gunn, and D. C. Tosteson. The transfer of monovalent inorganic anions through the red cell membranes. *In* "Erythrocytes, Thrombocytes, Leukocytes," (E. Gerlach, K. Moser, E. Deutsch, and W. Wilmanns, eds.), pp. 71-76. Georg Thieme Verlag, Stuttgart, 1973.

26. O. S. Andersen, A. Finkelstein, I. Katz, and A. Cass. Effect of phloretin on the permeability of thin lipid membranes. *J. Gen. Physiol.* *67*:749-771, 1976.

27. E. M. Wright and J. M. Diamond. Anion selectivity in biological systems. *Physiol. Rev.* *57*:109-156, 1977.

28. P. Lauger. Carrier-mediated ion transport. *Science 178*:24-30, 1972.

29. C. S. Patlak. Contributions to the theory of active transport. *Bull. Math. Biophys.* *18*:271-315, 1956.

30. E. I. Chow, E. D. Crandall, and R. E. Forster. Kinetics of bicarbonate-chloride exchange across the human red blood cell membrane. *J. Gen. Physiol.* *68*:633-652, 1976.

31. M. C. Blok, L. L. M. van Deenen, and J. De Gier. Effect of

the gel to liquid crystalline phase transition on the osmotic behavior of phosphatidylcholine liposomes. *Biochim. Biophys. Acta 433*:1-12, 1976.

32. J. W. Snow, J. F. Brandts, and P. S. Low. The effects of anion transport inhibitors on structural transitions in erythrocyte membranes. *Biochim. Biophys. Acta 512*:579-591, 1978.

33. A. L. Obaid and E. D. Crandall. $HCO_3^- - Cl^-$ exchange across the human erythrocyte membrane: Effects of pH and temperature. *J. Membr. Biol. 50*:23-41, 1979.

34. D. C. Tosteson. Halide transport in red blood cells. *Acta Physiol. Scand. 46*:19-41, 1959.

35. R. I. Sha'afi, C. M. Gary-Bobo, and A. K. Solomon. Permeability of red cell membranes to small hydrophilic and lipophilic solutes. *J. Gen. Physiol. 58*:238-258, 1971.

CHLORIDE TRANSPORT IN THE RENAL TUBULE

Maurice B. Burg

Laboratory of Kidney and Electrolyte Metabolism
National Heart, Lung, and Blood Institute
National Institutes of Health
Bethesda, Maryland

The renal tubule epithelium consists of a number of different types of cells arranged in series (Fig. 1). Each epithelial segment has its own characteristic cellular ultrastructure and specialized function. The purpose of this review is to summarize present knowledge of chloride transport in the different segments.

The proximal tubule consists of three segments which were originally classified according to their cellular ultrastructure (2). The segments are designated S1, S2, and S3, and are known to differ in function as well as structure. In addition, proximal tubules of superficial nephrons differ from those of deep nephrons in function (3), but not in cellular ultrastructure (2).

S1 proximal tubule is the initial part of the proximal convoluted tubule. The principal salt that normally is reabsorbed in this segment is sodium bicarbonate, not sodium chloride. Since there is relatively little chloride reabsorption, chloride concen-

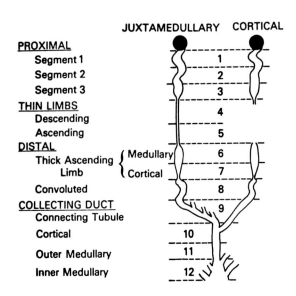

FIG. 1. *Diagram of nephron segments in rabbit kidneys* [*from* (1)].

tration rises in the tubule fluid as sodium bicarbonate and water are reabsorbed. Bicarbonate concentration falls, reciprocal to the increase in chloride.

S2 proximal tubule (Fig. 2) is the last part of the proximal convoluted tubule in all nephrons and includes the initial part of the proximal straight tubule in superficial cortical nephrons (2). A major part of the filtered chloride is reabsorbed in this segment. The nature of the chloride reabsorption depends on two factors: (1) the change in tubule fluid composition caused by selective transport in S1 (which affects salt and water reabsorption in the last part of S1 as well as in S2); and (2) the intrinsic function of S2 which differs from that of S1 (1). Since the observed transepithelial electrochemical difference is sufficient to drive its transport, chloride reabsorption is believed to be entirely passive in mammalian proximal tubules, including S2 (4 - 7). The passive chloride reabsorption occurs both by diffusion and by solvent drag. The high chloride concentration in the

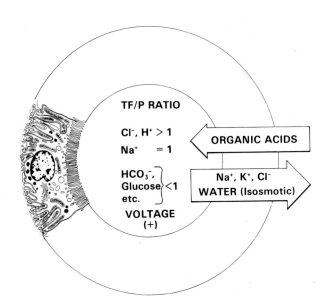

FIG. 2. Cellular ultrastructure and function of S2 proximal tubule. TF/P = tubule fluid/plasma concentration ratio [from (1)].

lumen provides the principal driving force for reabsorption of chloride by diffusion. A relatively small driving force suffices since chloride permeability is high (on the order of 11×10^{-5} cm/ sec in rats (7) and $6 - 7 \times 10^{-5}$ cm/sec in rabbits [Table I (8, 9)]. The chloride and bicarbonate concentration differences between the tubular and peritubular fluid also provide a driving force for re- absorption of fluid by osmosis (10). The reflection coefficient of bicarbonate across proximal tubules exceeds that of chloride. Therefore, despite equal osmolality of the tubular and peritubular fluid, the peritubular fluid has a higher effective osmotic pres- sure, because of its higher bicarbonate content. Reabsorption of fluid by osmosis results. The water reabsorbed by osmosis carries with it by solvent drag, solutes of low reflection coefficient, including chloride.

Although chloride reabsorption by mammalian proximal tubules apparently is passive, active chloride transport has been identi-

fied in *Necturus* tubules (11). In the *Necturus*, chloride is transported into the tubule cells from the lumen, linked to the entry
of sodium in an electrically neutral fashion. The primary active
transport process is extrusion of sodium from the cells at the
peritubular border, resulting in low intracellular sodium activity.
Sodium enters the cells from the lumen passively because of the
higher electrochemical activity of sodium in the lumen. Chloride,
on the other hand, enters the cells from the lumen against its
electrochemical activity difference, which indicates active transport. This is secondary active transport since it is linked to
dissipative sodium transport, rather than linked directly to a
source of metabolic energy. As a result of its secondary active
transport at the lumen border, chloride activity is maintained
high within the cells, and chloride diffuses passively out of the
cells across their peritubular border, completing its reabsorption.
In addition there is also completely passive reabsorption of chloride in *Necturus* proximal tubules, driven by the transepithelial
voltage, which is negative in the lumen. The passive chloride reabsorption passes between the cells through the tight junctions
and lateral intercellular spaces.

S3 proximal tubule is the medullary part of the mammalian
proximal straight tubule. Its individual function is not well
known since it generally has not been studied separately from S2.

Thin descending limbs of Henle's loop of rabbits were studied
directly by perfusion *in vitro*. When the perfusate and bath were
the same, there was neither a transepithelial voltage nor net
transport of salt or water, and thus there was no evidence of
active transport (12). The permeability to sodium chloride was
low and to water, was high. In the concentrating kidney, the
sodium chloride concentration in the tubule fluid at the bend of
Henle's loop is higher than that in the general circulation. The
high sodium chloride concentration results from equilibration of
the tubule fluid with the interstitium of the renal medulla which
is maintained at a high concentration by the renal countercurrent

system. Judging from the high water permeability and low sodium chloride permeability of rabbit thin descending limbs, equilibration in that species occurs by osmosis of water with little net sodium chloride movement. On the other hand, in rodents (such as rats, *Psammonys*, and hamsters), it was inferred that there is entry of sodium chloride into the thin descending limbs, as well as reabsorption of water (13). The sodium chloride presumably enters the thin descending limbs of the rodents by diffusion from the high concentration in the inner medulla.

Thin ascending limbs of Henle's loop in concentrating kidneys reabsorb chloride. In isolated perfused tubules there was no evidence of active chloride transport (14). Since the water permeability is low, and there is little net water absorption, chloride reabsorption results in a decrease in its concentration in the tubule fluid. The chloride transport is believed to be passive, driven by the concentration difference for chloride (14). Chloride was observed to be higher in the tubule fluid than in the contiguous capillaries (15). The permeability to chloride was observed to be extremely high in this segment [approximately 100×10^{-5} cm/sec (16)]. There is kinetic evidence that the high chloride permeability in thin ascending limbs results from facilitated transport. Thus, the chloride flux ratio (lumen to bath/bath to lumen) was significantly less than the value predicted theoretically for simple passive diffusion (16). Also, the chloride flux was a saturating function of chloride concentration, and was significantly inhibited by bromide (Fig. 3).

Cortical and medullary thick ascending limbs of Henle's loop (Fig. 4) reabsorb chloride by active transport (17). The active chloride transport results in a transepithelial voltage oriented positive in the tubule lumen. Since these segments, like the thin ascending limb, are virtually impermeable to water, the reabsorption of chloride results in a further decrease in its concentration in the tubule fluid. A number of diuretic drugs, such as furosemide, inhibit the active chloride transport. The mechanism of

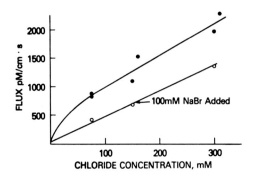

FIG. 3. Kinetics of chloride flux across isolated perfused
rabbit thin ascending limbs of Henle's loop. ^{36}Cl flux was
measured from lumen to bath at different concentrations of NaCl.
Note that chloride flux was a saturating function of concentration
and that addition of sodium bromide inhibited the chloride flux.
The high chloride permeability in this segment is believed to be
due to a facilitated passive transport system. [From the data in
(16).]

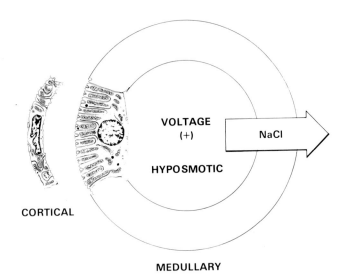

FIG. 4. Cellular ultrastructure and function of the thick
ascending limb of Henle's loop [from (1)].

the active chloride transport has not been resolved. The chloride

transport is dependent on sodium, and potassium-activated adeno-

sine triphosphatase and was observed to be inhibited under conditions known to inhibit that transport protein [i.e., removal of sodium or potassium or addition of ouabain (Fig. 5)]. Because of these findings, it was proposed that the active chloride transport is in fact secondary to primary active sodium transport, as is the case in the *Necturus* proximal tubule, but direct evidence for this mechanism in the thick ascending limb is lacking.

The cortical thick ascending limb differs from the medullary part in its cellular ultrastructure (2) and in the hormonal sensitivity of its adenylate cyclase (e.g., parathyroid hormone stimulates the enzyme in the cortical but not in the medullary segment). There are also differences in sodium chloride transport. The rate of sodium chloride absorption was much greater in the medullary than in the cortical portion when both segments were perfused at rapid rates with a solution containing a relatively high sodium chloride concentration. At slow perfusion rates, however, the

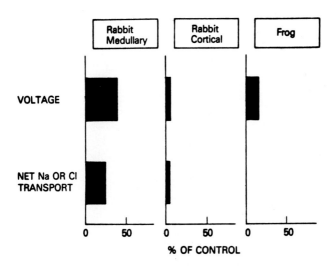

FIG. 5. *Ouabain inhibited NaCl absorption and the voltage across rabbit cortical and medullary thick ascending limbs and frog diluting segments. (The diluting segment in the frog is analogous to the thick ascending limb in the rabbit.) This result is significant since active chloride and passive sodium transport are present in these segments, yet ouabain is believed to be a specific inhibitor of active sodium (not active chloride) transport. The explanation remains speculative (see text)* [from (17)].

cortical portion reduced the sodium chloride concentration in the tubule fluid to a much lower level than did the medullary part (17).

Both segments of the thick ascending limb have a high permeability to salt. Thus, the electrical resistance of the cortical portion was low (25 Ω cm^2) and chloride permeability was high (2.5 × 10^{-5} cm/sec). The permability to sodium exceeded that to chloride in both the medullary and cortical portion. Because of these permeability characteristics, a dilution potential developed when the sodium chloride concentration was lowered in the lumen by reabsorption of salt. This dilution potential added to the transport potential caused by the active chloride transport.

The distal convoluted tubule (Fig. 6) has been principally studied by micropuncture in rats. It is defined anatomically by its specific cellular ultrastructure. The segment defined in this manner extends over only a part of the "distal tubule" available on the kidney surface for micropuncture. The remainder of the "distal tubule" actually consists of connecting tubule and cortical collecting duct. Since the type of cells at the site of micropuncture was not identified in most studies of "distal tubules," it is not feasible to ascribe the observed function to a particular cell type. Keeping this reservation in mind, the results of the micropuncture studies can be summarized as follows (18). The transepithelial voltage across "distal tubules" was oriented negative in the lumen, and increased from low values near the macula densa to approximately -40 mV in the "late distal tubule." The negative voltage is a consequence of active sodium transport and provides a driving force for passive chloride absorption (4). There may also be some active chloride absorption (4). When sulfate-containing solutions were infused into rats, the chloride concentration fell to levels in the "distal tubules" that were too low to be accounted for on the basis of the observed voltage. Also, the rate of chloride reabsorption was greater than that calculated for passive absorption, considering the chloride conductance and the voltage.

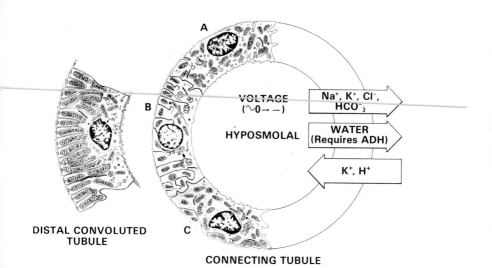

FIG. 6. Cellular ultrastructure and function of "distal tubules." The distal convoluted tubule, as strictly defined by its
cellular ultrastructure, comprises only a part of the so called
"distal tubule" that can be studied by micropuncture, since it is
accessible at the surface of the kidney. In addition to the distal
convoluted tubule, the "distal tubule" includes a short post-macula
densa segment of cortical thick ascending limb, the connecting tubule, and the initial part of the cortical collecting duct (1).
The functions outlined in the diagram were principally deduced from
the micropuncture studies. Since cellular ultrastructure generally
was not identified in the micropuncture studies, any given function
cannot be assigned with certainty to the distal convoluted tubule
or any of the other parts of the "distal tubule" [from (1)].

The collecting duct segments which comprise the remainder of
the nephron will not be considered here since they are discussed
in another part of this symposium.

SUMMARY

Permeability to chloride and the extent of its transport
varies greatly between the different nephron segments. These differences are outlined and discussed in conjunction with current
theories of the mechanisms involved.

TABLE I. *Permeability (P) of rabbit proximal tubules.*

	P_{Cl} (cm/s×10^5)		P_{Na}/P_{Cl}	
	Superficial	Juxtamedullary	Superficial	Juxtamedullary
S1			1.6	2.0
S2	5.6 – 7.3		0.3 – 0.6	2.0
S3		2.1		1.3 – 2.0

(From 8 + 9)

REFERENCES

1. M. B. Burg. The renal handling of sodium chloride, water
    amino acids, and glucose. *In* "The Kidney," (B. Brenner and
    F. Rector, eds.), 1980. W. B. Saunders, Co., Philadelphia,
    Pa. Ch. 7.
2. B. Kaissling and W. Kriz. Structural analysis of the rabbit
    kidney. *In* "Advances in Anatomy, Embryology, and Cell
    Biology," 1978. Springer-Verlag, New York.
3. H. R. Jacobson. Characteristics of volume reabsorption in
    rabbit superficial and juxtamedullary convoluted tubules.
    J. Clin. Invest. *63*:410-418, 1979.
4. G. Giebisch and E. Windhager. Electrolyte transport across
    renal tubular membranes. *In* "Handbook of Physiology, Section
    8. "Renal Physiology," pp. 315-376. American Physiol. Soc.
    1973, Washington, D.C.
5. J. A. Schafer, S. L. Troutman, M. L. Watkins, and T. E.
    Andreoli. Volume absorption in the pars recta. I. "Simple"
    active Na^+ transport. *Am. J. Physiol.* *234*:F332-F339, 1978.
6. M. Sohtell. Electrochemical forces for chloride transport in
    the proximal tubules of the rat kidney. *Acta Physiol. Scand.*
    103:363-369, 1978.
7. E. Fromter, G. Rumrich, and K. J. Ullrich. Phenomenologic
    description of Na^+, Cl^-, and HCO_3^- absorption from proximal
    tubules of the rat kidney. *Pflugers Arch.* *343*:189-220, 1973.
8. J. Schafer, S. Troutman, and T. Andreoli. Volume reabsorp-
    tion, transepithelial potential differences, and ionic per-
    meability properties in mammalian superficial proximal tu-
    bules. *J. Gen. Physiol.* *64*:582-607, 1974.
9. S. Kawamura, M. Imai, D. Seldin, and J. Kokko. Characteris-
    tics of salt and water transport in superficial and juxta-
    medullary straight segments of proximal tubules. *J. Clin.*
    Invest. *55*:1269-1278, 1975.
10. K. H. Neumann and F. C. Rector, Jr. Mechanism of NaCl and

water reabsorption in the proximal convoluted tubule of rat kidney. *J. Clin. Invest. 58*:1110-1118, 1976.

11. K. R. Spring and G. Kimura. Chloride reabsorption by renal proximal tubules of *Necturus*. *J. Membr. Biol. 38*:233-254, 1978.

12. J. P. Kokko. Sodium chloride and water transport in the descending limb of Henle. *J. Clin. Invest. 49*:1838-1846, 1970.

13. C. deRouffignac. Physiological role of loop of Henle in urinary concentration. *Kidney Internat. 2*:297-303, 1972.

14. M. Imai and J. Kokko. Sodium, urea, and water transport in the thin ascending limb of Henle. Generation of osmotic gradients by passive diffusion of solutes. *J. Clin. Invest. 53*:393-402, 1974.

15. D. R. Gelbart, C. A. Battilana, J. Bhattacharya, F. B. Lacy, and R. L. Jamison. Transepithelial gradient and fractional delivery of chloride in thin loop of Henle. *Am. J. Physiol. 235*:F192-F198, 1978.

16. M. Imai and J. P. Kokko. Mechanism of sodium and chloride transport in the thin ascending limb of Henle. *J. Clin. Invest. 58*:1054-1060, 1976.

17. M. B. Burg and J. E. Bourdeau. Function of the thick ascending limb of Henle's loop. *In* "New Aspects of Renal Function," (H. G. Vogel and K. J. Ullrich, eds.), pp. 91-102, Excerpta Medica, Amsterdam-Oxford, 1978.

18. E. E. Windhager and L. S. Constanzo. Transport functions of the distal convoluted tubule. *In* "Physiology of Membrane Disorders," (T. E. Andreoli, J. F. Hoffman, and D. D. Faniste eds.), pp. 681-706, Plenum Medical Book Co., New York and London, 1978.

MECHANISM OF CHLORIDE TRANSPORT

ACROSS MAMMALIAN COLLECTING DUCT

Juha P. Kokko

Department of Internal Medicine
University of Texas Health Science Center
Southwestern Medical School
Dallas, Texas

I. INTRODUCTION

In vivo micropuncture and *in vitro* microperfusion studies are
the two principle experimental techniques which have been utilized
to delineate the characteristics of chloride transport across the
mammalian collecting duct epithelium. These techniques are
schematically represented in Fig. 1 and show that the *in vivo*
micropuncture studies have examined chloride transport across the
papillary collecting duct (the papillary collecting duct is the
only segment of the collecting duct accessible to conventional
micropuncture techniques), while the *in vitro* microperfusion
studies have examined most extensively the characteristics of
chloride transport across the cortical segment of the collecting
duct (technically it is quite difficult to dissect long segments
of the papillary collecting duct). Thus it is evident that the

CHLORIDE TRANSPORT
IN BIOLOGICAL MEMBRANES

123

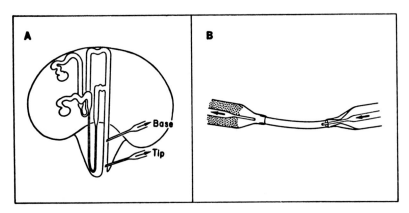

FIG. 1. *Schematics of the two principle techniques by which
to examine chloride transport across the collecting duct epithelium. Panel A demonstrates the in vivo micropuncture technique while panel B represents the in vitro microperfusion technique.*

direct *in vivo* micropuncture and *in vitro* microperfusion techniques
have not examined chloride transport in the same segment of the
collecting duct.   Also, the *in vitro* microperfusion techniques have
been limited to tubules obtained from the rabbit and the human
while the micropuncture techniques have examined chloride transport
across the papillary collecting ducts of rats and hamsters.   Therefore it is evident that the two techniques might not necessarily
yield information which is identical, however, it is interesting
to note that the conclusions reached from these two approaches are
quite complimentary and have not raised any major controversies.
This manuscript is divided in such a way that the *in vivo* micropuncture studies will be discussed first followed by a consideration of the *in vitro* microperfusion studies.

II.   MICROPUNCTURE STUDIES

It is well appreciated that the urinary chloride concentration is highly dependent on the physiological status of the animal.

Similarly, micropuncture studies have noted that the collecting
duct chloride concentration is quite variable and dependent on the
volume and dietary status of the animal (1 - 4). Also, it is of
importance to note that recent studies have demonstrated that fractional chloride reabsorption increases along the papillary collecting duct if the delivery of chloride is increased to this segment
by volume expansion (7). However, conclusions regarding the
mechanism of chloride reabsorption cannot be made without the
knowledge of the electrochemical driving forces that exist across
the collecting duct.

The measurement of the transepithelial potential difference
across the papillary collecting duct is associated with numerous
inherent uncertainties. First, it is not possible to be certain
of the magnitude of the liquid-junction potential which exists
between the exploring and reference electrodes since the concentration of the papillary solutes is highly variable. This source
of error can be minimized by filling the electrodes with 3 M KCl.
Second, it is not possible to control the degree of shunting of
the potential difference from the duct of Bellini. Third, it is
difficult to be certain whether shunting of the PD (potential
difference) occurs around the impaling pipette. Despite these
well recognized problems, the measured transepithelial PDs using
microelectrodes filled with 3 M KCl have reported values between
0 and -10 mV during hydropenia (1, 5) and values up to -34.2
(lumen negative) during sulfate diuresis (5). Potential difference values in our laboratory in hydropenic mutant Wistar rats
have been -2.5 ± 0.3 mV using 3 M KCl electrodes with OD of 6 μm
(15). Laurence and Marsh (5) and Diezi et al. (1) have also
measured simultaneously the chloride concentration of the collecting duct fluid and compared it to the adjacent vasa recta. Using
values derived from the measurement, they have calculated the magnitude of the equilibrium potential difference using the Nernst
equation and have concluded that it is likely that chloride is actively transported out of the collecting duct. However, these

authors also recognize the potential sources of error in the PD
measurements. Our own potential difference measurements and urinary chloride determinations are consistent with these earlier observations and support their conclusions.

Whether the papillary collecting duct regulates the urinary
chloride excretion is an unanswered question. Prostaglandins,
compounds which inhibit cortical collecting duct (6, 7) and thick
ascending limb of Henle salt reabsorption (6, 8), do not influence
chloride reabsorption across the papillary collecting duct (4).
Likewise it has been shown recently by micropuncture of the last
accessible superficial distal tubule and comparison to final urinary excretion, that chloride reabsorption in the collecting duct
is a function of load per se (2). It has also been shown by direct
collecting duct micropuncture studies that chloride reabsorption
increases with increasing delivery rates (3). This would suggest
that though the papillary collecting duct does participate in reabsorption of chloride, the main determinant of increased urinary
chloride excretion is the delivery of chloride to the base of the
papillary collecting duct. This can occur as a consequence of increased chloride delivery from the juxtamedullary nephron (3) or
inhibition of chloride reabsorption in any of the more proximal
segments.

III. MICROPERFUSION STUDIES

The technique whereby isolated segments of collecting ducts
are perfused *in vitro* is more suited for direct examination of the
mechanism of chloride transport than are the *in vivo* micropuncture
studies. The major advantages include: (i) direct control of
luminal and bathing constituents; (ii) accessibility of all segments of collecting ducts for study; and (iii) more exactness in
the measurement of transepithelial PD since complete electrical
seals are easy to accomplish and liquid junction PDs can be

measured or calculated.

The magnitude and polarity of the PD across the cortical collecting tubule of the rabbit is highly dependent on the mineralocorticoid state of the rabbit from which the tubule is dissected. Tubules obtained from rabbits with high serum mineralocorticoid levels have a negative PD while tubules obtained from rabbits with suppressed mineralocorticoids have a PD near zero or slightly positive (9, 10, 11). It is of interest that although the magnitude of the mean PD correlates nicely with the mineralocorticoid status of the rabbit, there nevertheless exists more intertubular baseline PD variation than in any other segment similarly studied. The origin of the variability has not been identified; however, it is not from technical artifacts since the same investigators do not see this much variation with other segments. Nevertheless, once the PD in a given tubule has been established, it remains reasonably constant (± several millivolts) for many hours. Current evidence is most consistent with the view that the lumen negative potential is secondary to lumen-to-blood sodium transport. However, the origin of the lumen positive PD has been more conjectural.

The origin of the lumen positive PD can be examined by various electrolyte substitution studies. Tubules with positive PD can be obtained by eliminating the negative PD caused by sodium transport: by addition of $10^{-5}M$ ouabain to the bath; or by replacement of sodium by choline in both the bath and perfusate; and/or by obtaining tubules from rabbits with suppressed mineralocorticoid status. If the chloride in the perfusate and bath is replaced by methyl sulfate then the previously recorded PD of +5.5 ± 1.5 mV was observed to decrease to +0.8 ± 1.4 mV, P < 0.0025 (12). Stoner *et al.* (13) have perfused cortical collecting tubules with solutions in which chloride was replaced by sulfate and found that the lumen positive PD did not decrease. We have repeated their protocol exactly and noted that when the perfusate and bath NaCl was replaced by $NaSO_4$ in the presence of luminal amiloride

the control PD fell from 7.4 ± 1.9 mV to -0.3 ± 0.2 mV (12). Since
the perfusate and bath contained all of the other major ionic species, and since the PD is not different from zero in the absence of
Cl and in the absence of Na transport, we therefore feel that Cl
ion is centrally important in the generation of a lumen-positive
potential.

The dependence of the positive PD on the Cl ion does not
necessarily mean that chloride is actively transported. To examine
this issue further we have conducted various isotopic, electrophysiological, and chemical studies.

The isotopic permeability of the rabbit cortical collecting
tubule is moderately high $[2-5 \times 10^{-5} \text{cm/sec}$ (13, 14)]. However,
Stoner *et al.* (13) have shown that the measured total conductance
of the CCT (cortical collecting tubule) is significantly lower
than would be calculated from the chloride flux. Furthermore,
their estimates of the maximum electrical conductance attributable
to chloride was only 15% of the partial chloride flux calculated
from their isotopic flux data (13). They, therefore, suggested
the possibility of the existence of an electrically silent exchange
diffusion process. The presence of exchange diffusion would sugges
some type of physiological significance. Since exchange diffusion
is not responsible for net chloride flux, it is attractive to postulate that this process is not specific only for the chloride to
chloride exchange but might reflect chloride to bicarbonate or some
other exchange. These possibilities have not been examined directly, but we have extended their suggestions testing for the possibility of chloride-for-chloride exchange diffusion by measuring
the isotopic chloride flux in the presence and absence of cold
chloride. When the bath chloride was replaced by methyl sulfate,
the apparent chloride lumen-to-bath permeability decreased from
2.41 ± 0.5 to 0.69 ± 0.08 $\times 10^{-5}$ cm/sec (14). Thus, these findings
confirmed the existence of a chloride-for-chloride exchange diffusion process and show that when exchange diffusion is eliminated
the isotopic permeability to chloride across the cortical collectir

tubule is quite low. This conclusion was further verified by the
observation that intraluminally applied current of sufficient magnitude to alter the transtubular PD by −35 ± 4 mV or +28 ± 2 mV,
did not significantly influence the isotopic flux of chloride (14).

Net chloride efflux from the cortical collecting tubule has
been examined isotopically and chemically in tubules obtained from
rabbits with suppressed mineralocorticoid status and rabbits which
had been treated with deoxycorticosterone, 5 mg/day, i.m. for 1
week prior to experimentation. The average flux ratio (lumen-to-
bath/bath-to-lumen) was 0.99 ± 0.04 across tubules obtained from
non-DOCA rabbits and 1.28 ± 0.09 in tubules from DOCA-treated rabbits (14). Though these numbers are not significantly different
from one, the DOCA-treated tubules, however, do have a higher permeability ratio than non-DOCA tubules (P < 0.01) (14). Thus, these
values suggest that DOCA treatment is responsible for stimulating
chloride efflux, but are not conclusive since the majority of each
unidirectional flux is composed of exchange diffusion. It is for
these reasons that we designed a series of experiments in which
net chloride movement was measured in the absence of antidiuretic
hormone.

The results of three representative experiments are depicted
in Fig. 2. In each case the bath is a solution with major electrolyte concentrations similar to serum and to which has been added
5% vol/vol fetal calf serum. The top tubule is from a mineralocorticoid suppressed rabbit. It was perfused with solution similar
to the bath except it did not contain fetal calf serum. This study
demonstrates that as perfusion rate is decreased the chloride concentration of the collected fluid remains unchanged. Six similar
experiments have shown that there is no change in either the chloride (−2.3 ± 1.4) or osmolal (−1 ± 2.0) concentration when tubules
are obtained from mineralocorticoid suppressed rabbits (14). On
the other hand the middle experiment in Fig. 2 was obtained from a
DOCA-treated rabbit. It shows that as the perfusion rate is decreased, there is a progressive decrease in the chloride concentra-

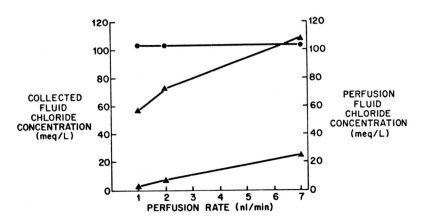

Fig. 2. *The results of collected tubular chloride measure-*
ment in three representative experiments (4) using cortical collect-
ing tubules obtained from corticoid suppressed (●) and DOCA treated
(▲) rabbits.

tion of the collected fluid. When six such tubules with lengths

ranging from 1.8 to 3.0 mm were perfused at rates of 2.0 to 3.5

ml/min, there was an average decrease in the chloride concentration

of −22 ± 5.5 mEq/L and osmolality of −32 ± 5.1 mOsm/L (14). Thus

the results demonstrate that pretreatment of the rabbit with miner-

alocorticoid stimulates net chloride efflux from the cortical col-

lecting tubule. The bottom experiment was designed to test whether

the ability to decrease the chloride concentration is capacity- or

gradient-limited. In this experiment the perfusion fluid initially

had a chloride concentration of 24 mEq/L. It can be readily ap-

preciated that the chloride concentration decreased to 1 mEq/L as

the perfusion rate was decreased to 1 nl/min. Thus it appears that

the transtubular chloride gradient is not a limiting factor in

transporting chloride out of the collecting duct.

The final group of studies was designed to compare the

measured chloride ion concentration distribution across cortical

collecting tubules obtained from DOCA-treated animals with the calculated distribution based on the measured potential and assuming
passive equilibration. The perfusion rate in these experiments
was 2-4 nl/min. The bath again was fluid simulating plasma with
5% vol/vol fetal calf serum. The mean perfusate chloride concentration was 7.5 mEq/L while the collected fluid chloride averaged
4.10 mEq/L. The mean observed PD was -63.6 mV (12). Using Nernst
equilibrium equation analysis, it can be shown that chloride transport occurs against an electrochemical gradient.

In summary, evidence has been put forth from both *in vivo*
micropuncture studies and *in vitro* microperfusion experiments that
both the cortical and papillary collecting tubules are able to reabsorb significant quantities of chloride. It is also concluded
that the chloride conductance of the membrane is quite low with
the apparent high isotopic permeability being principally due to an
exchange diffusion processes. Further, it was shown that under
certain circumstances, the data are most consistent with an active
processes for chloride efflux and that these active transport processes are in part regulated by mineralocorticoids. Since the collecting duct can modulate net sodium as well as chloride transport,
it then becomes apparent that the collecting duct participates in
volume homeostasis by having the ability to regulate net reabsorption of salt.

SUMMARY

A number of recent *in vivo* micropuncture and *in vitro* microperfusion studies have been designed to examine the nature of
chloride transport across the mammalian collecting duct epithelium.
These studies reveal that both the cortical and papillary collecting duct are able to reabsorb net quantities of chloride. Isotopic
chloride flux measurements indicate that the major quantity of unidirectional chloride flux is secondary to an exchange diffusion

process. When the exchange diffusion is eliminated, the isotopic
and electrophysiological measurements indicate that the cortical
collecting duct epithelium has a low partial ion conductance to
chloride. Electrophysiological ion substitution experiments
demonstrate that the net PD is a sum of a negative PD secondary
to active sodium transport and a positive PD which is dependent
on the presence of luminal chloride. Further electrophysiological
studies are consistent with the view that chloride is actively
transported out of the cortical and papillary collecting duct, and
that this active transport process, in part, can be modulated by
the mineralocorticoid status of the animal.

ACKNOWLEDGMENTS

    This work was supported in part by National Institutes of
Arthritis, Metabolism and Digestive Diseases Research Grant
1 RO1 AM14677 and National Institute of Arthritis, Metabolism,
and Digestive Diseases Training Grant 5 T32 AM07257.

REFERENCES

1. J. Diezi, P. Michoud, J. Aceves, and G. Giebisch. Micro-
    puncture study of electrolyte transport across papillary
    collecting duct of the rat. *Am. J. Physiol. 224*:623–634,
    1973.
2. T. D. DuBose, Jr., D. W. Seldin, and J. P. Kokko. Segmental
    chloride reabsorption in the rat nephron as a function of
    load. *Am. J. Physiol. 234*:F97–F105, 1978; or *Am. J. Physiol.:
    Renal Fluid Electrolyte Physiol. 3*:F97–F205, 1978.
3. E. Higashihara, T. D. DuBose, Jr., and J. P. Kokko. Direct
    examination of chloride transport across papillary collecting
    duct of the rat. *Am. J. Physiol. 235*:F219–F226, 1978; or

Am. J. Physiol.: Renal Fluid Electrolyte Physiol. 4:F219-F226, 1978.

4. E. Higashihara, J. B. Stokes, J. P. Kokko, W. B. Campbell, and T. D. DuBose, Jr. Cortical and papillary micropuncture examination of chloride transport in segments of the rat kidney during inhibition of prostaglandin production: A possible role for prostaglandins in the chloruresis of acute volume expansion. *J. Clin. Invest.* 64:1277-1287, 1979.

5. R. Laurence and D. J. Marsh. Effect of diuretic states on hamster collecting duct electrical potential differences. *Am. J. Physiol.* 220:1610-1616, 1971.

6. Y. Iino and M. Imai. Effects of prostaglandins on Na transport in isolated collecting tubules. *Pfleugers Arch.* 373: 125-132, 1978.

7. J. B. Stokes and J. P. Kokko. Inhibition of sodium transport by prostaglandin E$_2$ across the isolated, perfused rabbit collecting tubule. *J. Clin. Invest.* 49:1099-1104, 1977.

8. J. B. Stokes. Effect of prostaglandin E2 on chloride transport across the rabbit thick ascending limb of henle. *J. Clin. Invest.* 64:495-502, 1979.

9. J. B. Gross, M. Imai, and J. P. Kokko. A functional comparison of the cortical collecting tubule and the distal conoluted tubule. *J. Clin. Invest.* 55:1284-1294, 1975.

10. R. G. O'Neil and S. I. Helman. Transport characteristics of renal collecting tubules: Influences of DOCA and diet. *Am. J. Physiol.* 233:F544-F558, 1977; or *Am. J. Physiol.: Renal Fluid Electrolyte Physiol.* 2:F544-F558, 1977.

11. G. J. Schwartz and M. B. Burg. Mineralocorticoid effects on cation transport by cortical collecting tubules *in vitro*. *Am. J. Physiol.* 235:F576-F585, 1978; or *Am. J. Physiol.: Renal Fluid Electrolyte Physiol.* 4:F576-F585, 1978.

12. M. J. Hanley, J. P. Kokko, J. B. Gross, and H. R. Jacobson. An electrophysiologic study of the cortical collecting tubule of the rabbit. *Kidney Int.* 17:74-81, 1980.

13. L. C. Stoner, M. B. Burg, and J. Orloff. Ion transport in cortical collecting tubule; effect of amiloride. *Am. J. Physiol.* *227*:453-459, 1974.

14. M. J. Hanley and J. P. Kokko. Study of cloride transport across the rabbit cortical collecting tubule. *J. Clin. Invest.* *62*:39-44, 1978.

15. R. Hogg and J. P. Kokko. Unpublished observations.

BIDIRECTIONAL FLUID TRANSPORT IN THE PROXIMAL TUBULE

Jared Grantham
James Irish III
Daniel Terreros

Departments of Medicine and Physiology
University of Kansas School of Medicine
Kansas City, Kansas

Department of Physiology
School of Medicine
West Virginia University
Morgantown, West Virginia

In the normal course of urine formation mammalian nephrons absorb salts and water from the glomerular filtrate into the blood. The renal secretion of fluid seems to be a unique property of the proximal straight tubules (PST), specifically the S_2 segment (1, 2, 3). Proximal straight tubules probably do not secrete fluid under ordinary conditions *in vivo*; however, when the concentrations of aryl anions such as hippurate, benzoate, and paraamino hippurate (PAH) rise in the blood, fluid secretion potentially may occur in the PST of rabbits (2, 3), mice (J. Irish III and D. Hall, unpublished observations), and humans (4). The secretion of fluid in the PST is due to the intense transport of organic anions from the blood into urine. Only those anions

readily transported into the cells through the mechanism typified
by PAH appear capable of promoting the net movement of water and
salts across the epithelial layer in a direction opposite to the
normal absorptive flow (5).

The phenomenon of fluid secretion is best illustrated in
isolated tubules *in vitro* that are perfused. These tubules are
recovered from rabbit kidneys by a dissection technique (6),
placed in a bath of normal rabbit serum and held between two
pipets so as to completely obstruct the lumen at both ends of the
tubule. In this way an experiment is begun with the lumen tightly
collapsed--the only way fluid can enter the lumen is through the
tubule wall. Although many organic anions can cause fluid secre-
tions, we confine this discussion to PAH, the agent we know the
most about.

Fluid secretion in PST has a number of unusual characteristics. When PAH is placed in the rabbit serum surrounding a PST,
a lumen appears in a few minutes. The speed with which the lumen
appears is a function of the PAH transport rate, which in turn
depends in large measure on the region of the PST from which the
tubule is obtained (1). In our recent studies we found that the
initial portion of the S_2 segment forms a lumen within 1-2 minutes
after 1 mM PAH is placed in the bath, whereas terminal S_2 segments form lumens more slowly. This heterogeneity in the rate of
fluid secretion is in accord with the earlier finding that PAH
was accumulated in the initial portion of the S_2 segment to a
greater extent than in the terminal portions of this segment (2).
The S_3 segments, those portions of the PST just proximal to the
thin descending limb, and the S_1 segments, those portions of convoluted tubule nearest the glomerulus, both transport PAH at a
much slower rate than the S_2 segments. Neither S_1 nor S_3 portions
secrete fluid in response to PAH. The capability of S_2 segments
to secrete fluid is strongly dependent on the temperature at which
the tubules are incubated. Tubules secrete fluid at 37°C but not
at temperatures lower than about 28°C. The rate of fluid secre-

tion also depends on the concentration of PAH in the bath. The minimal concentration to cause secretion is about 50 µM; a maximum rate of fluid secretion of about 0.1 nl/min/mm, is seen with 1.3 mM/L PAH in the bath (7). Fluid secretion and PAH transport are blocked by probenecid, probably by competitive inhibition of PAH transport, and by ouabain, which inhibits the sodium pump (3). The proximal tubule mechanism of fluid secretion appears to be distinct from that mediated by cyclic AMP in other organs since cyclic AMP will not cause secretion by itself, nor does it accelerate the rate of fluid secretion caused by PAH (J. Grantham, unpublished observations).

A dynamic appreciation of the fluid secretion phenomenon can be gained by looking at the relative time course of net fluid transport, transtubule electrical potential difference and morphology of tubules subjected to PAH (1 mM in the bath (Fig. 1). In tubules perfused at one end with an isotonic Ringer's perfusate and obstructed at the other end, fluid absorption or secretion can be adduced from the movement of an oil drop in the shank of the perfusion pipet (7, 8). In tubules with open lumens to begin with, it takes about 7 minutes for PAH to reverse the flow of fluid from the usual absorptive to the secretory direction. It should be noted that the rate of fluid absorption slows shortly after PAH is added to the bath. In tubules perfused with Ringer's solution in the conventional fashion at relatively fast perfusion rates, PAH in the bath decreases fluid absorption to a small but significant extent; however, fluid secretion is not seen in this case (J. Irish III and J. Grantham, unpublished observations). In the nonperfused tubules fluid secretion, evinced by opening of lumens, is observed 2 - 5 min after adding PAH to the bath. The change in the size of tubule cells can be assessed by this method as well. In six of eight recent studies we observed a slight decrease in the size of tubule cells from S_2 segments as PAH and fluid entered the lumen.

The transtubule PD provides an additional clue to the trans-

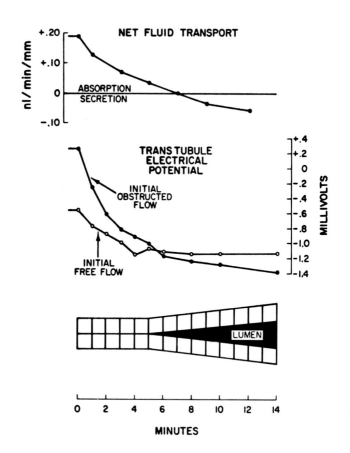

FIG. 1. Time course of PAH (1 mM) effect on S_2 PST. Net fluid absorption was measured in five tubules according to the method in ref. 7. Electrical potential measurements are from the same experiments reported in ref. 7. The schematic diagram is representative of the appearance and time course of nonperfused tubules after incubation in PAH.

port events going on in the tubules in association with PAH and fluid secretion. PAH was added to tubules perfused by the stationary method, and in the regular fashion at flow rates of about 10 nl/min (7). In both cases the PD increased shortly after PAH was added to the bath, and rose to final steady levels that were nearly the same. The increment in PD was greater for the "obstructed" tubules, since their initial PDs were slightly positive. The major point of these observations is that PAH had the same qualitative effect on the PD of tubules that ultimately secreted fluid (the "obstructed" tubules), as on those that did not (the "free-flow" tubules). Thus, it seems clear that the electrogenic effects of PAH secretion can be dissociated from the net movement of fluid through the epithelium.

The character of the secreted fluid is important to our understanding of the mechanism of fluid secretion (Table I). The secreted fluid is isosmotic to the bath, and the sodium concentrations of the two bulk phase solutions are virtually identical (7). The potassium concentration is slightly greater than in the bath. In recent studies we have found that chloride is accumulated in the secreted fluid, as well as sodium, potassium, and PAH. We have not analyzed chloride directly in the tubule fluid; instead, we have used an indirect approach to gain additional information about the amount of chloride in the cells, as well as in the urine (Table II). In these experiments segments of superficial S_2 PST were incubated in ^{36}Cl medium of high specific activity. The total chloride content in the tubules (cells plus lumens) was determined from the activity of ^{36}Cl in the tissue using a method developed previously for Na, K, PAH, and urea (3). In control nonperfused tubules in which the lumens are collapsed, the chlor-

TABLE I. *Approximate Composition of Secreted Fluid in Proximal Straight Tubules*

	Bath medium (mEq/liter)	Lumen fluid (mEq/liter)
Sodium	150	150
Potassium	5	7
Chloride	120	80
PAH	1	40

TABLE II. Chloride[a] Content of Nonperfused PST Treated with PAH
or Ouabain

Control medium	Expanded lumen due to PAH (1 mM)	Ouabain (10^{-4}M)
	$\times\ 10^{-9}$ mEq/mm tubule length	
18.4 ± 1.7	37.9 ± 7.2	84.9 ± 22.2
(6)	(4)	(4)

[a]Tubules were incubated in ^{36}Cl to constant specific activity.
They were analyzed for isotope by previous methods (3) in control
media, 20 minutes after addition of PAH and 30 minutes after addition of ouabain. Tubules were rinsed in ice-cold isotonic NaNO$_3$
to remove extracellular chloride. Mean ± SE.

ide is principally in the cells; the nominal concentration of
chloride in cell water is about 25 mEq/L. Twenty min. after PAH
was added (1 mM) the chloride content had increased 105% above
the control. In these segments all of the chloride in the tubules
in the presence of PAH (37.9 × 10^{-9} mEq/mm) could have been contained within the lumen if the diameter of the lumens was 20 μm
and the lumen concentration 120 mEq/L (equal to the bath). If
the lumen concentration were only 80 mEq/L, a value probably
closer to the truth when 40 mM/L PAH has been added to the urine,
a diameter of 24 μ would be needed to accommodate all of the tubule chloride in the urine. From previous studies we know that
the lumen diameter of nonperfused tubules secreting fluid in response to PAH is about 20-25 μ (7,8). Thus, it is reasonable to
suppose that during PAH-induced fluid secretion most of the tubular chloride may be within the lumens.

The effect of ouabain was tested for comparison since this
drug causes chloride to accumulate within tubule cells (Table II).
Ouabain also causes the cells to swell, to lose potassium and accumulate sodium (3). The increase in tissue chloride of 67 × 10^{-9}
mEq/mm caused by ouabain is reasonably close to the increase in
tubule (Na$^+$ + K$^+$) of 77 × 10^{-9} mEq/mm in previous studies (3).
This finding lends credence to our method for measuring tubule

chloride content.

In tubules undergoing net secretion, the most impressive constituent of the fluid is PAH, which rises to a concentration of about 40 mM/L. In our initial studies it seemed obvious that the PAH in the lumen was tied to the movement of fluid in the secretory direction; however, we did not have a clear idea about how PAH, salts, and water were coupled. On the basis of our more recent work, we have a better understanding of how the transport of PAH is tied to the movement of salt and water into the urine.

First we consider the initial movement of PAH from the blood into the cell. Earlier studies by many workers (9, 10, 11) indicated that under the usual *in vitro* experimental conditions, perfused tubules secreted PAH into the urine first by accumulating the anion in the cells, followed by the movement of PAH down an electro-chemical gradient into the urine. These studies were all performed under conditions in which net fluid secretion did not occur. There is reason to suspect that PAH may move across the luminal membrane by a facilitated or carrier-mediated mechanism in lower species (12). It was important, therefore, to determine the concentration profiles for PAH in the unusual case where relatively large amounts of PAH were being transported into the urine in high concentrations in association with net fluid secretion (Fig. 2). In our initial studies fluid secretion in nonperfused tubules the final urine had a PAH concentration of about 40 mM/L. The concentration of PAH in the cells was not measured, but in recent studies we have analyzed the PAH content of PST under the same conditions reported previously (3). In these experiments we measured the cell content of PAH isotopically 2 - 3 minutes after 1 mM was added to the bath of nonperfused tubules. In this way no lumens had yet been formed, and the PAH in the tubules may be presumed to be principally within the cells. The concentration of PAH in the intracellular fluid was consistently greater than that in the urine. The absolute gradient between cells and urine is not known for sure since these latter experi-

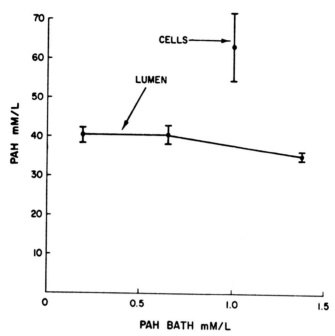

FIG. 2. *The intracellular and luminal concentrations of PAH*
in tubules secreting fluid in response to PAH in the bath (1 mM).
Measurements of lumen PAH are taken from reference 7. The con-
centration of PAH in the cells was determined in four experiments
2 - 3 min. after putting PAH in the bath and just before lumens
appeared. The mean tubule content of PAH was 46.3 ± 6.9 mEq/mm
tubule length. The value in the figure was corrected for a tubule
water content of 0.73 nl/mm.

ments were performed in different animals. Nonetheless, it is
clear that in tubules secreting fluid, the PAH concentration of
the cells is very high.

The potential energy needed to move PAH against such a steep
gradient is remarkable (Table III). We have calculated the E_{PAH}
in the usual way from the Nernst equation. It is interesting to
compare this value with that of other ions in cell water taken
from our earlier studies (Table IV). The transmembrane electrical
potential is about -27 mV (13). The E_K is -89 mV and the E_{Na} is
+30 mV, suggesting active accumulation of potassium and active ex-

TABLE III.   *Chemical Potential of PAH across Peritubular Membrane*

Bath PAH (mM/L)	Lumen (cell[a]) PAH (mM/L)	Chemical potential (mV)
0.24	41	-137
0.65	41	-110
1.37	35	-87
Em (resting)		-27
E_K		-89
E_{Na}		+30
E_{Cl}		-42

[a]*These data were taken from ref. 7 in which secreted fluid was analyzed for PAH content. On the basis of the results in Fig. 2, we assume that for each concentration of PAH in the bath the intracellular PAH concentration was at least as high as that in the urine.*

TABLE IV.   *Tissue Electrolyte Content of Nonperfused Proximal Straight Tubules*

Solute	Tissue content (mEq/mm × 10^{-9})	Tissue concentration[a] (mEq/l)
Potassium[b]	102	139
Sodium+	36	49
Chloride[c]	18	25

[a]*Estimated from 0.73 nl cell water per mm length*
[b]*From AJP 232:F42-49, 1977.*
[c]*New Data.*

trusion of sodium by the cation pump. E_{Cl} is -42 mV, indicating that chloride may be accumulated in the cell by a nonpassive mechanism. While there is probably enough potential energy in the sodium gradient to drive the entry of chloride into the cells across the peritubular membrane, the chemical potential for PAH appears to exceed the capability for a symport with sodium. This consideration, together with the nonspecific response of the PAH transport system to sodium gradients across the peritubular membrane (14) suggests that the anion enters the cell via a unique pump.

   The chemical nature of PAH in the cell is of some importance.
The general correspondence between the levels in the cells and the
levels appearing in the urine suggests that cell PAH is ionic and
not bound. To obtain additional information about the chemical
state of PAH in cells, in recent studies we have determined the
effect of PAH on the size of the cells in PST. If PAH is pumped
into the cell as an anion, electroneutrality can be preserved if
a cation accompanies PAH, in which case the cells should swell.
Alternatively, if chloride or another permeable anion is extruded
from the cell as PAH is accumulated, the cell volume might not
change appreciably. In six of eight experiments we found the tubule cells shrank slightly as PAH was accumulated by the cells,
suggesting that in the course of transport, chloride was extruded
from the cell. Our studies of the cell chloride content in the
early period after PAH was added to the bath suggest that the
chloride content of the cells decrease, but this portion of the
study is incomplete. Thus, although we have not yet made direct
measurements, it appears that as PAH is accumulated by the cells
intracellular chloride decreases to very low levels. In this way
the cells do not swell as they fill with an anion to which the
peritubular membrane is relatively impermeant.

   An increase in the electrochemical gradient against which
chloride moves out of the cell is one important consequence of
loading cells with an anion that does not penetrate the peritubular
membrane. The increased energy barrier against which chloride
must leave the cell probably decreases the net extrusion of sodium
chloride and, consequently, decreases the absorption of salt.
This effect of PAH as an impermeant intracellular anion undoubtedly
potentiates its effectiveness to cause fluid secretion.

   As PAH is accumulated in the cell, a steep electrochemical
gradient develops favoring diffusion of the anion into the urine.
The lumenal membrane is highly permeable to PAH, perhaps 16 times
more permeable than the peritubular membrane (9). We do not know

the relative permeabilities to PAH and Cl of the lumenal membrane, but it is not unreasonable to suppose that they are nearly the same. As PAH moves across the lumenal membrane, it obligates the extrusion from the lumen of an anion, e.g., Cl^- or HCO_3^-, or the addition of a cation. In tubules with collapsed lumens there is little Na^+, Cl^-, or HCO_3^- in the lumen, so more than likely the deposition of PAH first causes some expulsion of Cl^- until a limiting gradient is achieved. Then sodium is attracted into the urine and water along with it. In view of the fact that PAH secretion is associated with increased lumenal negativity, more than likely extracellular cations enter the lumen down an electrical gradient. Sodium might enter the urine through either a transcellular or a paracellular route. The transcellular route seems unlikely since the electrochemical forces favor movement of Na^+ into the cells at both the lumenal and antilumenal membranes; this could not lead to net sodium movement into the urine. It is more reasonable to conclude that Na^+ enters the urine via the rather permeable paracellular channel.

When sodium enters the urine to correct the charge imbalance caused by PAH, the stage is set for further fluid and salt movement into the urine. Since the tight junction is relatively impermeant to PAH and highly permeable to Na^+ and Cl^-, the deposition of impermeant anion into the urine favors the movement of solute and water into the urine, probably via the paracellular pathway. Indeed, in preliminary studies reported by Irish and Grantham (15) we found that PAH in the urine exerted a major influence on the movement of water from the bath into the urine. For these studies we perfused tubules with different isosmotic saline solutions in which PAH was substituted for chloride to a variable extent. In these studies net water movement changed from absorption to secretion when the concentration of PAH in the perfusate was increased to 40 - 60 mM/L, values in reasonable correspondence to the 40 mM concentration for PAH found in the fluid of tubules undergoing secretion driven by PAH transport.

From this experiment it seems clear that PAH has to be in the
urine to cause secretion, and more specifically the anion has to
exceed some limiting concentration. One factor determining the
magnitude of the limiting concentration of PAH needed to cause
fluid secretion was revealed by studies of the effect of ouabain
on PST. In tubules perfused with PAH substituted for chloride,
ouabain caused net fluid secretion at all concentrations of PAH
in the urine, indicating that relief of the force promoting absorption of salt and water enhanced the effectiveness of PAH to
cause fluid secretion. It will be interesting to learn if pro-
portionate, rather than complete inhibition of the sodium pump
can augment the secretion of fluid allied to PAH transport.

On the basis of these studies we propose a model of the
events that transpire as a PST with a collapsed lumen secretes
fluid in response to PAH (Fig. 3):

(1) PAH is actively transported into the cells to be "exchanged" for chloride.

(2) PAH enters the lumen down an electrochemical gradient and
is balanced by Na^+ entering via the paracellular shunt pathway.

(3) Owing to the steep gradient against chloride movement
out of the cell, the transport of NaCl is diminished, thereby
enhancing the effectiveness of PAH in the lumen to attract solute
and water into the lumen probably via the paracellular pathways.

By this account the net secretion of Na, K, and Cl is viewed as
secondary to the active transport of PAH into the urine. More
than likely the fluid secretion caused by hippurate, carbenicil-
lin, cephalothin, benzoic acid, and aryl anions of uremia is due
to an identical mechanism (5, 8, 11).

Finally, the principles of fluid secretion, established in
the proximal straight tubule, may be applicable to liver where
bile is secreted in association with organic anions (bile salts)
that are actively secreted by the hepatocytes (16).

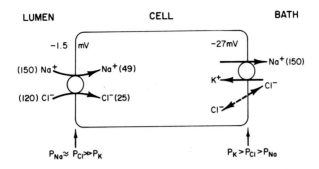

A. ORDINARY PST IN SYMMETRICAL BATHING MEDIA.

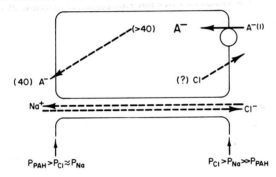

B. ORDINARY PST IN BATHING MEDIUM CONTAINING PAH.

*FIG. 3. Model of fluid secretion in PST driven by PAH trans-
port. (A) An ordinary PST perfused with isotonic medium in an external bath of rabbit serum. The principal features include:
(i) in the peritubular membrane, a sodium-potassium exchange pump
(Na-K, ATPase), a chloride translocator (perhaps ionic and symport
mechanisms operating in parallel), and relative permeabilities
favoring the passive movement of K^+ through the membrane. (ii) in
the lumen membrane, Na^+ and Cl^- enter the cell via a symport mechanism, and the membrane is more permeable to Na^+ and Cl^- than to K^+.
The electrolyte concentrations are taken from Tables I and IV.
(B) A nonperfused tubule exposed to PAH (1 mM) in the bath. There
is electrogenic pumping of PAH into the cell leading to the exit
of chloride across the peritubular membrane which is highly impermeant to PAH. The reduction of cell chloride limits the net movement of NaCl out of the cell, and net fluid absorption is slowed.
PAH moves into the lumen and initially causes the expulsion of
chloride; when the concentration of PAH rises sufficiently (and
lumen Cl^- decreases) there is movement of Na^+ into the lumen to
balance the continued entry of PAH. Ultimately, the accumulation
in the urine of the impermeant anion causes the net movement of salt
and water into the lumen by solvent drag, probably via the paracellular pathway.*

SUMMARY

The straight portion of the rabbit proximal tubule (PST) or-
dinarily absorbs salts and water from the lumen into the blood.
This segment can also secrete fluid into the urine in the presence
of relatively high concentrations of aryl anions (PAH is the proto-
type) especially when tubular perfusion (glomerular filtration)
approaches stopped-flow. Fluid secretion is due to the uphill
transport of PAH by a specific mechanism located in the peritubular
membrane, an interface that is also highly impermeant to the organic
anion. Because of the accumulation of high concentrations of PAH
in the cells net fluid absorption driven by the sodium pump is decreased. PAH moves into the urine readily across the lumenal membrane down an electrochemical gradient. The deposition of impermeant anions into the urine ultimately leads to filtration of extracellular solutes and water into the lumen, probably via paracellular pathways. The PAH transport system offers a unique way to
alter the intracellular ionic environment of renal cells. Moreover, the fluid secretion model provides a reasonably clear view
of how the active transport of anions (PAH in this case) can lead
to the secondary net movement of salts and water in a direction
opposite to the normal flow.

ACKNOWLEDGMENTS

We thank Joan Al-Attar for secretarial assistance. Portions
of this work were supported by a grant from the USPHS AM 13476.

REFERENCES

1. P. B. Woodhall, C. C. Tisher, C. A. Simonton, and R. R.
    Robinson. Relationship between para-aminohippurate secretion

and cellular morphology in rabbit proximal tubules. *J. Clin. Invest.* *51*: 1320-1329, 1978.

2. J. Grantham, J. M. Irish III. Organic acid and fluid secretion in the pars recta (PST) of the proximal tubule. *In* New Aspects of Renal Function," (H. G. Vogel and K. J. Ullrich, eds.). International Congress Series, No. 422, Excerpta Medica Amsterdam, 1978.

3. J. Grantham, C. M. Lowe, M. Dellasega, and B. R. Cole. Effect of hypotonic medium on K and Na content of proximal renal tubules. *Am. J. Physiol.* *232*:F42-F49, 1977.

4. F. B. Stapelton and M. A. Linshaw. Regulation of cell volume in a human proximal straight tubule. *Renal Physiol.* *1*:334-337, 1978.

5. R. D. Porter, W. F. Cathcart-Rake, S. H. Wan, F. C. Whittier, and J. J. Grantham. Secretory activity and aryl acid content of serum, urine, and cerebrospinal fluid in normal and uremic man. *J. Lab. Clin. Med.* *85*:723-733, 1975.

6. M. Burg, J. Grantham, M. Abramow, and J. Orloff. Preparation and study of fragments of single rabbit nephrons. *Am. J. Physiol.* *210*:1293-1298, 1966.

7. J. Grantham, P. B. Qualizza, and R. L. Irwin. Net fluid secretion in proximal straight renal tubules *in vitro*: Role of PAH. *Am. J. Physiol.* *226*:191-197, 1974.

8. J. Grantham, R. L. Irwin, P. B. Qualizza, D. R. Tucker, and F. C. Whittier. Fluid Secretion in Isolated Proximal Renal Tubules. *J. Clin. Invest.* *52*:2441-2450, 1973.

9. B. M. Tune, M. B. Burg, and C. S. Patlak. Characteristics of p-aminohippurate transport in proximal renal tubules. *Am. J. Physiol.* *217*:1057-1063, 1969.

10. J. M. Irish and W. H. Dantzler. PAH transport and fluid absorption by isolated perfused frog proximal renal tubules. *Am. J. Physiol.* *230*:1509-1516, 1976.

11. J. Grantham. Fluid secretion in the nephron: Relation to renal failure. *Physiol. Rev.* *56*:248-258, 1976.

12. W. B. Kinter. Chlorphenol red influx and efflux: Micro-
     spectrophotometry of flounder kidney tubules. *Am. J. Physiol.*
     211:1152-1164, 1966.

13. D. Terreros, J. A. Grantham, M. Tarr, and J. J. Grantham.
     Axial heterogeneity of transmembrane electrical potential in
     isolated proximal renal tubules. *Kidney Int*. December 1980.

14. K. J. Ullrich. Renal tubular mechanisms of organic solute
     transport. *Kidney Int*. 9:134-148, 1976.

15. J. M. Irish and J. Grantham. Role of luminal para-amino hip-
     purate in fluid secretion in proximal straight tubules. *Fed.*
     Proc. *36*:592, 1977.

16. J. L. Boyer. New concepts of mechanisms of hepatocyte bile
     formation. *Physiol. Rev*. *60*:303-326, 1980.

CHLORIDE TRANSPORT BY GASTRIC MUCOSA AND
MEMBRANES ISOLATED FROM GASTRIC OXYNTIC CELLS

John G. Forte

Hon Cheung Lee

Haim Breitbart

Department of Physiology-Anatomy
University of California
Berkeley, California

I. INTRODUCTION

Gastric mucosa is capable of actively transporting both H^+
and Cl^- from blood to lumen. This conclusion is firmly established
both from analysis of the transmucosal electrochemical potential
gradient for each ion *in vivo* (1), and from ion flux measurements
carried out on isolated, shortcircuited, mucosal preparations (2).
However, a great deal of controversy has surrounded the question of
the nature and possible interrelationship between the transport
mechanisms for these two ionic species. At the one extreme (3)
there exists the postulate of biochemically separate and independent functional transport mechanisms (electrogenic pumps) with the
proviso of simple electrical coupling (e.g. through membrane potential and ionic conductance pathways). At the other extreme tight
coupling between the transport processes has been suggested (4, 5).

151

Copyright © 1982 by Academic Press, Inc.
All rights of reproduction in any form reserved.
ISBN 0-12-775280-3

The Cl^- requirement for gastric H^+ secretion is well known
(6,7). Replacement of Cl^- with impermeant anions, such as SO_4^{-2}
or isethionate, depresses H^+ transport by 75% or more. However,
more permeable anions will support H^+ secretion, with an order of
selectivity being $Br^->Cl^->I^-$ (4). It is also clear from a number
of studies that Cl^- secretion by intact gastric mucosa is stimulated when H^+ secretion is stimulated, and inhibited when H^+ secretion is similarly reduced (8,9). Such interdependencies between
H^+ and Cl^- secretion have suggested functional commonalities in
the transport process, but it has not been possible to unequivocally distinguish between a direct biochemical coupling and indirect
electrical coupling.

Attempts to identify gastric cell fractions and membrane components with direct anionic involvement in enzymatic or transport
function has met with limited success. Interest in the HCO_3^--stim-
ulated, SCN^--inhibited ATPase discovered by Kasbekar and Durbin
(10) has waned since this system appears to be a general mitochon-
drial enzyme with no specific role in H^+ and/or Cl^- transport at
the apical plasma membrane of the acid secreting cell (11). Our
appreciation of gastric ion transport events at the subcellular
level has been considerably reoriented by the discovery of a unique
K^+-stimulated ATPase activity (8,11,12). This gastric K^+-stimula-
ted ATPase is associated with microsomal vesicles isolated from
acid secreting cells (tubulovesicle membranes and apical plasma
membranes), and has the interesting property of being further
stimulated by K^+ ionophores (12,13,14). Peter Scholes' group in
England was the first to show that the gastric microsomal vesicles
can transport (accumulate) H^+ in the presence of K^+, Mg^{+2}, and ATP
(14). George Sachs' laboratory has provided extensive documenta-
tion on the operation of gastric vesicular H^+ transport which he
has suggested to be an electroneutral H^+/K^+ exchange pump (15).

In this report we present a survey of our recent studies on
the influence and involvement of anions in the ATPase H^+ transport
activities of gastric microsomal vesicles (16). We demonstrate

an important modulating influence for anions in vesicular H^+ transport; however, the role appears to be a permissive one, that is by virtue of the passive permeability characteristics of the anion. Thus the "ATP-driven" anion transport in isolated microsomal vesicles is coupled to H^+ and K^+ transport. The observations, along with recent results from Machen's laboratory (17.18) on the involvement of Na^+ in gastric Cl^- flux, will be incorporated into a comprehensive picture of gastric Cl^- transport.

II.   METHODS FOR MEASURING GASTRIC VESICULAR H^+ TRANSPORT AND
      ATPase ACTIVITIES

Gastric microsomal vesicles were isolated from hog fundic mucosal homogenates by the differential and density gradient centrifugation procedures previously described (12). Assay of ATPase activity was by the conventional means of measuring the rate of inorganic phosphate production from ATP (8,11,12) in the presence of various activators, anions, and ionophores, as indicated for the individual experiments.

H^+ transport was measured using the principle of weak base accumulation in response to a pH gradient across a membrane system. In particular, we measured accumulation of fluorescent amines by the gastric vesicles, using the quench of signal determined spectrofluorometrically as an index of proton accumulation (19). Depending on the particular dye used, this method has been shown to provide sensitive qualitative (e.g., acridine orange) or quantitative (e.g., 9-aminoacridine) assessment of vesicular pH gradient formation (19,20). Wavelengths used for spectrofluorometry were 493→530 nm (excitation→emission) for acridine orange and 422→455 nm for 9-aminoacridine.

III.   ACTIVATION OF ATPase AND H[+] TRANSPORT

Earlier studies (11,13,21) have established the characteristics of K[+]-activation, effects of ionophores, and selectivity or various cations in stimulating gastric vesicular ATPase. The effects of KCl on hog gastric microsomal ATPase activity and the further activating effects of K[+] ionophores are summarized in Table I. Stimulation by valinomycin strongly suggests a limitation for the rate of K[+] accessibility to an activating site; this is further supported by an equivalent degree of enzyme activation when microsomal vesicles were ruptured by lyophilization or mild detergent treatment (13).

TABLE I. ATPase Activities of Gastric Microsomes[a]

	ATPase activity $(\mu molmg^{-1} hr^{-1}$	
	21^0-24^0C (n = 10)	37^0C (n = 6)
Mg^{2+} only	2.7+0.3	6.0+0.5
K^+-stimulated	4.8+0.5	17.6+2.4
K^+ plus val-stimulated	14.7+1.3	54.7+7.7
Val-stimulated	9.7+0.3	38.0+5.1

[a] *Values are shown for rate of ATP hydorlysis at room temperature and 37^0C. Basal ATPase activity for preparation is that measured in the presence of 1 mM $MgSO_4$. The additional stimulation of ATPase produced by 104 mM KCl (K^+-stimulated) and 140 mM KCl plus 10 μM valinomycin (K^+ plus val-stimulated) is also shown. The enchancement of K^+-ATPase specifically produced by valinomycin is shown as val-stimulated activity. (Data adapted from ref. 16).*

Gastric K^+-stimulated ATPase is the pump enzyme responsible for the transport of H^+ into the vesicles in exchange for K^+ (15, 19). This vesicular H^+ accumulation, like the ATPase, is also limited by K^+ accessibility as demonstrated by studies using uptake of fluorescent amines, such as shown in Fig. 1. In Fig. 1A it can be seen that the rate and degree of H^+ accumulation is en-

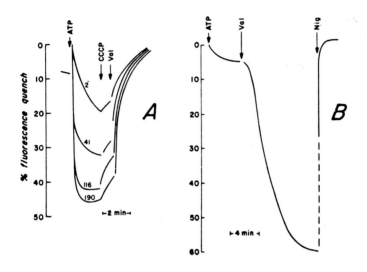

FIG. 1. *Influence of K^+ on H^+ transport activity of gastric microsomal vesicles. H^+ transport was monitored by assaying the degree of fluorescence quenching resulting from vesicular accumulation of fluorescent amines in response to a transmembrane pH gradient (see ref. 19). A. Gastric vesicles were preincubated with 150 mM KCl for various times (in min) shown on trace. Acridine orange (5 μM) was then added to vesicles and fluorescence monitored. The degree of quenching, after addition of ATP, is related to time of KCl preincubation. Protonophores (CCCP, 10^{-5}M) and valinomycin (val, 10^{-5}M) were added to dissipate the H^+ gradient. B. Gastric vesicles were placed into 150mM KCl and 10 μM 9-aminoacridine with no prior incubation. Addition of ATP produced only a small quenching of fluorescence. Subsequent addition of 10^{-5}M valinomycin (val) produced a large H^+ uptake (signal quench). H^+ gradient was dissipated by addition of 10^{-6} M nigericin (nig). Simultaneous measurements of ATPase activity before and after addition of valinomycin gave values of 6.8 and 13.8 μmol/mg protein/hr, respectively.*

hanced by a time of preincubation of vesicles in KCl. In Fig. 1B
with short periods of K^+ preincubation, it is apparent that addition of ATP produced only minimal H^+ transport; however, subsequent
addition of valinomycin stimulates an effective H^+ accumulation.
Thus K^+ accessibility to some intravesicular or intramembrane site
is important for optimal function of the gastric vesicular ATPase
and H^+ transport.

   The influence of varying concentrations of K^+ in stimulating
ATPase activity of hog gastric microsomes in the presence and absence of valinomycin is shown in Fig. 2. The apparent K_A for K^+
is about 4 mM without K^+ ionophore. On the other hand, the valinomycin-stimulated K^+-ATPase activity does not show clear saturation kinetics, but continues to increase even at 140 mM K^+. In
fact, the augmentation in enzymatic activity by valinomycin, i.e.,
rate in K^++valinomycin minus rate with K^+ alone, is nearly a linear
function of $[K^+]$ above 10 mM. Moreover, the ATP-dependent accumulation of H^+, as judged by acridine orange uptake, is also a linear
function of $[K^+]$ in these experiments in which valinomycin was used
as an activator (Fig. 2). Other cations will substitute for K^+ in
stimulating gastric ATPase with a selectivity $Tl^+>K^+>Rb^+>Cs^+>NH_4^+>$
$>Na^+$ (21), thus the system has some analogy with the K^+-dependent
component of the (Na^++K^+)-ATPase.

IV. ROLE OF ANIONS IN GASTRIC VESICULAR ATPase AND H^+ TRANSPORT

   Initial experiments to define the role of anions in this system involved measurement of ATPase at a constant $[K^+]$ of 140 meq/L
with reciprocal variation in the concentration of two anions, for
example, Cl^- and isethionate as shown in Fig. 3. Note that in the
absence of added ionophores K^+-stimulated ATPase was very little
affected over a wide range of variation in the two anions. However,
the degree of stimulation by valinomycin was markedly influenced by
the nature of the anion; thus the presence of an ionophore, which

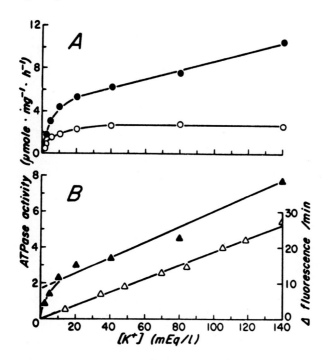

FIG. 2. The effect of $[K^+]$ on gastric vesicle ATPase activity and H^+ transport as measured by the initial rate of acridine orange fluorescence quenching. A. ATPase activity is shown in the absence (O) and present (O) of 10 μM valinomycin. B. The val-stimulated K^+-ATPase (Δ) calculated as the difference between ATPase activity measured in the presence and absence of valinomycin and shown along with parallel measurements for the initial rate of acridine orange fluorescence quenching (Δ). The ionic strength and $[Cl^-]$ of the reaction medium was maintained constant with varying amounts of NaCl as the K^+ concentration was changed, as indicated (reproduced from ref. 22).

specifically increased permeability to K^+, the K^+-ATPase was more markedly stimulated by Cl^- than by isethionate. When vesicular permeability to both K^+ and H^+ was increased (e.g., addition of nigericin or valinomycin plus a protonophore), there was a marked stimulation of ATPase in the isethionate solutions and the differential effects between Cl^- and isethionate became less pronounced. Similar experiments with reciprocal variations of other anions gave results which were more or less exaggerated, depending on the

John G. Forte *et al.*

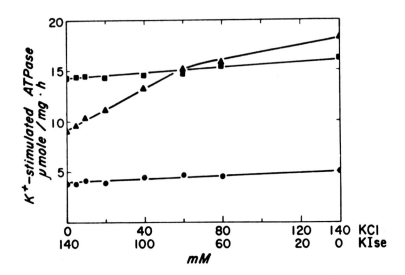

FIG. 3. K^+-stimulated ATPase activity of gastric vesicles
under conditions where Cl^- and isethionate were reciprocally varied
at a constant $[K^+]$ of 140 meq/L. ATPase activity is shown without
ionophore (o), with 10^{-5} M valinomycin (▲), and with 10^{-6} M nigericin (■).

two anions (16).

A summary of the effects of various anions on K^+-stimulated
ATPase activity is shown in Fig. 4. In the absence of ionophore
there were relatively small differences among various anions in
stimulating K^+-ATPase. However, the differences, or sensitivity
to various anions, became most striking when K^+ permeability was
specifically increased by valinomycin. For KBr, KNO_3, and KCl,
valinomycin produced a 200–300% stimulation of K^+-ATPase. The
rather small stimulation of K^+-ATPase by valinomycin (50–100%) for
larger anions like acetate, isethionate, and sulfate (not shown)
may be due to limited permeability of the anions. Thus, the flux
of K^+ into some activating site within the vesicles would be limited by counter ion movement, either the influx of an accompanying
anion or the efflux of the other principal cation, H^+. This interpretation is supported by the experiment where vesicular permea-

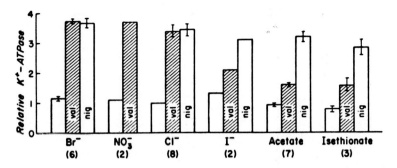

FIG. 4. Effect of various anions on K^+-stimulated ATPase
activity. Each of the indicated anions was present as the K^+ salt
at 140 mM. Results of a number of experiments (shown in parentheses) on different preparations have been summarized as the relative K^+-ATPase activity, taking the activity measured in 140 mM
KCl, with no ionophore, as 1.0 for any given experiment. Where
indicated, ionophore concentrations were 10^{-5} valinomycin (val) and
10^{-6} M nigericin (nig).

bility to both K^+ and H^+ was increased by the addition of nigericin
(Fig. 4) or valinomycin plus a protonophore (16).

In order to evaluate the influence of anions on vesicular H^+
accumulation, the uptake of acridine orange (or 9-aminoacridine)
was measured under conditions analogous to those described above
for K^+-ATPase. A comparison of various anions in supporting H^+
(dye) uptake is shown in Fig. 5. The less permeable anions-acetate,
isethionate, and sulfate-supported little or no net H^+ transport,
while the order of selectivity for the other anions tested was
$NO_3^->Br^->Cl^-$. (I^- is probably somewhat less effective than Cl^-, but
since I^- has independent quenching effects on the dyes used in this
study, the data are quantitatively unreliable.) Addition of nigericin led to immediate and complete dissipation of all gradients.
A progressive effect of concentration of the permeable anion species
on H^+ accumulation is shown in Fig. 6 where reciprocal variation of
Cl^- and isethionate were made at constant K^+.

Another anion of some interest was also tested, SCN^-, the well-
known inhibitor of gastric HCl secretion. In the range of 1-10 mM,

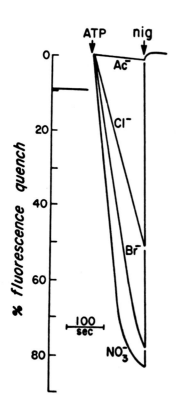

FIG. 5. The effect of several anions on acridine orange uptake by gastric vesicles. In addition to 150 mM of the K^+ salts of the indicated anions, the incubation medium contained 5 µM acridine orange, 10 mM Pipes buffer (pH 6.5), 1 MM $MgSO_4$, 2.5 µM, valinomycin, and 60 µg/ml of membrane protein. The H^+ transport reaction, as shown by the rate of fluorescence quenching, was initiated with 1 mM na-ATP at arrow. H^+ gradient was completely dissipated, as shown, by the addition of 10^{-6} M nigericin (nig). Experiments with K_2SO_4 and potassium isethionate gave identical results to potassium acetate, i.e., essentially no H^+ uptake (reproduced from ref. 16).

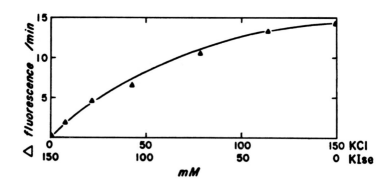

FIG. 6. Effect of Cl^- concentration on the initial rate of acridine orange fluorescence quenching at a constant $[K^+]$ of 150 meq/l. The Cl^- concentration was increased in substitution for isethionate (ISE). The reaction medium was otherwise identical to that of Fig. 4. After addition of 1 mM Na-ATP, the initial rate of fluorescence quenching was measured.

this lipid permeable anion was the most effective of all the anions tested (16) in stimulating valinomycin-stimulated K^+-ATPase activity and the rate of H^+ accumulation. However, concentrations of SCN^- above 10 mM were definitely inhibitory to both systems (16); the mechanism of this inhibition remains to be established.

V. A PUMP-LEAK MODEL FOR GASTRIC VESICULAR H^+ TRANSPORT

These studies show that there are clear anion effects on both the gastric K^+-stimulated ATPase activity and vesicular proton accumulation. The data are best explained on the basis of permeability limitations and not on the basis of direct enzymatic effects. We have summarized the ATPase activity and ion transport properties of gastric microsomal vesicles in the pump-leak model shown in Fig. 7. The first element in the model is the ATP-driven K^+/H^+-exchange

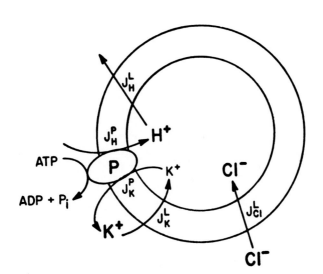

FIG. 7. Schematic accounting of ion movements across gastric microsomal vesicles. The J's are ion fluxes with the superscript designating pump flux (P) or leak pathway (L). The model consists of an ATP-driven H^+/K^+ exchange pump, in accord with earlier work (e.g. refs. (14,15,19) and the passive leak pathways for the principal ions K^+, H^+, and Cl^- (reproduced from ref. 16).

pump (14,15), which transports n_H^K potassium ions out of the gastric vesicles in exchange for each H^+ transported in. This pump can be either electroneutral ($n_H^K=1$) or electrogenic ($n_H^K{\neq}1$). The other elements are the leak, or conductance pathways for H^+, K^+, and anions. Because the site for K^+ activation is internal, any process which increases the internal $[K^+]$ or the inflow of K^+ should activate the pump and stimulate H^+ uptake. This has been shown to be true either by long term preincubation of the vesicles in K^+-containing media or by the use of valinomycin to increase membrane permeability to K^+. However, from the requirements for macroscopic electroneutrality, K^+ can enter the vesicles only when accompanied by a counter ion: either a cation gets out in exchange for K^+, or anions follow the K^+ influx. An observed stimulation of H^+ accumulation by valinomycin indicates that anion influx must accompany the K^+ influx. Thus, by electrical coupling through the membrane potential, the anions can influence the permeation of K^+ to the internal pump site; the more permeable the anion, the higher the permeation of K^+ in the presence of valinomycin, and therefore the larger the activation of the pump. On the other hand, if H^+ permeability is increased at the same time that K^+ permeability is increased (e.g., nigericin), the resulting passive H^+/K^+ exchange process should tend to lower or eliminate the H^+ uptake at the same time that ATPase activity is stimulated. A more quantitative expression of this gastric vesicular pump-leak model has been given elsewhere (16).

VI. APPLICATION TO CL^- TRANSPORT BY INTACT GASTRIC EPITHELIUM

One lesson to be drawn from these studies is that the role for Cl^- in H^+ transport by the vesicular K^+-ATPase system is clearly passive. On the basis of their permselectivity and ability to support the gastric transport system, we can arrange the anions tested in the following order: $SCN^->NO_3^->Br^->Cl^->I^->$isethionate $\simeq$

SO_4^-. If we attempt to extrapolate these results to intact tissue we find some agreement on the selectivity of anions in promoting H^+ secretion, e.g. $Br^- > Cl^- > I^- >$ isethionate $\approx SO_4^{-2}$. However, NO_3^- is a poor substitute for Cl^- in terms of supporting H^+ secretion by intact gastric mucosa (23), and SCN^- is, in fact, an inhibitor of H^+ secretion. Such difficulties do not completely invalidate the proposed extrapolation, but they must eventually be explained. It is clear that the gastric HCl secretory process *in situ* has more facets and limitations (e.g., anion permeation at the serosal membrane, HCO_3^-/anion exchange, membrane transformations, etc.) than those of the fundamental H^+ pump mechanism, and these may be the site of action of SCN^- and/or NO_3^-. But it is also conceivable that the negative effects of these anions *in vivo* may be related to their apparently high membrane permeability. Rehm has presented evidence to suggest that SCN^- may "uncouple" the gastric H^+ pump, with the putative HSCN complex recycling within the membrane nearly as fast as it is formed (24).

In any event a specific Cl^- transport enzyme has not yet been identified in gastric cell fractions. This is by no means conclusive evidence, but we might examine the possibility that active Cl^- transport by gastric mucosa is coupled to other ion transport processes as has been suggested at this symposium for so many other tissues. A large part of gastric Cl^- movement is closely associated with H^+ transport. For simiplicity, this has been termed the "acidic" Cl^- transport (5,25), and in general, the observations reported here for anion-dependent effects on vesicular K^+-ATPase and H^+ transport would fit these conditions. However, there is a component of active Cl^- transport which occurs independent of H^+ secretion. Recent studies by Machen and his collaborators have demonstrated that Cl^- transport by frog gastric mucosa is dependent upon the presence and concentration of Na^+ in the serosal solution (17,18). This, coupled with the reaffirmation that the $[Cl^-]$ of gastric epithelial cells (30-40 mM, ref. (26) is greater than could be accounted for by an equilibrium distribution across the serosal surface, prompted Machen's suggestion that a coupled NaCl transport system may be driven by the electrochemical gradient for Na^+ at the serosal membrane. This system might be analogous to the coupled NaCl system that Frizzell *et al.* (27) have suggested for flounder intestine and several other epithelia.

A scheme for various proposed modes of Cl^- movement across gastric mucosa is shown in Fig. 8. The primary H^+ pump is located at the mucosal surface; the system operates as a H^+/K^+ exchange mechanism. In this particular scheme the ATP-driven exchange occurs within the inner membrane aspect, allowing K^+ to recycle from the cytoplasm to the activating site and back again. This mode of K^+ recycling has been suggested to account for several details regarding K^+-site availability (28); it differs from the pump-leak model of Fig. 6 in that the K^+ activating site is within a deeply embedded, and possibly hydrophobic, phase of the membrane rather than within the vesicle. The net effect of this mode of operation (i.e. K^+ recycling) would be an e.m.f. (electromotove force) associated with H^+ secretion, hence an electrogenic pump. Movement of Cl^- would be passively responsive to the H^+ e.m.f. and to the driving force for K^+ entry. Whether this latter pathway involves a coupled carrier-mechanism or simple electrical coupling is uncertain. Wolosin and Forte have recently shown that distinctive membrane fractions can be isolated from rabbit stomach during physiological stages of rest vs. HCl secretion (29,30). Membrane preparations from resting stomachs have all the characteristics described for the vesicular model described above, while the larger membrane vesicles from histamine-stimulated stomachs have, in addition, a KCl entry pathway, thus they can transport protons via the K^+-ATPase without the requirement for valinomycin or preincubation. Important questions to pursue will be whether the KCl entry port is (i) a neutral or electrically coupled cotransport system, and (ii) activated by direct cytoplasmic reactions or the resulting fusion of cytoplasmic membranes with the apical surface at the time of stimulation (e.g., see Ref. 31).

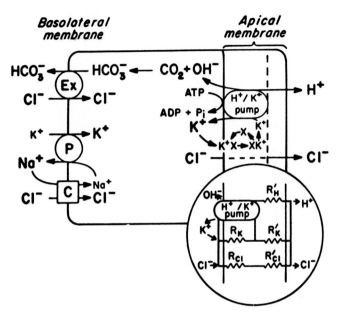

FIG. 8. Schematic representation of gastric epithelial transport processes. An ATP-driven H^+/K^+ exchange pump is shown at the
apical membrane. In order to fit with observed conditions for K^+
activation in both intact preparations and isolated vesicles a
modification from Fig. 6 is proposed: that K^+ is largely recycled
within the membrane phase. An ad hoc postulate is introduced whereby K^+ movement with the membrane phase is facilitated by a specific
channel or carrier (X), which has not yet been identified. (Perhaps
it has been removed by vesicle preparation or requires a particular
activating reaction--e.g. pump activation.) The equivalent circuit,
in which resistance to K^+ movement might be different in the inner
and outer membrane region $(R'_K > R_K)$, is shown in the inset. The recycling of K^+, coupled to the H^+/K^+ exchange, would provide the
basis for an electrogenic pump that would be especially apparent in
Cl^--free solutions. Cl^- at the apical surface would provide the
return limb of the circuit, thus permitting the net flow of HCl,
which appear tightly coupled on open circuit. Under conditions of
active HCl secretion Cl^- entry into the oxyntic cell would be provided by the neutral Cl^-/HCO_3^- exchanger (Ex) at the basolateral
surface. The Na^+ pump (P) at that same surface would serve to maintain the cellular ionic milieu and K^+ balance. The Na^+ gradient,
serosa to cell, could provide the driving force for moving Cl^-, via
a coupled NaCl entry (C), into the cell against its electrochemical
potential gradient. If the permeability of the apical membrane to
Na^+ and K^+ were very low, we would not expect this system to provide
significant net transepithelial ion flow; however, under short-
circuit conditions there would be an associated Cl^- current. This
same serosal Na^+ pump system could provide for net K^+ secretion or

*net Na$^+$ absorption, depending upon the permeability of the apical
membrane to these ions. At rest the membrane potential would
largely be the summation of a K$^+$ concentration emf across the serosal surface and a Cl$^-$ concentration emf across the mucosal surface.
When secreting HCl there would be the additional emf associated
with current flow through the H$^+$ and K$^+$ limbs at the apical surface.
This highly simplified representation does not account for the different cell types of the gastric epithelium, nor for the membrane
transformation known to be associated with secretion (reproduced
from ref. 31).*

Cl$^-$ entry from the serosal solution into the epithelial cells
would occur by three pathways: (i) Cl$^-$/HCO$_3^-$ exchange; (ii) Coupled
NaCl entry, and (iii) A Cl$^-$ conductive pathway (not shown). These
would not necessarily all be on the same cell type of the histolo-
gically complex gastric epithelium. Rehm (32) has offered convincing evidence for the operation of neutral Cl$^-$/HCO$_3^-$ exchange across
the serosal surface of oxyntic cells which would operate to eliminate cellular HCO$_3^-$ (or OH$^-$) and to provide an enhanced source of
Cl$^-$, both associated with H$^+$ secretion. A Cl$^-$ conductance at the
serosal surface has been demonstrated (7,33), but this is less than
K$^+$ conductance at that site (33).

The proposed coupled NaCl entry is consistent with Machen's
observations (17,18), and could represent an "active" force for
driving Cl$^-$ into the cell. The energy would of course come from the
Na$^+$ pump and the Na$^+$ gradient. When the tissue is not secreting
HCl and on open circuit, a steady-state could be maintained with no
flow of solute if permeability to Na$^+$ and K$^+$ at the mucosal surface
were very low. (The short circuited tissue would deliver a current
of Cl$^-$ from serosal to mucosa.) If mucosal surface permeability to
K$^+$ were increased there would be a net secretion of KCl (e.g., ref.
34); while an increase in mucosal surface permeability to Na$^+$ would
be associated with net absorption of NaCl (e.g., ref. 5,9,31). At
rest, the membrane potential would largely be the summation of a K$^+$
concentration e.m.f. across the serosal surface and a Cl$^-$ concentration e.m.f. across the mucosal surface, the net polarity being

mucosal-side negative. When secreting HCl there would be the additional e.m.f. associated with electrogenic H^+ transport.

VII. SUMMARY

Gastric mucosa actively transports both H^+ and Cl^- from blood
to lumen. Numerous experiments have shown interdependencies as
well as independencies for these in transport processes. Recent
studies have demonstrated K^+-stimulated ATPase and H^+ transport
activities in membrane vesicles isolated from gastric mucosa, and
these activities have been implicated as the gastric H^+ pump.
Characteristics of vesicular H^+ transport are as follows: (i) Requires K^+ at some intravesicular or intramembranous site (H^+/K^+
exchange pump); (ii) Anions are not involved in direct enzymic
stimulation but "regulate" pumping rate by virtue of their permeability, characteristics with a sequence, $NO_3^->Br^->Cl^->I^->$acetate $\simeq$
isethionate$\simeq SO_4^{-2}$; (iii) Pump is "uncoupled" from anion-control by
treatments which increase both H^+ and K^+ permeability.
    These observations plus recent studies showing Na^+-dependent
transport of Cl^- by gastric mucosa form the basis for the suggestion that gastric Cl^- transport is principally coupled to transport
of other ions. Thus at the apical cell surface Cl^- would be coupled
to H^+ transport via the H^+/K^+ pump. At the basolateral surface Cl^-
movement is coupled to HCO_3^- (Cl^-/HCO_3^- exchange) and to Na^+ (NaCl
cotransport). For Cl^- transport associated with HCl secretion the
driving forces would be provided by the H^+ pump and the Cl^-/HCO_3^-
exchange at their respective surfaces. For the nonsecreting
stomach, Cl^- transport would be driven by the Na^+ gradient at the
serosal surface.

ACKNOWLEDGMENTS

This work was supported in part by a grant from the U.S. Public Health Service, AM10141.

REFERENCES

1. W. S. Rehm. A theory of the formation of HCl by the stomach. *Gastroenterology 14:* 401-417, 1950.

2. C. A. M. Hogben. The chloride transport system of the gastric mucosa. *Proc. Nat'l. Acad. Sci.,* Wash. 37: 393-395, 1951.

3. W. S. Rehm. Hydrochloric acid secretion, ion gradients and the gastric potential. In *"The Cellular Functions of Membrane Transport"* (J.F. Hoffman, ed.), Prentice-Hall, New Jersey, 1964, pp. 231-250.

4. R. P. Durbin. Anion requirements for gastric acid secretion. *J. Gen. Physiol. 4:* 735-748, 1964.

5. T. E. Machen and J. G. Forte. Gastric Secretion. In *"Transport Across Biological Membranes"* V. 4 (G. Giebisch, ed.), Springer Verlag, Heidelberg, 1979, pp. 693-747.

6. E. Heinz and R. P. Durbin. Evidence for an independent H^+ pump in the stomach. *Biochim. Biophys. Acta 31:* 246-247, 1959.

7. J. G. Forte, P. H. Adams, and R. E. Davies. The source of the gastric mucosal potential difference. *Nature, Lond. 1974:* 874-876, 1963.

8. J. G. Forte. Three components of Cl^- flux across bullfrog gastric mucosa. *Am. J. Physiol. 216:* 167-174, 1969.

9. J. G. Forte and T. E. Machen. Transport and electrical phenomena in resting and secreting piglet gastric mucosa. *J. Physiol. 224:* 33-51, 1975.

10. D. K. Kasbekar and R. P. Durbin. An ATPase from frog gastric mucosa. *Biochim. Biophys. Acta 105:* 472-482, 1965.

11. A. L. Ganser and J. G. Forte. K^+-stimulated ATPase in puri-

fied microsomes of bullfrog oxyntic cells. *Biochim. Biophys. Acta 387:* 169-180, 1973.

12. J. G. Forte, A. L. Ganser, R. Beesley, and T. M. Forte. Unique enzyme of purified microsomes from pig fundic mucosa. *Gastroenterology 69:* 175-189, 1975.

13. A. L. Ganser and J. G. Forte. Ionophoretic stimulation of K^+-ATPase of oxyntic cell microsomes. *Biochem. Biophys. Res. Comm. 54:* 690-696, 1973.

14. J. Lee, G. Simpson, and P. Scholes. An ATPase from dog gastric mucosa: Changes of outer pH in suspension of membrane vesicles accompanying ATP hydrolysis. *Biochem. Biophys. Res. Comm. 60:* 825-832, 1974.

15. G. Sachs, H. H. Chang, E. Rabon, R. Schackmann, M. Lewin, and G. Saccomani. A non-electrogenic H^+ pump in plasma membranes of hog stomach. *J. Biol. Chem. 251:* 7690-7698, 1976.

16. H. C. Lee, H. Breitbart, M. Berman, and J. G. Forte. Potassium stimulated ATPase activity and H^+ transport in gastric microsomal vesicles. *Biochim. Biophys. Acta 553:* 107-131, 1979.

17. W. L. McLennan, T. E. Machen, and T. Zeuthen. Ba^{2+} inhibition of electrogenic Cl^- secretion in vitro frog and piglet gastric mucosa. *Am. J. Physiol. 238:* G403-G413, 1980.

18. T. E. Machen and W. L. McLennan. Na^+-dependent H^+ and Cl^- transport in vitro frog gastric mucosa. *Am. J. Physiol. 239:* G151-G160, 1980.

19. H. C. Lee and J. G. Forte. A study of H^+ transport in gastric microsomal vesicles using fluorescent probes. *Biochim. Biophys. Acta 508:* 339-356, 1978.

20. D. W. Deamer, R. C. Prince, and A. R. Crofts. The response of fluorescent amines to pH gradient across liposome membrane. *Biochim. Biophys. Acta 274:* 323-335, 1972.

21. J. G. Forte, A. L. Ganser, and T. K. Ray. The K^+-stimulated ATPase from oxyntic glands of gastric mucosa. In *"Gastric Hydrogen Ion Secretion"* (D. Kasbekar, W. R. Rehm, and G. Sachs, eds.), Marcel Dekker, New York, 1976, pp. 302-330.

22. H. C. Lee, H. Breitbart, and J. G. Forte. The functional role of K^+-ATPase in proton transport by gastric microsomal vesicles. *Annals N.Y. Acad. Sci. 341:* 297-311, 1980.

23. C. A. M. Hogben. Ultrastructure and transport across epithelial membranes. *Circulation 26:* 1179-1188, 1962.

24. W. S. Rehm. Electrogenicity of gastric proton mechanism in light of effects of thiocyanate and weak bases. *Proc. of XXVI Int. Cong. Physio. Sci.*, Budapest. Vol. XIV, p. 660.

25. R. P. Durbin and E. Heinz. Electromotive chloride transport and gastric acid secretion in the frog. *J. Gen. Physiol. 41:* 1035-1047, 1958.

26. H. W. Davenport and F. Alzamora. Sodium, potassium, chloride and water in frog gastric mucosa. *Am. J. Physiol. 202:* 711-715, 1962.

27. R. A. Frizzell, M. Field, and S. G. Schultz. Sodium-coupled chloride transport by epithelial tissue. *Am. J. Physiol. 236:* F1-F8, 1979.

28. H. C. Lee. H^+ transport system of microsomes isolated from pi gastric mucosa. Ph.D. Thesis, Univ. of California, Berkeley, CA, 1978.

29. J. M. Wolosin and J. G. Forte. Changes in the membrane envirc ment of the (K^++H^+)-ATPase following stimulation of the gastri oxyntic cell. *J. Biol. Chem.*, 1981, in press.

30. J. M. Wolosin and J. G. Forte. Functional differences betweer K^+-ATPase rich membranes isolated from resting or stimulated rabbit fundic mucosa. *FEBS Letters 125*(2): 208-212, 1981.

31. J. G. Forte, T. E. Machen, and K. J. Obrink. Mechanisms of gastric H^+ and Cl^- transport. *Physiol. Rev. 42:* 111-126, 1980.

32. W. S. Rehm. Ion permeability and electrical resistance of the frog's gastric mucosa. *Fed. Proc. 26:* 1303-1313, 1967.

33. J. B. Harris and I. S. Edelman. Chemical concentration gradients and electrical properties of gastric mucosa. *Am. J. Physiol. 206:* 769-782, 1964.

34. J. B. Harris and I. S. Edelman. Transport of potassium by
     the gastric mucosa of the frog. *Am. J. Physiol.* *198*: 280-
     284, 1980.

THE ROLE OF SODIUM IN ANION TRANSPORT

ACROSS RENAL AND SMALL INTESTINAL CELLS:

STUDIES WITH ISOLATED PLASMA MEMBRANE VESICLES

Heini Murer

Physiologisches Institut der Universität
Zürich, Switzerland

Rolf Kinne

Department of Physiology
Albert Einstein College of Medicine
Bronx, New York

INTRODUCTION

The renal and intestinal epithelium transports a great number of anions, inorganic as well as organic. Thus in the kidney
the proximal tubule reabsorbs chloride, bicarbonate, phosphate,
and sulfate, and extracts the acidic amino acids, tricarboxylic
acid cycle intermediates, lactate, pyruvate and taurocholate from
the primary urine. Additionally, organic acids are secreted by
the tubule. In the intestine similar transport processes are
observed.

Considering the driving forces for the movement of these
anions across the epithelium, the general thermodynamical descrip-

173

tion has to be applied, which is given in Eq. (1).

$$J_i = J_i act + (1-\sigma_i)\bar{c}_i J_v + P_i\left(\Delta c_i + \frac{z_i F}{RT}\bar{c}_i \Delta\phi\right) \tag{1}$$

where J_i is the transepithelial net transport of solute; $J_i act$ the active transport component; σ_i the reflection coefficient of the substances; $\bar{c}_i$ the mean concentration across the membrane; Δc_i the transepithelial concentration difference; J_v the volume flux; P_i the permeability coefficient; z_i the charge and valency of charged species; F the Faraday number; and $\Delta\phi$ the transepithelial electrical potential difference. This equation shows that for the transepithelial transport, active $(J_i act)$, as well as passive, fluxes have to be taken into account. These passive fluxes comprise the transport by solvent drag $[(1-\sigma_i)\bar{c}_i J_v]$ and diffusion $[P_i(\Delta c_i + (z_i F/RT)\bar{c}_i \Delta\phi]$, and occur especially in low-resistance epithelia predominantly through the leaky paracellular shunt pathways. Using this type of phenomenological description in the rat proximal tubule, it was found that chloride transport in this epithelium is mostly passive, whereas the reabsorption of bicarbonate is the result of active absorption and passive backflux (1). For weak organic acids also, the nonionic diffusion governed by pH differences has to be considered. A detailed treatise of this problem is given by Weiner and Maffey in a recent review (2).

For the transport of an anion across the luminal and contraluminal border of the epithelial cell the following conditions exist. Since the interior of the cell is charged negatively compared to the extracellular fluid, the entry of an anion into the cell requires energy, unless the intracellular concentration is so low, that the chemical concentration difference is high enough to overcome the opposing electrical potential difference. If the intracellular concentration of the transported species is higher than predicted from equilibrium distribution, and reabsorption is to occur, the active entry step can be expected to be located in the luminal cell border, or in the instance of secretion, the active entry step is present in the contraluminal cell border. Exit

of an anion from the cell at the opposing cell border might proceed passively, driven by the electrochemical potential difference.

The present contribution summarizes the knowledge obtained
by using isolated plasma membrane vesicles derived from the luminal or contraluminal cell pole on the properties of the transport
systems, located in the two cell borders. A major emphasis is
placed on the role of the sodium gradient as a driving force in
the small intestinal and renal proximal tubular anion transport,
since flux measurements in intact epithelia have shown that the
majority of anion transport processes are dependent on the presence
of sodium at the cell side where the active uptake of the anion
into the cell is thought to take place (1, 3).

I. SYSTEMS INVOLVED IN ABSORPTION

A. *Transport Systems in the Luminal Membrane*

1. *Absorption of Organic Acids.* Two examples will be given
of the mechanisms involved in the absorption of organic acids in
kidney and intestine: the transport of L-lactate and the transport of L-glutamate by rat renal proximal tubule plasma membranes.
As shown in Fig. 1A, the uptake of L-lactate by isolated brush
border vesicles is stimulated by the presence of a concentration
difference for sodium (sodium gradient) across the membrane. An
overshoot phenomenon is observed, which indicates that lactate is
transiently accumulated inside the membrane vesicles. If sodium
is replaced by choline the uptake is much slower and the overshoot
is absent. D-lactate inhibits L-lactate uptake only in the
presence of sodium. Stimulation of lactate flux, or in general of
any anion flux, by the sodium gradient does not necessarily imply
that a sodium-anion cotransport system, similar to the systems observed for D-glucose and other neutral sugars and amino acids is
present in the membrane (4). Indirect coupling, e.g., through

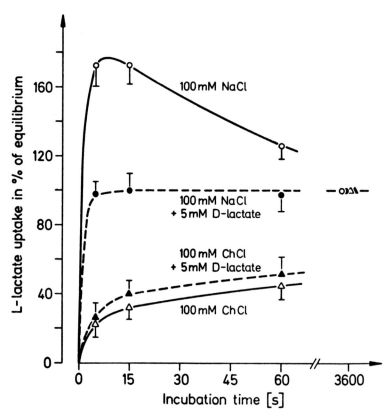

FIG. 1A

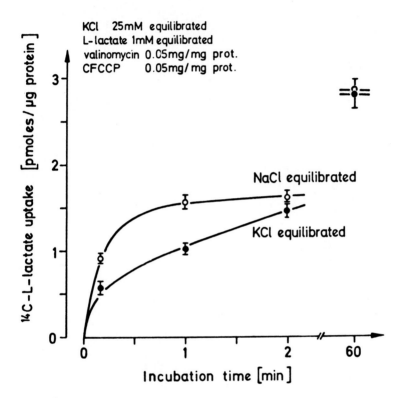

FIG. 1B

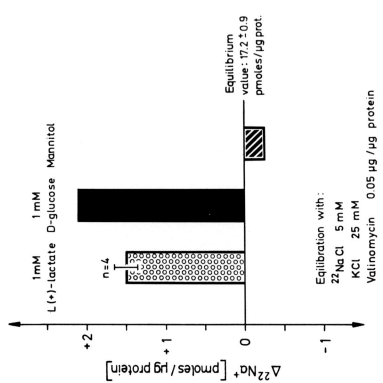

FIG. 1. Lactate sodium cotransport in rat renal brush border membranes: Brush border membranes from kidney cortex were prepared by the method of Evers et al. (48). Membranes were treated with valinomycin (0.05 µg/µg membrane protein) and loaded with 25 mmol/liter KCl, 100 mmol/liter mannitol and 20 mmol/liter HEPES-Tris, and incubated at 25°C in solutions containing in addition 1 mmol/liter ^{14}C-lactate, 100 mmol/liter NaCl or choline chloride with or without 5 mmol/liter D-lactate (Fig. 1A). For the experiments given in Fig. 1B the vesicles were loaded with either 75 mmol/liter NaCl and 25 mmol/liter KCl or with 100 mmol/liter KCl in addition to the mannitol and HEPES-Tris; incubation was carried out in identical solutions containing 1 mmol/liter ^{14}C-L-lactate (absence of sodium or potassium gradients). For the study of L-lactate-induced sodium tracer fluxes, concentrated solutions of L-lactate (or D-glucose) were given to membrane vesicles which were preloaded with 25 mmol/liter KCl and 5 mmol/liter $^{22}NaCl$ in addition to mannitol and HEPES-Tris; in Fig. 1C the uptake value after 5 s is given as the difference between control (no L-lactate or D-glucose) and L-lactate or D-glucose-containing media.

FIG. 1C

generation of a vesicle-inside- positive diffusion potential, can take place. Furthermore, through the action of the sodium-proton exchange system known to be present in the brush border membrane, the inside of the vesicle can become more alkaline and entry of the weak acid by nonionic diffusion might be facilitated (5). Therefore, experiments excluding such indirect coupling phenomena were conducted. As shown in Fig. 1B, L-lactate influx into brush border membrane vesicles is also stimulated under conditions where the sodium gradient has been abolished by preincubation of the vesicles in a sodium containing buffer. It could also be demonstrated that a L-lactate gradient can drive the uptake of sodium into the brush border vesicles even if the membrane potential had been short-circuited by potassium and valinomycin (Fig. 1C). These three observations taken together provide strong evidence for the existence of a L-lactate-sodium cotransport system in the brush border membrane (16). Similar observations, although in some instances obtained under different experimental conditions, have been made for the transport of lactate in rat intestinal brush border membranes, for various trixarboxylic acid cycle intermediates in rabbit and rat renal brush borders, for taurocholate in rat intestinal brush border membranes and ketone bodies in rat renal brush border membranes (6 - 15).

All the characteristics mentioned above for L-lactate have been observed in the uptake of L-glutamate by the renal brush border membrane vesicles. Both systems differ, however, in two aspects. First, the sodium-dependent L-lactate transport in renal membrane is rheogenic. More than one sodium ion is transported together with the anion lactate, therefore L-lactate uptake is stimulated when the inside of the brush border vesicle is hyperpolarized by super-imposing a vesicle-inside-negative diffusion potential (16). The same maneuver leaves the transport of L-glutamate unaltered; at least in the complete absence of internal potassium. Second, as shown in Fig. 2, the presence of potassium inside the brush border vesicle stimulates the uptake of L-gluta-

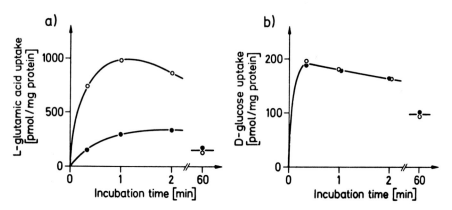

FIG. 2. Effect of intravesicular potassium on the sodium-
dependent uptake of L-glutamate (left diagram) and D-glucose
(right diagram) into rat renal brush border membrane vesicles.
The uptake was determined under two different conditions: (●)
no potassium present inside and outside the vesicles (K_i = K_O = 0);
(○) potassium gradient (K_i = 50 mmol/liter, K_O = 8.3 mmol/liter).
Vesicles were loaded with 200 mmol/liter mannitol, 20 mmol/liter
HEPES/Tris, pH 7.4, 50 mmol/liter choline chloride (●) or KCl (o).
The incubation media contained 100 mmol/liter mannitol, 50 mmol/
liter NaCl, 50 mmol/liter choline chloride and 20 mmol/liter
HEPES-Tris, pH 7.4.

mate markedly. This stimulation seems not to be related to a
change in the membrane potential, since potassium preloading has
no effect on the rheogenic transport of D-glucose (4). The mechanism of this effect of potassium is not yet elucidated, but one could
envisage a model in which the uptake of glutamate in the presence
of a potassium gradient is coupled to an entry of two or more
sodium ions and an exit of one potassium ion. In studies with rat
renal brush border vesicles in the presence of potassium, sodium-
dependent glutamate uptake is potential sensitive, whereas in
studies with rabbit membrane preparations, transport seems to occur electroneutrally (17 - 20). Interestingly, similar effects
have been already described in other plasma membranes (e.g., 21).

2. *Absorption of Inorganic Anions.* The best studied examples of inorganic anion transport in the renal proximal tubule and in the small intestine are the phosphate transport and the sulfate transport. In rat and rabbit kidney and small intestine, the brush border membrane contains a sodium-phosphate cotransport system (22 - 24). Sulfate, which has been studied more recently, is taken up by renal and intestinal brush border membrane vesicles also in a sodium-dependent manner. As illustrated in Fig. 3, evidence for a transport system for sulfate can be obtained only in the presence of sodium, where the sulfate transport is saturated with increasing concentrations of sulfate in the incubation medium (25, 26).

Studies with respect to the mode of chloride transfer utilizing isolated plasma membrane vesicles have been hampered by the fact that experiments with chloride isotopes are very difficult, mainly because the more stable isotope ^{36}Cl has a specific activity which is extremely low for vesicle studies. Therefore, conclusions on chloride movements have been derived until recently, from two types of indirect experiments, one employing pH measurements and testing for the presence of a chloride/hydroxylion exchange system, and the other, investigating the effect of chloride on sodium movements across the membranes, thereby looking for sodium-chloride cotransport systems. Concerning the former type of experiments (Cl/OH-exchange), Liedke and Hopfer could demonstrate that in rat small intestinal brush border membrane vesicles a proton gradient across the membrane was dissipated much more rapidly in the presence of chloride than in the presence of nitrate or thiocyanate (27). These authors provided evidence for the presence of an OH/Cl-exchange system in these membranes by demonstrating in addition that the proton conductance of the membrane is very low and therefore the conductance of the membrane for the counterion is rate limiting for the proton transfer (27). In similar experiments performed in our laboratory with rat renal brush border membranes no such preferential role of chloride could be observed (M. Barac-

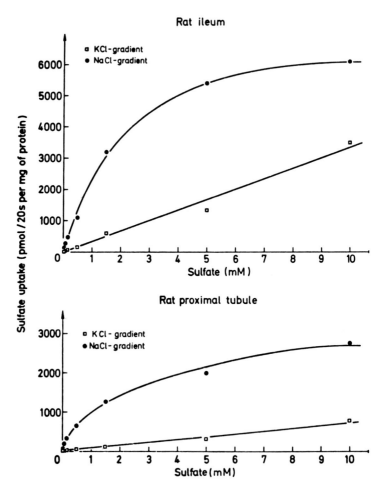

FIG. 3. Saturation of SO_4^{2-} uptake by rat renal and ileal
brush border vesicles: Brush border membranes from rat proximal
tubules were isolated as described by Evers et al. (48); those
from rat ileum as described by Murer et al. (5). Membrane
vesicles were loaded with 100 mmol/liter mannitol and 20 mmol/
liter HEPES-Tris (pH 7.4) and incubated in a medium containing in
addition 100 mmol/liter salt and the sulfate concentrations as indicated on the figures.

Nieto, H. Murer, and R. Kinne, unpublished observations). More
recently, tracer flux experiments using [36]Cl or [82]Br isotopes,
respectively, provided firm evidence for the existence of a

Cl^-/OH-exchanger in small intestinal brush border membranes (28, 29). Similar experiments were reported by Warnock and Yee for rabbit renal brush border membrane vesicles (30). In these experiments on chloride tracer flux, it was possible to demonstrate that an inwardly directed proton gradient drives chloride uptake into the vesicles in the complete absence of sodium.

From electrophysiological experiments as well as from tracer flux experiments in the intact epithelium, a model for electroneutral sodium chloride uptake into the internal cell has been proposed, in which a chloride/hydroxyl (bicarbonate) exchange system and a sodium/proton exchange system are coupled functionally (31). Indeed, a sodium/proton exchange system has been demonstrated in renal as well as in intestinal brush border membranes from rat, rabbit, and flounder (5, 32). The operation of this system is illustrated in Fig. 4. A transmembranal sodium gradient elicits an extrusion of protons from the vesicle. A proton gradient directed from the inside or the outside stimulates sodium uptake and maintains a transient intravesicular accumulation of sodium (overshoot). Thus, the possible elements (Na/H exchange and Cl/OH-exchange) for the proposed model of electroneutral sodium chloride transport have been identified, however, the demonstration of their combined function in the vesicle has been very difficult, if not impossible (32, 33). This might be due to an uncoupling of these systems when studied in the isolated membrane system. Liedke and Hopfer recently presented kinetic as well as experimental arguments against the existence of a Na-Cl cotransport system in the intestinal brush border membranes (34).

The sodium/proton exchange system in rat renal brush border membranes is involved in the bicarbonate reabsorption by the proximal tubule. According to the current concepts on bicarbonate transport in the proximal tubule, the acidification of the urine leads to an increased formation of H_2CO_3 and CO_2, as these two enter the tubular cell. At this point it should be mentioned also that the possibility exists that bicarbonate transport might not

Effect of H⁺/OH⁻ on sodium flux

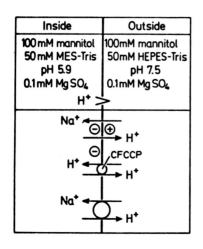

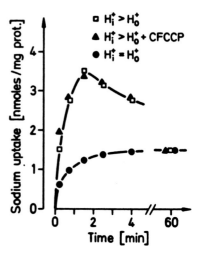

Effect of sodium on H⁺/OH⁻ flux

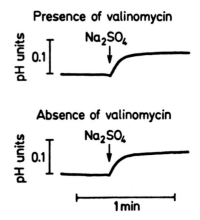

FIG. 4. *Coupling between sodium and H^+/OH^- flux across brush border membranes (kidney). Experimental conditions are given in scheme. CFCCP (=carbonyl cyanide p-trifluoromethoxy-phenylhydrazone) and valinomycin were added to incubation medium as ethanolic solutions (final concentration: 0.05% ethanol; 10 µg CFCCP mg protein^{-1}; 10 µg valinomycin mg protein^{-1}). Sodium concentration in H^+ efflux experiments was 100 mmol/liter. Sodium concentration in sodium uptake experiments was 1 mmol/liter. Results from Murer et al. (5).*

only be driven by the sodium gradient but might also be energized by ATP. Thus, a nonmitochondrial Mg-ATPase stimulated by bicarbonate has been found to be located in the brush border membranes of the proximal tubular cells (35). Intestinal brush border membranes seem to contain a chloride/bicarbonate stimulated ATPase (36). Recently, evidence has been presented that this nonmitochondrial Mg-ATPase present in renal brush border membranes is able to catalyze proton transfer across the brush border membrane (29).

B. Transport Systems in the Contraluminal Membranes

Only little is known hitherto on the mechanisms involved in the exit of anions from the absorptive cells. This is mainly due to the difficulties encountered in isolating pure basal-lateral plasma membranes in reasonable yields. For L-lactate, it has been found recently that basal-lateral plasma membranes from rat proximal tubule contain a facilitated diffusion system which is sodium-independent (14). As shown in Fig. 5, transstimulation of L-lactate uptake can be observed in the basal-lateral plasma membranes, which in contrast to the brush border membranes, is also observed in the absence of sodium. [Note: Similar findings were recently obtained in studies on lactate transport with intestinal basal-lateral membrane preparations (37).] The transport of phosphate and sulfate across the basal-lateral plasma membranes of renal proximal tubular and small intestinal epithelial cells is also sodium-independent (H. Murer, H. Lücke, and R. Kinne, unpublished observations). Physiological observations on the transport of sulfate and inorganic phosphate across the intact epithelium suggest that the exit step might involve an anion/hydroxylion exchange mechanism. The transfer of L-glutamate across the basal-lateral plasma membranes involves probably also sodium and potassium-dependent transport systems. This finding can explain the very high accumulation of this amino acid in the proximal tubular cell (38).

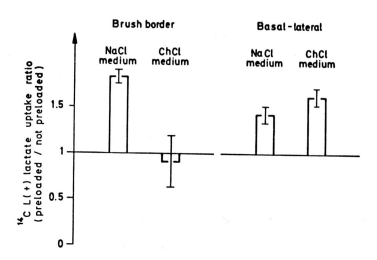

FIG. 5. Effect of sodium on the tracer L-lactate uptake after preloading with cold L-lactate: Basal-lateral membranes and brush border membranes from rat kidney cortex were isolated by the free flow electrophoresis method (49). Membranes were loaded with 100 mmol/liter mannitol, 25 mmol/liter KCl and 75 mmol/liter NaCl or choline chloride, 20 mmol/liter HEPES-Tris, pH 7.4. For lactate preloading 1 mmol/liter cold L-lactate was also included in this buffer. Tracer L-lactate uptake was studied in the corresponding media containing 1 mmol/liter ^{14}C-L-lactate. Valinomycin at a concentration of 0.05 µg/µg protein was present during the incubations.

II.   SYSTEMS INVOLVED IN SECRETION

A.   *Contraluminal Systems*

1.   Organic Acids. One of the most prominent examples for studies of secretory processes of anions is the secretion of p-aminohippuric acid (PAH) by the renal proximal tubule. Studies with isolated basal-lateral plasma membranes derived from rat kidney cortex revealed that the membranes contain a probenecid-sensitive transport system for PAH. The finding with isolated membrane vesicles, that PAH crosses the membrane as an anion, does not provide any information on the driving force for the entry of the organic acid at the contraluminal cell pole (39).

2. *Inorganic Anions.* Sodium-dependent secretion of chloride can occur in the small intestine, a tissue that normally absorbs chloride in a sodium-dependent manner (3). Evidence suggests that in the small intestine, absorption of chloride proceeds in the mature enterocytes, whereas secretion proceeds in the crypt region of the folded epithelial layer. To omit the problems related to the inhomogeneity of the epithelium, model studies on secretory processes have been performed with membrane vesicles isolated from the rectal gland of the dogfish, an epithelium with uniform cell population. The rectal gland of the dogfish (*Squalus acanthias*) has a very potent excretory system for chloride. The excretion is dependent on the presence of sodium and is inhibited by ouabain (40). In membrane vesicles derived predominantly from the basal-lateral plasma membranes of the gland, a sodium-chloride cotransport system could be demonstrated (41). One piece of evidence for the presence of this system is shown in Fig. 6, where the tracer replacement for sodium (which is equivalent to a saturation phenomenon) was studied in an incubation medium containing chloride or nitrate. Saturability of sodium uptake into the rectal gland plasma membrane vesicles could only be observed in the presence of chloride, indicating that a limited number of transport sites are present in the membrane which require the presence of chloride for their interaction with sodium. Furthermore, the chloride-dependent sodium flux is inhibited by furosemide which also inhibits chloride secretion in the intact gland (41).

B. *Luminal Systems*

1. *Organic Acids.* For a long time it has been postulated by Foulkes that in rabbit proximal tubule a luminal transport system is present for PAH. In their studies with rat brush border membranes, Berner and Kinne were not able to demonstrate such a system (39). However, recent experiments with flounder kidney brush border membranes provided unequivocal evidence for a luminal

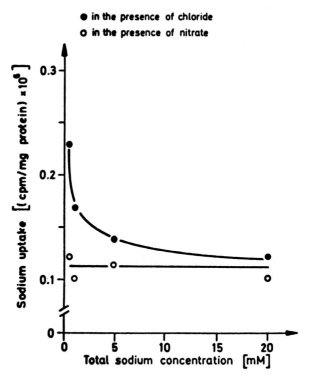

● in the presence of chloride
○ in the presence of nitrate

FIG. 6. Effect of chloride on tracer replacement of $^{22}Na^+$ uptake into rectal gland plasma membrane vesicles: The membranes were prepared from dogfish rectal gland (41) and loaded with 100 mmol/liter mannitol, 1.2 mmol/liter MgNO₃, 20 mmol/liter Tris-HEPES, pH 7.6, and 100 mmol/liter KCl (●) or KNO₃ (○). The incubation medium contained increasing concentration of NaCl (●) or NaNO₃ (○) as given in the figure. Total salt concentration in the incubation was also 100 mmol/liter by either adding KCl (●) or KNO₃ (○). Membranes were treated with valinomycin (around 10 µg/mg protein).

transport system for PAH which facilitates the transfer of the organic acid as an anion across the membrane (42). Similar findings have also been obtained for rabbit renal brush border vesicles (43). Thus, the exit of PAH from the cell at the luminal side could be driven by the electrical potential and by the chemical concentration difference of the anion (42). More recently, Blomstedt and Aronson provided evidence for a PAH/OH⁻ exchange mechanism to be located in the brush border membrane of rabbit proximal

tubule, rendering PAH transport electroneutral and driven by transmembrane pH-differences (44). It is difficult to visualize the
physiological importance of such a mechanism in tubular PAH secretion. However, since this mechanism also accepts urate as a substrate, it may play a significant role in effecting uphill urate
reabsorption. This elegant study further demonstrated that secondary active solute transport in epithelial membranes may be coupled
to the electrochemical gradient of an ion other than Na^+. Since
the establishment of a transmembrane proton-gradient is itself
sodium gradient-dependent via the operation of the Na^+/H^+ exchange
system and is therefore secondary active, one could use the term
"tertiary active" for proton coupled transport of a solute across
the luminal membrane.

   2. *Inorganic Anions.* The luminal exit steps for the secretion of inorganic anions by epithelial cells are only poorly understood hitherto. These systems seem to be, however, quite important for the regulation of secretory processes. Indirect evidence, for example, from the rectal gland stimulated by cyclic
AMP suggests that the permeability of the luminal membranes to
chloride is the rate-limiting step for chloride secretion that is
increased, in order to establish the higher rate of secretion (45).
Evidence for a conductive chloride pathway was recently obtained
in studies with rat intestinal brush border vesicles (28).

III. SYNOPSIS

   From the results reviewed above, some general principles on
the role of sodium-dependent anion transport systems in the epithelial absorption and secretion of anions can be delineated. All
epithelial cells - secretory as well as absorptive - have a low
concentration of intracellular sodium and a high concentration of
intracellular potassium. The gradients between the intra- and
extracellular fluid are maintained by the action of the Na^+-K^+

ATPase. With the possible exception of the choroid plexus, this
enzyme is located at the side of the cell which is in close contact
with the blood stream (46). Given this situation, sodium-anion
cotransport systems involved in secretion are located in the basal-
lateral membrane of the cell, and sodium-anion cotransport systems
involved in absorption are located in the luminal membrane of the
cell (see scheme in Fig. 7). Thereby the sodium gradient is used
as a driving force for the entry of the anion into the cell and
consecutively, since the exit of the anion out of the cell can follow the electrochemical gradients passively, the sodium gradient
acts as driving force for the whole transcellular transport of the
anion. As can be seen from the examples given above, the sodium-
anion cotransport systems are already relatively well understood,
the knowledge about the systems facilitating the exit of the anions
are, however, relatively scarce. In future, this area of research
should receive special attention. In this context it should be
emphasized that it is not at all certain that the exit mechanism
is always represented by a conductive pathway. Evidence is now
accumulating from studies using intact epithelial preparations
that anion-exchange mechanisms could be involved, e.g., recently

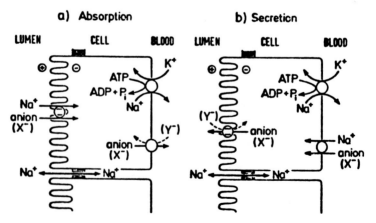

FIG. 7. Schematic representation of the polar distribution
of transport systems possibly involved in the epithelial absorption or secretion of anions.

it was demonstrated that proximal tubular transport of sulfate is dependent on HCO_3^- (K. J. Ullrich, unpublished observations). It has also been demonstrated that the exit step for chloride in the rabbit gallbladder epithelial cell is nonconductive (47). Finally, such anion-exchange mechanisms might also be involved in the cellular uptake of organic acids in the secretory process.

SUMMARY

Membrane vesicles isolated mainly from small intestinal or renal proximal tubular epithelia have repeatedly been used for the elucidation of the mechanisms involved in secretion or absorption of anions. From the results, some general principles on the role of sodium-dependent anion transport systems can be delineated. As examples lactate- and glutamate-transport for organic anion absorptive transport are discussed, and as examples for inorganic anion transport sulphate- and chloride transport are discussed in more detail. Anion secretory mechanisms are exemplified by experiments on sodium-chloride transport performed with membrane vesicles isolated from the dogfish rectal gland.

ACKNOWLEDGMENTS

The experiments presented and discussed in this article were in part supported by the Swiss National Science Foundation (Grant No. 3.451.079).

REFERENCES

1. E. Frömter, G. Rumrich, and K. J. Ullrich. Phenomenologic
    description of Na$^+$, Cl$^-$, and HCO$_3^-$-absorption. *Pflügers Arch.*
    343:189-220, 1973.

2. M. W. Weiner and R. H. Maffey. The provision of cellular
    metabolic energy for active ion transport. *In* "Physiology
    of Membrane Disorders," (T. E. Andreoli, J. F. Hoffmann, and
    D. D. Fanestil, eds.), New York and London, Plenum Medical
    Book Company, 1978, pp. 287-308.

3. R. A. Frizzell, M. Field, and S. G. Schultz. Sodium coupled
    chloride transport by epithelial tissues. *Am. J. Physiol.*
    236:F1-F8, 1979.

4. H. Murer and R. Kinne. The use of isolated membrane vesicles
    to study epithelial transport processes. *J. Membrane Biol.*
    55:81-95, 1980.

5. H. Murer, U. Hopfer, and R. Kinne. Sodium/proton antiport in
    brush border vesicles isolated from rat small intestine and
    kidney. *Biochem. J. 154*:597-604, 1977.

6. M. L. Garcia, J. Benavides, and F. M. Valdivieso. Ketone
    body transport in renal brush border vesicles. *Biochem.*
    Biophys. Acta 600:922-930, 1980.

7. J. Kippen, B. Hirayama, J. R. Klinenberg, and E. M. Wright.
    Transport of tricarboxylic acid cycle intermediates by mem-
    brane vesicles from renal brush border. *Proc. Natl. Acad.*
    Sci. USA 76:3397-3400, 1979.

8. M. S. Medow, S. B. Baruch, O. Gutienez, V. F. King, and
    E. L. Pinto. Transport of citric acid by luminal and contra-
    luminal membrane vesicles of dog renal cortex. *Fed. Proc. 37*:
    1347, 1978.

9. F. A. Wilson, G. Burckhardt, H. Murer, G. Rumrich, and K. J.
    Ullrich. Sodium coupled taurocholate transport in the proxi-
    mal convolution of the rat kidney *in vivo* and *in vitro*.
    J. Clin. Invest. 67:1141-1150, 1981.

10. S. H. Wright, J. Kippen, J. R. Klinenberg, and E. M. Wright. Specificity of the transport system for tricarboxylic acid cycle intermediates in renal brush borders. *J. Membrane Biol.* *57*:73-82, 1980.

11. R. C. Beesley and R. G. Faust. Sodium ion coupled uptake of taurocholate by intestinal brush border membrane vesicles. *Biochem. J.* *178*:299-303, 1979.

12. B. Hildmann, C. Storelli, W. Haase, M. Barac-Nieto, and H. Murer. Sodium ion/L-lactate cotransport in rabbit small intestinal brush border membrane vesicles. *Biochem. J.* *186*: 169-176, 1980.

13. H. Lücke, G. Stange, R. Kinne, and H. Murer. Taurocholate-sodium cotransport by brush border vesicles isolated from rat ileum. *Biochem. J.* *174*:951-958, 1978.

14. H. Murer, M. Barac-Nieto, K. J. Ullrich, and R. Kinne. Renal transport of lactate. *In* "Renal transport of organic substances," (P. Deetjen, S. Silbernagl, F. C. Lang, and R. Greger, eds.), Springer Vergag, Berlin-Heidelberg-New York, in press.

15. F. A. Wilson and L. L. Treanor. Glycodeoxycholate transport in brush border membrane vesicles isolated from rat jejunum and ileum. *Biochem. Biophys. Acta* *554*:430-440, 1979.

16. M. Barac-Nieto, H. Murer, and R. Kinne. Lactate-sodium cotransport in renal brush border membranes. *Am. J. Physiol.* *239*:F496-F506, 1980.

17. G. Burckhardt, R. Kinne, G. Stange, and H. Murer. The effects of potassium and membrane potential on sodium dependent glutamic acid uptake. *Biochem. Biophys. Acta* *599*:191-201, 1980.

18. E. G. Schneider, M. R. Hammermann, and B. Sacktor. Sodium gradient dependent L-glutamate transport in renal brush border vesicles. *J. Biol. Chem.* 2557645-7649, 1980.

19. E. G. Schneider and B. Sacktor. Sodium gradient dependent L-glutamate transport in renal brush border vesicles. *J. Biol. Chem.* *255*:7650-7656, 1980.

20. J. Seeger, G. Stange, H. Murer, and R. Kinne. Transport of L-glutamate by brush border membrane vesicles isolated from rat kidney cortex. *Pflugers Arch. 379*:R. 18, 1979.

21. G. Rudnick and P. J. Nelson. Platelet 5-hydroxytryptamine transport, an electroneutral mechanism coupled to potassium. *Biochemistry 17*:4739-4742, 1978.

22. H. Murer and B. Hildmann. Transcellular transport of calcium and inorganic phosphate in the small intestinal epithelium. *Amer. J. Physiol. 240*, g409-g416, 1981.

23. H. Murer, H. Stern, G. Burckhardt, C. Storelli, and R. Kinne. Sodium dependent transport of inorganic phosphate across the renal brush border membrane. *In* "Phosphate and Minerals in Health and Disease," (S. G. Massry, E. Ritz, and H. Jahn, eds.), Plenum Press, New York and London, 1980, pp. 11-23.

24. L. Cheng and B. Sacktor. Sodium gradient dependent phosphate transport in renal brush border membrane vesicles. *J. Biol. Chem. 256*, 8080-8084, 1981.

25. H. Lücke, G. Stange, and H. Murer. Sulfate sodium cotransport by brush border membrane vesicles isolated from rat ileum. *Gastroenterology 80*:22-30 (1981).

26. H. Lucke, G. Stange, and H. Murer. Sulphate-ion/sodium-ion cotransport by brush border membrane vesicles isolated from rat kidney cortex. *Biochem. J. 182*:223-229, 1979.

27. L. M. Liedke and U. Hopfer. Anion transport in brush border membranes isolated from rat small intestine. *Biochem. Biophys. Res. Commun. 76*:579-585, 1977.

28. C. M. Liedke and U. Hopfer. Mechanisms of Cl^- translocation across the small intestinal brush border membrane. II. Demonstration of Cl^-/OH^- exchange and Cl^- conductance. *Am. J. Physiol.*, in press.

29. H. Murer, E. Kinne-Saffran, R. Beauwens, and R. Kinne. Proton fluxes in isolated renal and intestinal brush border membranes. *In* "Hydrogen Ion Transport in Epithelia,"

(J. Schultz, ed.). Elsevier/Excerpta Medica/North Holland, Amsterdam, 1980, pp. 267-276.

30. D. G. Warnock and V. J. Yee. Chloride uptake by brush border membrane vesicles isolated from rabbit renal cortex. *J. Clin. Invest.* 67:103-115, 1981.

31. A. F. Bieberdorf, Ph. Gorden, and J. S. Fordtran. Pathogenesis of congenital alkalosis with diarrhea. *J. Clin. Invest.* 51:1958-1968, 1972.

32. H. Murer, J. Eveloff, R. Kinne, W. B. Kinter, and M. Field. Transport of sodium into brush border membrane vesicles isolated from flounder intestine and flounder kidney tubules. *MOJBL Bulletin* 17:52-56, 1980.

33. J. Eveloff, M. Field, R. Kinne, and H. Murer. Sodium cotransport systems in intestine and kidney of the winter flounder. *J. Comp. Physiol.* 135:175-182, 1980.

34. C. M. Liedke and U. Hopfer. Mechanisms of Cl^- translocation across the small intestinal brush border membrane. I. Absence of Na-Cl cotransport. *Am. J. Physiol.,* in press.

35. E. Kinne-Saffran and R. Kinne. Further evidence for the existence of an intrinsic bicarbonate stimulated Mg^{2+}-ATPase in brush border membranes isolated from rat kidney. *J. Membrane Biol.* 49:235-251, 1979.

36. M. H. Humphreys and L. J. N. Chou. Anion stimulated ATPase activity of brush border from rat small intestine. *Am. J. Physiol.* 236:E70-76, 1979.

37. C. Storelli, A. Corcelli, G. Cassano, B. Hildmann, H. Murer, and C. Lippe. Polar distribution of sodium dependent and sodium independent transport systems for L-lactate in the plasma membrane of rat enterocytes. *Pflügers Arch.* 388:11-16, 1980.

38. B. Sacktor, J. L. Rosenbloom, C. T. Liang, and L. Cheng. Sodium gradient- and sodium plus potassium-gradient dependent L-glutamate uptake in renal basolateral membrane vesicles. *J. Membrane Biol.* 60, 63-71, 1981.

39. W. Berner and R. Kinne. Transport of p-aminohippuric acid
     by plasma membrane vesicles isolated from rat kidney cortex.
     Pflügers Arch. 361:269–277, 1976.

40. P. Silva, J. Stoff, M. Field, L. Fine, J. N. Forrest, and
     F. H. Epstein. Mechanisms of active chloride secretion by
     shark rectal gland: role of Na^+-K^+-ATPase in chloride trans-
     port. *Am. J. Physiol. 233*:F298–F306, 1977.

41. J. Eveloff, R. Kinne, E. Kinne-Saffran, H. Murer, P. Silva,
     F. H. Epstein, J. Stoff, and W. B. Kinter. Coupled sodium
     and chloride transport into plasma membrane vesicles pre-
     pared from dogfish rectal gland. *Pflügers Arch. 378*:87–92,
     1978.

42. J. Eveloff, R. Kinne, and W. B. Kinter. Transport of glu-
     cose and PAH into plasma membrane vesicles prepared from
     flounder kidney tubules. *Am. J. Physiol. 237*:F291–298,
     1979.

43. J. Kippen, B. Hirayama, J. R. Klinenberg, and E. M. Wright.
     Transport of p-aminohippuric acid, uric acid and glucose in
     highly purified rabbit renal brush border membranes. *Bio-
     chem. Biophys. Acta 556*:161–174, 1979.

44. J. W. Blomstedt and P. S. Aronson. PH-gradient stimulated
     transpo-t of urate and p-aminohippurate in dog renal micro-
     villous membrane vesicles. *J. Clin. Invest. 65*:931–934,
     1980.

45. J. S. Stoff, R. Rosa, R. Hallac, P. Silva and F. H. Epstein:
     Hormonal regulation of active chloride transport in the dogfis
     rectal gland. *Emer. J. Physiol. 237*, F138–F144, 1979.

46. J. M. Tormey. Anatomical methods for studying transport
     across epithelia. *In* "Water Relations in Membrane Transport
     in Plants and Animals," (A. Jungreis, J. K. Hodges, A. Klein-
     zeller, and S. G. Schultz, eds.), New York and London, Aca-
     demic Press, 1977, pp. 233–248.

47. R. A. Frizzell, M. Dugas, and S. G. Schultz. Sodium chloride
     transport by rabbit gallbladder: Direct evidence for a coupled
     NaCl influx process. *J. Gen. Physiol.* *65*:769-795, 1975.

48. C. Evers, W. Haase, H. Murer, and R. Kinne. Properties of
     brush border vesicles isolated from rat kidney cortex by
     calcium precipitation. *Membrane Biochem.* *1*:203-219, 1978.

49. R. Kinne, H. Murer, E. Kinne-Saffran, M. Thees, and G. Sachs.
     Sugar transport by renal plasma membrane vesicles. *J. Mem-
     brane Biol.* *21*:375-395, 1975.

Cl TRANSPORT IN RABBIT CORNEA*

Stephen D. Klyce[+]

Department of Surgery, Division of Ophthalmology
Stanford University School of Medicine
Stanford, California

Control of cell and/or tissue hydration is a primary function
of certain biomembrane-resident ion pumps that balance the swelling
tendency (colloid osmotic pressure) of membrane-bound aqueous
phases (cytoplasm, connective tissue, etc.). In thermodynamic
terms, the maintenance of steady-state in living systems results
from a balance between dissipative and accretive ion flows, a
situation of minimum entropy production. The demonstration of
metabolism-linked vectorial solute transport by any single membrane
or membrane system is necessary to provide the driving force for
the maintenance of the nonequilibrium status of biosystems.

*Supported by USPHS grant EY 03311 from the National Eye In-
stitute.
+Present address: Louisiana State University Eye Center,
Louisiana State University Medical Center School of Medicine, New
Orleans, Louisiana 70112.

The control of corneal transparency is a good example of balance between passive (stromal swelling) and active (ion transport) processes. Corneal epithelium and endothelium are distinctly different general types of fluid transporting layers, yet each has resident ion pumps that participate synergistically in the regulation of stromal hydration. In this brief review the Cl transport system in the rabbit corneal epithelium will be discussed, beginning with its characterization at the tissue level, with membrane level studies following, and concluding with an analysis of the regulation of corneal hydration that deals with flows across both membrane and gel phases of tissues.

I. FLUXES, PHARMACOLOGY AND NEURAL CONTROL

The demonstration of secretory (stroma to tears) active Cl transport across the rabbit corneal epithelium depended in large measure upon the development of improved isolation procedures and incubation techniques since the mammalian cornea proved to be very sensitive to isolation damage--wrinkling, etc. It became evident that the inward epithelial Na transport discovered by Donn et al. (1) accounted for only half of the measured short-circuit current: the remaining current matched the net secretory flux of Cl (2, 3). The existence of two ion transport systems of similar magnitude with opposing charge and moving ions in opposite directions required further study for functional analysis, since in the presence of normal resting potential, electroneutrality must be obeyed. Hence, Cl and Na fluxes were measured at open circuit to determine the direction of net salt transport in this tissue (4). Under control conditions there was no significant salt transport at open circuit. However, with dibutyryl cyclic-AMP stimulation, Cl transport dominated at short circuit, and at open circuit 20 - 40 nmol/cm^2hr of NaCl moved from stroma to tears (Table I). This provides a possible driving force for epithelial fluid secretion.

TABLE I. Ion Transport across $10^{-3}M$ Dibutyryl Cyclic AMP-Treated Rabbit Corneal Epithelium

	Resting potential (mV)	Short-circuit current	Resistance $(k\Omega cm^2)$	J_{Cl}^{net}	J_{Na}^{net}
At short circuit (3)	--	255 ± 3	3.32 ± 0.08	$+194 \pm 6$	-39 ± 10^a
At resting potential (4)	27.7 ± 1.3	--	$3.13 \pm .14$	$+29 \pm 9$	$+36 \pm 9$

[a]Obtained from separate study. Negatively signed flux indicates tears to aqueous direction.
Units for current and net fluxes are $nEq/cm^2 hr$.

Many of the pharmacological features of Cl transport by amphibian and mammalian corneal epithelia are similar, and these are presented in the preceding article. Summarizing, Cl transport in rabbit cornea appears to be under control of cell cyclic AMP (which is found there in relative abundance), as the transport is stimulated by catecholamines, methyl xanthines, and dibutyryl cyclic AMP (2, 3). Prostaglandins (5) and ascorbate (6) also stimulate Cl transport. Finally, ouabain appears to inhibit Cl transport.

The presence of catecholamine receptors linked to the modulation of epithelial transport led to the speculation (3) that ocular sympathetics might exert some influence on transport in addition to their previously proposed antimitogenic influence. However, evidence for the presence of sympathetic nerves in the mammalian cornea has only recently been supported by continuing histochemical studies (7, 8).

Direct evidence for the presence and source of monoaminergic fibers can be obtained with neuroanatomical mapping procedures (9). To demonstrate the source of corneal sympathetic terminals, a 1-3 μl bleb of 2% (w/v) solution of horseradish peroxidase (HRP)

was formed in rabbit corneal stroma by injection. One day to 36
hr later, examination of serial frozen sections of entire superior
cervical ganglia (SCG) revealed up to 20 cell bodies in the ipsilateral ganglion containing HRP (Fig. 1). Hence, sympathetic nerve
terminals within or near the cornea originate in the SCG.

An animal model with which to study the possible influence of
corneal sympathetics on corneal electrophysiology has been developed (10). In these studies the stimulation of the pregang-
lionic trunk of the ipsilateral SCG was found to evoke the expected
hyperpolarization of epithelial potential when the tear surface was
bathed in Cl-free medium (to amplify the response to catecholamines,
cf., ref. 11). This effect was blocked by propranolol. These
findings suggest that corneal Cl transport may be modulated by sympathetic fibers from the SCG.

II. INTRACELLULAR MEASUREMENTS AND MEMBRANE MODEL

Understanding the role of individual membranes in the corneal
epithelium is complicated by the fact that it is composed of
several layers of cells. On the other hand, at a given depth in
the epithelium, there is only a single cell type. The electrical
potential and transverse membrane resistance profiles across the
rabbit corneal epithelium have been measured with microelectrodes
(11, 12). On the average, the potential profile occurred in three
reasonably discrete steps, each associated with a discrete change
in transverse membrane resistance. Correlation of these steps
with structure was confirmed by histologic demonstration of cell
type following intracellular injection of dye.

While making microelectrode penetrations from the solution
bathing the tear side of the epithelium, it was found that the
outer membrane of the squamous cell and the inner membrane of the
basal cell enclosed two regions of stable potential separated by
a small central potential step; the intracellular potential of all

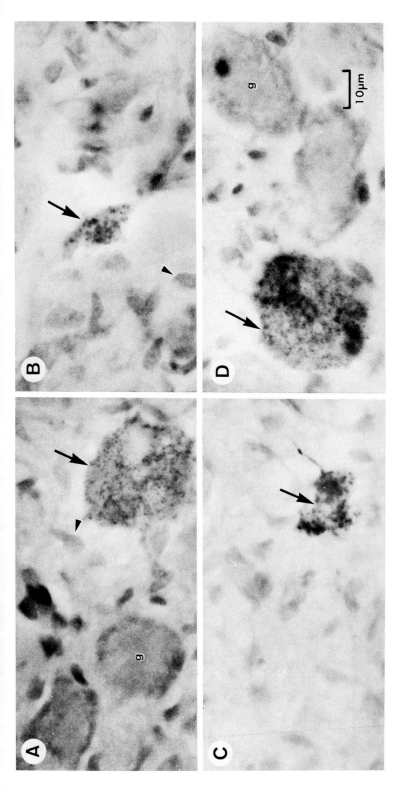

FIG. 1. Light micrographs of 20 – 30 μm thick frozen sections of superior cervical ganglion 24 hr after intracorneal injection of HRP. Arrows indicate HRP-labeled ganglion cell bodies. HRP was enhanced following the procedures given in ref. 9. Sections were counterstained with cresyl violet acetate. In A and D unlabeled ganglion cell bodies are indicated (g). Arrow heads in A and B point to the nuclei of satellite cells.

cells was negative with respect to both sides. The most recent measurements for control rabbit corneas (10) give the following values for outer membrane, central step, inner membrane, respectively, -26.2 ± 1.1 mV, -6.5 ± 0.7 mV, $+72.3 \pm 0.8$ mV (transepithelial potential = 39.6 ± 1.6 mV). Transverse membrane resistances associated with these potentials were (in the same order) 5.7 ± 0.3 kΩcm^2, 1.1 ± 0.1 kΩcm^2, and 0.8 kΩcm^2 (total epithelial resistance = 9.1 ± 0.5 kΩcm). Most of the residual resistance (16% of the total) occurred in the wing or intermediate cell region, which was essentially isopotential.

Early studies of corneal physiology noted the low permeability of the epithelium to Na and other ions and ascribed a passive barrier function to that layer with respect to corneal hydration. The microelectrode experiment indicated that the location of at least the major portion of that barrier was at the epithelial surface, probably represented by the combined resistance of the squamous cell outer membrane and the peripheral tight junctions.

Circuit analysis of the simplest Thevenin equivalent for this epithelium (12) showed that the 'shunt' resistance paralleling the outer membrane (presumably combined tight junction and edge damage resistance) averaged 4.5 kΩcm^2 compared to a cellular pathway resistance of 16 kΩcm^2. Recently, these data have been refined in tissue with minimal edge damage. Using two different methods in the determination values of 25.2 ± 2.8 kΩcm^2 for the unstimulated cellular pathway and 13.5 ± 1.5 kΩcm^2 for the paracellular shunt have been found (13).

Microelectrode penetrations were used to examine the mode of action of epinephrine in the stimulation of epithelial Cl transport (11). Pulses of epinephrine delivered to either medium bathing the isolated cornea transiently decreased epithelial resistance (Fig. 2). Analysis of membrane resistance changes indicated that epinephrine rather specifically produced a Cl-dependent decrease in the resistance of the outer cellular membrane. The resistance change was generally accompanied by a small increase in corneal

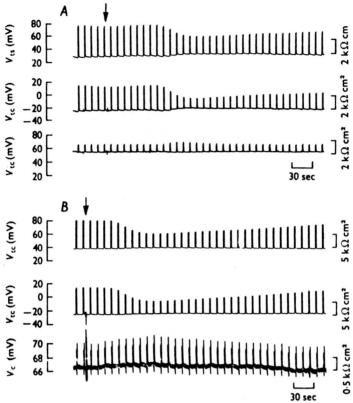

FIG. 2. *Response of epithelial (ts) and the outer (tc) and inner (cs) barrier potentials (V) and transverse resistance to epinephrine added to the tear side bathing solution at ↓ (final concentration = 5 × 10⁻¹⁰M).* Resistances were determined from the deflections produced by applying 1 sec constant current pulses every 10 sec. (A) Record made from a wing cell. The decrease in epithelial resistance was matched by changes at the outer barrier. The small increase in inner barrier <u>apparent</u> resistance was the consequence of the paracellular pathway. (B) Similar observations were made recording from a basal cell. (Reprinted from ref. 11.)

potential due to outer membrane depolarization. The sign and magnitude of the outer membrane polarization to epinephrine pulses were found to be dependent upon the Cl activity gradient across the outer cellular membrane. Lowering tear solution [Cl] led to larger depolarizing responses across the outer membrane, whereas depletion of cell Cl from the stromal side produced hyperpolarizing responses.

With current clamp techniques, the reversal potential for the epinephrine response of the outer membrane was measured as a func-

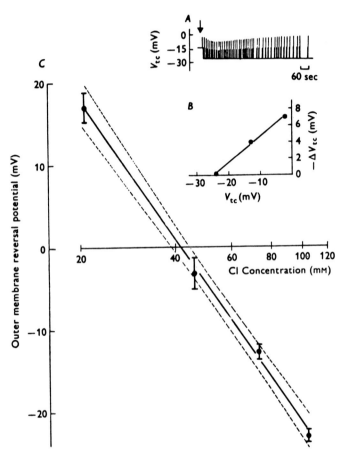

FIG. 3.   *The relationship between epinephrine-induced outer membrane reversal potential and tear side Cl concentration.   A. Outer membrane was polarized with three levels of constant current while recording its potential and response to epinephrine added at* ↓ .   *B.   At the peak response to epinephrine, the drug-induced voltage transients are plotted to determine the reversal potential.   C.   The slope of the relationship was 56 mV/decade tear* [Cl].   *The concentration at which the reversal potential is 0* (I_{Cl} = 0) *should be equivalent to intracellular* [Cl].   *(Reprinted from ref. 11.)*

tion of tear side Cl concentration (Fig. 3). The reversal potential for the epinephrine response matched the predictions of the
Nernst diffusion equation, indicating that the mechanism of stimulation was an increase in the Cl conductance of the outer barrier.
Furthermore, since there was such excellent Cl electrode behavior
exhibited in this type of experiment, cell Cl concentration was
taken as 41.5 mM, the concentration of Cl in the tears when the
reversal potential was zero.

Studies on the inner barrier showed that cell Cl could be de-
pleted slowly (over some hours) by bathing the inner corneal surface with Cl-free medium. This was reversible except when ouabain
(10^{-5} M) was present, which suggests that the entrance of Cl to
the epithelial cells is coupled in some fashion to Na transport.

The model currently proposed for rabbit corneal epithelial
transport at the membrane level is drawn in Fig. 4. The entrance
of Cl into the cells would be against a 43 mV electrochemical potential gradient. Its exit into the tears is down a small gradient
across the outer barrier whose Cl permeability is modulated by
catecholamines via cyclic AMP.

III. COUPLING OF Cl TRANSPORT TO FLUID TRANSPORT

Ultimately it is desirable to compare vectorial solute transport to epithelial fluid transport, since these are coupled processes. In corneal experiments volume flows have been measured
with several different techniques. Probably the most satisfactory
of these is the direct measurement of stromal thickness, since it
is related to hydration which, in turn, is the parameter solute
pumps must control for normal corneal function. For this purpose
we (14, 15) have automated the Maurice specular microscope pacho-
meter so that averaged and scaled readings of stromal and epithelial thicknesses can be obtained conveniently over tens of
hours with an overall accuracy better than 1%.

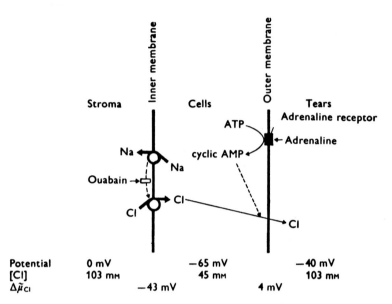

FIG. 4. Membrane model proposed for the action of epinephrine (adrenaline) in rabbit corneal epithelial Cl transport.
Potentials and Cl for each compartment have been determined experimentally from which the Cl electrochemical potential gradients
$(\Delta\mu_{Cl})$ were calculated. (Reprinted from ref. 11.)

The contribution of the epithelium alone to stromal hydration
was assessed in isolated rabbit corneas with the inner endothelial
layer scraped off and the surface blocked with silicone oil (4,
16). In control experiments, stromal thickness was nearly constant
over several hours of incubation *in vitro*, which is consistent with
fluxes measured at open circuit (*vide supra*). Stimulation of Cl
transport with the phosphodiesterase-inhibitor theophylline caused
epithelia to thin previously swollen stromas at a steady rate
(6 hr or more) somewhat greater than 1 μm/hr. This rate is consistent with isotonic fluid transport from the closed stromal compartment based on fluxes measured at resting potential. Hence, it was
concluded that the epithelium may participate synergistically along
with the endothelium in the prevention of stromal edema.

However, assessing the functional significance of the epithelial transport system with the intact and functional *endothelium* (now thought to transport HCO_3; ref. 17) was not a straightforward matter, due to controversies in the literature concerning the values of membrane phenomenological coefficients in the cornea. Furthermore, theoretical approaches modeling fluid transport in cornea and other epithelial tissues did not deal simultaneously with both solute transporting membranes (across which fluid flows are predominantly controlled by concentration gradients) and the gel properties of connective tissue or other polyelectrolyte gel phase (through which fluid flows are driven primarily by local colloid osmotic and hydrostatic pressure gradients). Below a method is presented to incorporate in a single model the flow properties of both membranes and gel (18, 19).

Coupled flows across membranes can be described with the steady state linear coupled flow equations based on nonequilibrium thermodynamics as most usefully developed by Kedem and Katchalsky (20). The linear equations can be integrated analytically to obtain a kinetic analysis of flows across membranes and membrane systems. The problem encountered with the application of these equations to continuous phases such as connective tissue, cytoplasm, etc. is the time dependence of the local transport coefficients (i.e., hydraulic conductivity and solute permeability are functions of time and space within gels). This makes the governing equations nonlinear, and therefore defies, at least at present, analytical solution for transient moving boundary problems.

Nevertheless, a numerical solution is possible as follows: Using the technique of finite element analysis, a tissue system may be divided into sufficiently small pieces so that the transport coefficients governing the rates of flows into and out of the element are time independent. A system modeled with this approach can be as complex as will be allowed by the availability of accurate, high speed digital computation so long as experimental perturbations (e.g., "osmotic shocks"), which may be used to charac-

terize system boundary coefficients, do not drive a portion of the
system so far from equilibrium that its macroscopic behavior is no
longer described by the governing equations (cf. ref. 35). The application of this approach to dealing with nonlinear irreversible
thermodynamics for the cornea is illustrated below (19).

The cornea was modeled as a series array of $n + 1$ planar membranes bounding n concatenated stromal compartments. The central
stromal membrane transport coefficients were defined by functions
that were empirical fits to experiment (*vide infra*). The limiting
membranes were given the attributes of corneal epithelium and endothelium. Each was characterized by time independent reflection
coefficient (σ), a hydraulic conductivity (Lp), a solute permeability (ωRT), and an active solute pump (Ja).

The general coupled flow equations (20) were expanded into
the set:

$$Jv_{i,t} = Lp_{i,t}(\Delta P_{i,t} - \sigma_i RT\Delta C_{i,t})$$

$$Js_{i,t} = (1-\sigma_i)\overline{C}_{i,t}Jv_{i,t} + \omega_{i,t}RT\Delta C_{i,t} + Ja_i$$

in which solute concentrations C were corrected for nonideal solution behavior. Coefficients Lp and ω were assumed constant for the
limiting membranes and variable for the stromal "membranes." The
latter have been previously characterized as follows: $Lp = \rho H^4$
(22), where ρ is a constant and H is the local hydration of the
corneal stroma; the stromal solute permeability can be calculated
directly from the stromal solute diffusion coefficient. Local
stromal pressure P was evaluated from the sum of intraocular
pressure IOP, and swelling pressure SP; P = IOP − SP, where
$SP = \gamma e^{-\overline{H}}$. Since corneal hydration is a linear (and known) function of stromal thickness, expansion of these relationships can be
used to calculate pressure, solute concentration, and volume
(thickness) in the stroma.

This approach was used to redetermine σ, Lp, ωRT, and Ja for
both the corneal epithelium and endothelium by time sequence

analysis of osmotic perturbation experiments as illustrated in
Fig. 5. First the coefficients for the epithelium were determined
by perturbing the cornea with the endothelial surface blocked with
oil. Then, with these coefficients known, similar experiments were
repeated with whole cornea fitting thickness changes to determine
the endothelial coefficients. Hypotonic and hypertonic perturba-

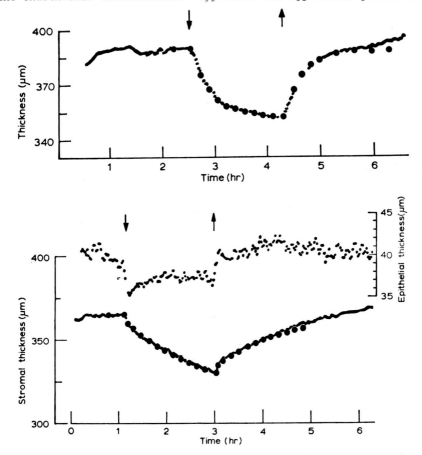

FIG. 5. *Influence of a 30 mOsm hypertonic perturbation on
stromal thickness. Epithelial perfusate osmolarity was increased
at ↓ and returned to normal at ↑ . The calculated fit (●) is
superimposed on the experimental data. (A) Response with the endothelial surface scraped away for the determination of epithelial coefficients. (B) Response perfusing both surfaces with Ringer's solution. The influence on epithelial thickness is shown as well,
primarily due to indicate the resolution of the measuring system.
(Reprinted from ref. 19.)*

tions were found to yield the same membrane phenomenological coefficients. Hence, it was argued that the experimental procedure did
not significantly alter the measured parameters. The results of
this study are listed in Table II.

The values obtained were quite dissimilar from previous measurements (23, 25). It is therefore valid to ask whether the model
can be of any general use in the analysis of corneal hydration dynamics, especially since the model was parameterized to fit the
theory to a single type of experimental observation. Briefly, the
parameterized model above was found to predict reasonably accurately the results of many other well documented corneal experiments. First of all, the model required the presence of a solute
pump in the endothelium to maintain normal stromal hydration equal
in magnitude to the HCO_3 transport measured recently (17). A good
match was found for the temperature reversal phenomenon (26) (time
course of corneal thinning at $37°C$ following overnight cold
swelling). It predicted the rate at which ouabain caused stromal
swelling through the inhibition of endothelial solute transport
(24). Several other predictions [viz.: the effect of intraocular
pressure changes (28), the effect of evaporation from the tear
film (23, 29), the rate of stromal swelling following destruction
of a limiting membrane *in vivo* (30), the influence of hibernation
on stromal hydration (31)] either matched experimental observation
or provided explanations where none had previously existed. Finally, the model predicted that the stimulated Cl transport system
of the rabbit corneal epithelium should thin a swollen stroma at
the rate of 1.28 µm/hr, which is identical to the experimental value
(16). Hence, the usefulness and validity of the model with its
parameterization seems apparent.

Finally, one might ask why this method for obtaining membrane
phenomenological coefficients gave answers so distinctly different
from other procedures such as the zero-time extrapolation method
used to determine hydraulic conductivity. The answer was not obvious without further analysis, although support for the approach

TABLE II. Corneal Phenomenological Coefficients and Solute Transport Rates (ref. 19)

	$\Delta\pi_{NaCl}$ ($\times 10^{-6} osm/cm^3$)	σ_{NaCl}	Lp ($\times 10^{-12} cm^3/dyn\ sec$)	$\omega_{NaCl}RT$ ($\times 10^{-6} cm/sec$)	Ja ($\times 10^{-10} mol/cm^2\ sec$)
Epithelial values with endothelium blocked[a]					
	+30.0	0.78 ± .07	5.8 ± 0.7	0.15 ± .01	-.120 ± .064
	-23.3	0.80 ± .06	6.4 ± 1.5	0.23 ± .02	+.038 ± .031
Endothelial values using whole cornea[b]					
	+30.0	0.45(.45-.60)	53(43-55)	80(73-80)	4.8(3.1-4.8)
	-23.3	0.45(.43-.50)	32(24-53)	80(90-100)	4.6(4.6-5.5)

[a] Means ± SEM. Number of observations = 5.
[b] Values fitting median response. The range of values determined by fit to minimum and maximum response to osmotic perturbation. Number of observations = 6.

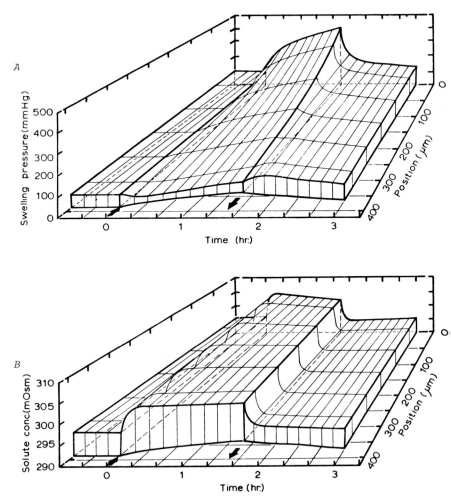

FIG. 6. *Stromal gradients as a function of time and space
for the experiment shown in the preceding figure. Epithelium lies
at position 0, endothelium at forward surface. Arrows indicate
application and removal times for the 30 mOsm hypertonic perturbation. (A) Solution for swelling pressure which shows increased
swelling pressure in the anterior stroma, decreased in posterior.
If the stroma were well-stirred, these gradients would not exist.
(B) Solution for stromal solute concentration. Note the rapidity
with which the concentration changes occur following the perturbation. (Drawn from data calculated in ref. 19.)*

seemed adequate, since the coefficients could be used as a set to predict accurately corneal hydration dynamics for a variety of observations. Plotted in Fig. 6A,B are the solutions for stromal swelling pressure and solute concentration as a function of time and space for the experiment fitted in Fig. 5. Earlier theories (e.g., refs. 32,33) assumed that the central gel phases between biomembranes were well-stirred compartments. With respect to solute concentration in this experiment, this assumption was apparently justified although additional problems arise on inspection. The measurement of volume flows (or thickness) requires a finite amount of time. Should zero-time extrapolation procedures be used to estimate membrane coefficients, they will underestimate Lp due to the rapidity with which solute concentration changes within the gel phase. Furthermore, pressure gradients arise across gel phases primarily as a consequence of viscosity effects (as noted above, stromal hydraulic conductivity varied with the fourth power of local hydration while the driving force increased exponentially with hydration). These factors can cause significant colloid osmotic pressure gradients to arise as shown in the illustrations.

Even more complex pressure gradients can arise within the corneal stroma if the same osmotic perturbation was applied to the endothelial surface (Fig. 7A,B), which is a common procedure used to obtain endothelial hydraulic conductivity (25). In this situation the stroma first thinned then swelled due to the asymmetrical nature of corneal membrane permeabilities. While swelling pressure increased in the posterior stroma, it decreased anteriorly as the imposed osmotic perturbation rapidly increased the subepithelial solute concentration.

Hence, it was concluded that the gel properties of aqueous tissue phases should and can be incorporated into models dealing with coupled flows across epithelia. It was noted that the simple method applied here to the cornea did not deal with the issue of paracellular pathway influence, since doing so would have trebled the number of unknowns to be determined. Nor was it

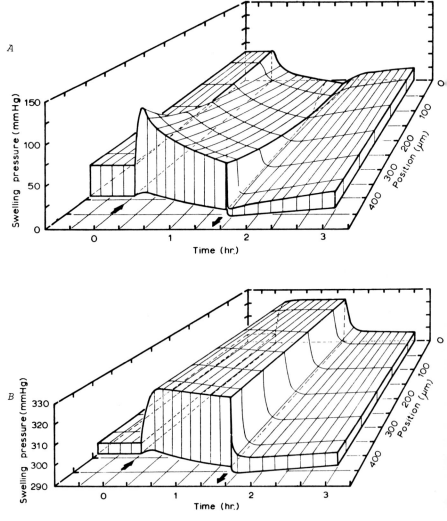

FIG. 7. As in previous figure, but applying the 30 mOsm hypertonic perturbation to the endothelial surface. (A) Solution for swelling pressure. (B) Solution for stromal solute concentration (Drawn from data calculated in ref. 19.).

necessary to do so since the model adequately predicted corneal hydration dynamics for a variety of observations, the stroma apparently provided the necessary compartmentalization. The model has been scaled down to the dimensions of a hypothetical epithelium

with a good possibility that the general approach can be used to
describe quantitatively the transport of fluid across cytoplasmic
phases while dealing simultaneously with the problems of cell
volume regulation and transepithelial fluid transport (18, 34).
In this way the functional significance of Cl and other ion transport systems might be better understood.

SUMMARY

    Cyclic AMP mediates the active secretory transport of Cl by
the rabbit corneal epithelium. Neural modulation of this transport is possible by sympathetic fibers from the superior cervical
ganglion. A model for the transport system at the membrane level
proposes that catecholamines stimulate Cl transport by causing a
Cl-specific increase in the Cl conductance of the outer epithelial
cellular membrane. Thermodynamic considerations place the active
step at the inner epithelial barrier. As the reuptake of Cl across
the inner membrane is blocked by ouabain, coupling to Na transport
is probable. The epithelial Cl transport system is capable of
thinning previously swollen corneal stromas, albeit at a rate 30
times less than that of the endothelium. The nature of the coupling
between the ion transport and fluid transport was explored with
open circuit flux measurements and the development of a new model
with which to deal quantitatively with flows across both membrane
and gel phases of tissues. The latter was used to redetermine the
phenomenological coefficients of the epithelium and endothelium,
which allowed the successful prediction of many features of corneal
hydration dynamics.

ACKNOWLEDGMENTS

The author is indebted to his collaborators for their active
role in some of the studies reviewed here: W. S. Marshall, D. M.
Maurice, A. H. Neufeld, S. R. Russell, R. K. S. Wong, and J. A.
Zadunaisky. Gratitude is expressed to the Radiological Physics
Section at Stanford Medical School for use of their computer fa-
cility to provide the background graphics for Figs. 6 and 7, and to
O. Bernegger and T. Chang for their excellent technical assistance.

REFERENCES

1. A. Donn, D. Maurice, and N. Mills. Studies on the living
    cornea *in vitro*. II. The active transport of sodium across
    the epithelium. *Arch. Ophthalmol.* *62*:748-757, 1959.

2. S. D. Klyce. Microelectrode and tracer analysis of ion trans-
    port in the rabbit corneal epithelium. *Abstracts of the IV
    International Biophysics Congress, Moscow 3*:106, 1972.

3. S. D. Klyce, A. H. Neufeld, and J. A. Zadunaisky. The acti-
    vation of chloride transport by epinephrine and Db cyclic-AMP
    in the cornea of the rabbit. *Invest. Ophthal.* *12*:127-139,
    1973.

4. S. D. Klyce. Transport of Na, Cl and water by the rabbit
    corneal epithelium at resting potential. *Am. J. Physiol.*
    228:1446-1452, 1975.

5. B. R. Beitch, I. Beitch, and J. A. Zadunaisky. The stimula-
    tion of Cl transport by prostaglandins and their interaction
    with epinephrine, theophylline, and cyclic AMP in the corneal
    epithelium. *J. Membr. Biol.* *19*:381-396, 1974.

6. M. G. Buck and J. A. Zadunaisky. Stimulation of ion transport
    by ascorbic acid through epithelium and other tissues.
    Biochim. Biophys. Acta 389:251-260, 1975.

7. B. Ehinger. Distribution of adrenergic nerves to orbital

structures. *Acta Physiol. Scand.* *62*:291-292, 1964.

8. A. Laties and D. Jacobinowitz. A histochemical study of the adrenergic and cholinergic innervation of the anterior segment of the rabbit eye. *Invest. Ophthalmol.* *3*:592-600, 1964.

9. J. H. LaVail and M. M. LaVail. The retrograde intraaxonal transport of horseradish peroxidase in the chick visual system: A light and electron microscopic study. *J. Comp. Neurol.* *157*:303-358, 1974.

10. S. D. Klyce and R. K. S. Wong. Sympathetic influence on corneal epithelial electrophysiology. *Invest. Ophthalmol. Vis. Sci. Suppl.* *190abs.*; 1978.

11. S. D. Klyce and R. K. S. Wong. Site and mode of adrenaline action on Cl transport across the rabbit corneal epithelium. *J. Physiol.* *266*:777-799, 1977.

12. S. D. Klyce. Electrical profiles in the corneal epithelium. *J. Physiol.* *226*:407-429, 1972.

13. W. S. Marshall and S. D. Klyce. Membrane resistances in rabbit corneal epithelium. *Federation Proc.* *40*:370, 1981.

14. S. D. Klyce and D. M. Maurice. Automatic recording of corneal thickness *in vitro*. *Invest. Ophthalmol.* *15*:550-553, 1976.

15. S. D. Klyce and S. R. Russell. System for monitoring the thickness of transparent layered structures. *Rev. Scient. Instrum.* *49*:1318-1321, 1978.

16. S. D. Klyce. Enhancing fluid secretion by the corneal epithelium. *Invest. Ophthalmol. Vis. Sci.* *16*:968-073, 1977.

17. S. Hodson and F. Miller. The bicarbonate ion pump in the endothelium which regulates the hydration of rabbit cornea. *J. Physiol.* *263*:563-577, 1976.

18. S. D. Klyce and S. R. Russell. Cytoplasmic standing gradients could link epithelial Jv to solute transport. *Biophys. J.* *25*:95a, 1979.

19. S. D. Klyce and S. R. Russell. Numerical solution of coupled transport equations applied to corneal hydration dynamics.

J. Physiol. 292:107-134, 1979.

20. O. Kedem and A. Katchalsky. Thermodynamic analysis of the permeability of biological membranes to nonelectrolytes. *Biochim. Biophys. Acta* 27:229-246, 1958.

21. S. Mishima and D. M. Maurice. The effect of normal evaporation on the eye. *Exp. Eye Res.* 1:46-52, 1961.

22. I. Fatt and T. K. Goldstick. Dynamics of water transport in swelling membranes. *J. Colloid Sci.* 20:434-460, 1965.

23. S. Mishima and B. O. Hedbys. The permeability of the corneal epithelium and endothelium to water. *Exp. Eye Res.* 1: 39-45, 1967.

24. S. M. Trenberth and S. Mishima. The effect of ouabain on the rabbit corneal endothelium. *Invest. Ophthalmol.* 7:44-52, 1968.

25. K. Green and M. A. Green. The permeability to water of rabbit corneal membranes. *Am. J. Physiol.* 217:634-641, 1969.

26. H. Davson. The hydration of the cornea. *Biochem. J.* 59:24-28, 1955.

27. J. Fischbarg, C. R. Warshavsky, and J. J. Lim. Pathways for hydraulically and osmotically-induced water flows across epithelia. *Nature* (London) 266:71-74, 1977.

28. S. Hodson. The regulation of corneal hydration by a salt pump requiring the presence of sodium and bicarbonate ions. *J. Physiol.* 236:271-302, 1974.

29. S. Mishima and D. M. Maurice. The oily layer of the tear film and evaporation from the corneal surface. *Exp. Eye Res.* 1:39-45, 1961.

30. D. M. Maurice and A. A. Giardini. Swelling of the cornea *in vivo* after the destruction of its limiting layers. *Br. J. Ophthalmol.* 35:791-797, 1951.

31. L. Z. Bito, J. C. Roberts, and S. Saraf. Maintenance of normal corneal thickness in the cold *in vivo* (hibernation) as opposed to *in vitro*. *J. Physiol.* 231:71-86, 1973.

32. M. H. Friedman. A quantitative description of equilibrium

and homeostatic thickness regulation in the *in vivo* cornea.
I. Normal cornea. *Biophys. J. 12*:648-665, 1972.

33. M. H. Friedman. A quantitative description of equilibrium
and homeostatic thickness regulation in the *in vivo* cornea.
II. Variations from the normal state. *Biophys. J. 12*:666-682,
1972.

34. S. D. Klyce and S. R. Russell. The viscous flow theory:
An adjunct to the standing gradient hypothesis. *Federation
Proc. 39*:378, 1980.

35. I. Prigogine. Time, structure, and fluctuations. *Science
201*:777-785, 1978.

THE ACTIVE TRANSLOCATION OF Cl AND Na

BY THE FROG CORNEAL EPITHELIUM:

COTRANSPORT OR SEPARATE PUMPS?

Oscar A. Candia

Departments of Ophthalmology, and Physiology and Biophysics
Mount Sinai School of Medicine of the City University of New York
New York, New York

Active Cl transport has been described in a large variety of
epithelial membranes including those from mammals, amphibia, and
fish (1 - 7). One typical characteristic of most Cl transporting
epithelia is that Cl is seldom the only ion transported by the
tissue. Active H, Na, or HCO_3 transport are usually associated
with the transport of Cl. It has also become apparent that not
all Cl transport mechanisms are the same. Watlington et al. (8),
for example, divided the Cl transport mechanisms in two main
categories: those which are inhibited by acetazolamide and those
which are inhibited by ouabain. In a more recent review, Frizzell
et al. (9) divided active Cl transport into three categories: an
electrically silent $Cl-HCO_3$ exchange; Cl absorption, and Cl secretion. They concluded that both Cl secretion and Cl absorption are
coupled to the Na movement produced by the Na gradient created by
the Na pump. It is also worth noting that net active Na transport

223

usually occurs in only one direction (towards the blood side); whereas active Cl transport can occur in either direction.

In defining active transport, one can use the Rosenberg criterion (10): a net movement against an electrochemical gradient; or the Kedem criterion (11): a direct coupling of the ion flow to the flow of a metabolic reaction. In using the Rosenberg criterion, the so-called cotransport or "secondary active transport" processes may be included. In using the Kedem criterion, only "primary active transport" processes will be included. If active Cl transport was the only active solute translocation present in an epithelial membrane, its transport against an electrochemical gradient and its inhibition by a metabolic inhibitor would certainly prove its status of "primary active Cl transport." Initially it appeared that the frog corneal epithelium only actively transported Cl and therefore constituted a "primary active Cl transport system." However, it was later shown by Candia and Askew (12) and Candia *et al.* (13) that the frog cornea was capable of a net Na transport which was particularly evident when the cornea was treated with amphotericin B. More recently we have shown that the epithelial "Na pump" is actually a coupled "Na:K pump" (14).

In this paper we examine ionic transport by the frog corneal epithelium, namely, Cl, Na, and K. Published information and new experimental data are analyzed in an effort to determine whether active Cl transport in the cornea is a primary process or secondary to an active Na:K transport.

I. THE CORNEAL EPITHELIUM

The bullfrog cornea is a transparent membrane about 0.2 mm thick and up to 1 cm in diamter. It consists of three distinct layers: the endothelium, stroma, and epithelium. The epithelium can not be successfully separated from the stroma for electro-physiological studies and the results described here were generally

obtained from the epithelial-stroma preparation. The stroma,
basically a connective tissue, does not contribute to the electrical and ionic transport properties. The epithelium is a multicel-
lular layer that can be compared with the frog skin epithelium consisting of basal cells, intermediary cells, and squamous cells.
There are desmosomes connecting the basal and intermediary cells,
with tight junctions occurring in the most superficial cell layer.

II. THE Cl TRANSPORT

     Zadunaisky first showed in 1966 (5, 15) that the frog cornea
isolated in an Ussing-type chamber develops a PD in the range of
12 - 40 mV, the tear side being negative with respect to the stromal side. When the PD was short-circuited and unidirectional Cl
fluxes measured, the net Cl flux matched the SCC within 10%. In
addition, when Cl was replaced by SO_4 in the bathing solutions,
the PD declined to slightly above zero. The first clue of an interdependence between Cl transport and Na was also given by
Zadunaisky who showed that when Na was removed from the bathing
solutions, the Cl-originated SCC rapidly declined to zero (5).

     Sodium was required on the stromal side. Removal of Na from
the tear bathing solution did not affect the Cl-originated SCC
(5). It was later shown by Candia (16) how the lack of Na in the
medium affected the unidirectional Cl fluxes. The effect was bi-
phasic. During the first 90 min in Na-free Ringer, Cl transport
was inhibited and the stroma to tear Cl flux diminished. After
90 min, although the SCC remained at zero, both unidirectional Cl
fluxes steadily increased to values three to five times larger
than the control (see Table I). Sodium is necessary not only to
sustain the Cl transport but also to maintain the normal permeability of the corneal epithelium.

     Zadunaisky (17) studied the relationship between Cl trans-
port and Na concentration on the stromal side. There was no

TABLE I.   *Unidirectional Chloride Fluxes across the Short-Circuited
Frog Cornea Bathed in Na-Rich and Na-Free Solutions*[a]

	Stroma to Tear (n=6)	Tear to Stroma (n=6)
Control (Na-rich)	0.69 ± 0.03	0.29 ± 0.03
Exp (60 min Na-free)	0.48 ± 0.05	0.56 ± 0.08
Exp (120 min Na-free)	2.40 ± 0.72	2.42 ± 1.01

[a]*Values are means ± SE in* $\mu Eq/hr\ cm^2$.

change in the SCC when the Na concentration was reduced from 100 mM
to about 40 mM.   Only when the Na concentration was below 30 mM did
the SCC rapidly decline.   In interpreting Zadunaisky's data, it
seems that 5 mM Na was enough to maintain about 50% of the control
SCC.   This implies that if Cl was driven by a Na gradient, the cellular Na concentration was in this condition less than 5 mM.   A logical extension of these studies was to test the effect of ouabain on
Cl transport.   Table II (modified from Candia, ref. 2) shows that
ouabain $10^{-4}M$ has a clear inhibitory effect on the Cl-originated SCC
and net Cl transport.   The effect is seen as a reduction on the
stromal to tear flux and as a slight increase of the opposite flux.
Unlike the Na-free effect, there was not a second phase of increase
in Cl permeability.

    Recently we have found that a K-free medium on the stromal side
also reduces the Cl-originated SCC (unpublished observations) as

TABLE II.   *Effect of Ouabain* $10^{-4}M$ *on Unidirectional Chloride Fluxes
and Short-Circuit Current Across the Bullfrog Cornea*[a]

	Stroma to Tear (n=10)	Tear to Stroma (n=6)	SCC (n=10)
Control	0.76 ± 0.06	0.34 ± 0.02	0.61 ± 0.04
Ouabain	0.54 ± 0.05	0.44 ± 0.04	0.14 ± 0.01

[a]*Values are means ± SE in* $\mu Eq/hr\ cm^2$.

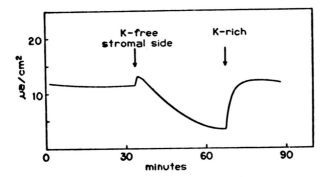

FIG. 1.   Effect of K-free medium on the stromal side on the
Cl-originated SCC across the corneal epithelium.   Addition of K
restores the SCC.

shown in Fig. 1.

These results clearly demonstreated that Na was required for
the transport of Cl and that the enzyme Na-K ATPase was involved in
the active translocation of Cl.   At this point an important distinction should be made in regard to the Na-dependence.   One can
speculate that Na in the stromal medium is necessary to maintain
an effective cellular Na concentration.   Cellular Na and external
K are required to activate the Na-K ATPase that would provide (by
ATP hydrolysis) the energy for the Cl transport.   Two possible
couplings are shown in Fig. 2.   In one scheme, the coupling of
metabolism to the Cl flow, although undefined, is independent of
Na movement.   The other alternative suggests that the Na-K ATPase
energizes a Na:K transport, which maintains a Na gradient.   This
gradient provides, by means of a Na-Cl cotransport system, the
driving force for the Cl transport.

In differentiating between these alternatives, it was important to clearly characterize the Na transport in the corneal epithelium.   This will be described next.

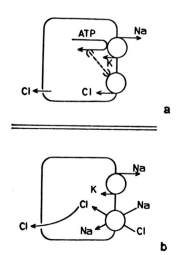

FIG. 2. *Models of active Cl transport coupling. (a) Direct coupling to metabolic flow. (b) Coupling to the Na movement across the basolateral membrane.*

III. THE Na AND K TRANSPORT

When the cornea is bathed in Cl-free Na_2SO_4 Ringer, the PD, in many cases, does not drop to zero and remains at a 1 to 3 mV level. Unidirectional Na fluxes can be extremely variable across different corneas. Even in two electrically well-matched corneas from the same frog, Na fluxes could be different, and because of this, a small net difference between the two opposite Na fluxes is difficult to prove. Candia and Askew (12) resolved this problem by measuring both unidirectional Na fluxes in the same cornea simultaneously with ^{22}Na and ^{24}Na. The results are shown in Table III.

In paired measurements, the tear to stroma flux was always larger, both in Cl-rich and Cl-free short-circuited preparations. Furthermore, in Cl-free Ringer, the small net Na transport closely matched the small SCC.

The existence of an active net Na transport was unequivocally shown when the corneas were treated with 10^{-5} M amphotericin B on

TABLE III. *Simultaneous Unidirectional Sodium Fluxes and Short-Circuit Current across the Isolated Bullfrog Cornea[a]*

	Sodium Fluxes		Net	SCC
	Tear to Stroma	Stroma to Tear	Net	SCC
NaCl Solutions (15)	0.64	0.54	0.10[b]	1.10[c]
Na$_2$SO$_4$ Solutions (17)	0.62	0.48	0.14[b]	0.16

[a]*Values are means in µEq/hr cm^2*
[b]*The difference was significantly larger than zero; P < 0.005.*
[c]*Indicates both chloride and sodium transport.*

the tear side solution. As shown in Fig. 3, the SCC is immediately and greatly stimulated in Na$_2$SO$_4$ Ringer.

Measurement of unidirectional Na fluxes shows a substantial net Na transport in the tear to stroma direction. The results are summarized in Tables IV and V. Ouabain reduced considerably the net Na transport and the SCC. The following points are worth noticing for a later discussion: (a) The stroma to tear Na flux was significantly larger in NaCl Ringer than in Na$_2$SO$_4$ Ringer. (b) After ouabain, when both unidirectional Na fluxes were similar, those in

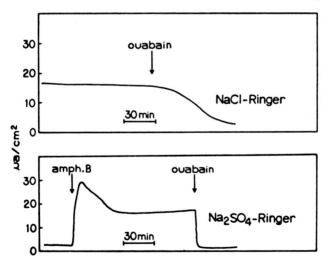

FIG. 3. *Effects of ouabain on the SCC across the corneal epithelium. Top panel: SCC represents Cl transport. Lower panel: SCC represents Na transport stimulated by amphotericin B.*

TABLE IV. *Unidirectional Sodium Fluxes across the Short-Circuited Bullfrog Cornea Bathed in NaCl Ringer. Effects of 10^{-5} M Amphotericin B, and 10^{-4} M Ouabain.[a]*

| | Sodium Fluxes | | SCC[b] |
	Tear to Stroma	Stroma to Tear	
Control	0.32 ± 0.04	0.28 ± 0.04	0.38 ± 0.03
Amphotericin B	1.70 ± 0.17	0.86 ± 0.15	0.78 ± 0.09
+ Ouabain	1.15 ± 0.19	1.07 ± 0.13	0.08 ± 0.02

[a]*Values are means $\pm$ SE in $\mu Eq/hr\ cm^2$.*
[b]*SCC includes Cl transport.*

TABLE V. *Unidirectional Sodium Fluxes across the Short-Circuited Bullfrog Cornea Bathed in Na_2SO_4 Ringer. Effects of 10^{-5} M Amphotericin B, and 10^{-4} M Ouabain.[a]*

| | Sodium Fluxes | | SCC |
	Tear to Stroma	Stroma to Tear	
Control	0.31 ± 0.03	0.29 ± 0.03	0.02 ± 0.01
Amphotericin B	1.38 ± 0.12	0.39 ± 0.05	0.31 ± 0.03
+ Ouabain	0.63 ± 0.09	0.56 ± 0.10	0.05 ± 0.01

[a]*Values are means $\pm$ SE in $\mu Eq/hr\ cm^2$.*

NaCl Ringer were significantly larger than those in Na_2SO_4 Ringer.

Comparison of the net Na flux with the SCC in amphotericin B-treated corneas bathed in Na_2SO_4 Ringer shows a discrepancy. Thus the possibility of an active K transport was investigated. Unidirectional K fluxes were then measured under short-circuited conditions in amphotericin B treated corneas, and a net K transport in the stroma to tear direction was found. These results are shown in Table VI. It is also shown that ouabain inhibits the net K transport.

TABLE VI. Unidirectional Potassium Fluxes across the Short-
Circuited Bullfrog Cornea Bathed in Na_2SO_4 Ringer. Effects of
10^{-5} M Amphotericin B and 10^{-4} Ouabain.[a]

| | Potassium Fluxes | | |
	Stroma to Tear	Tear to Stroma	SCC[b]
Control	0.006 ± 0.001	0.007 ± 0.001	0.04 ± 0.01
Amphotericin B	0.410 ± 0.034	0.026 ± 0.004	0.80 ± 0.07
+ Ouabain	0.037 ± 0.004	0.034 ± 0.004	0.02 ± 0.01

[a]Values are means $\pm$ SE in $\mu Eq/hr\ cm^2$.
[b]SCC is the difference between net Na minus net K transport.

IV. THE LOCATION OF THE Na:K AND Cl PUMPS

In considering the possibility of a Na-Cl cotransport system,
the location of the Na:K and Cl pumps in either the basolateral
membrane or apical membrane of the epithelium is an important consideration. Since Na and Cl are actively transported in opposite
directions, the location of the Na and Cl pumps on opposite cell
sides would exclude the possibility of cotransport. Sodium:po-
tassium pumps have been demonstrated to be located in the basolateral membranes of a large variety of epithelia. This location
is associated with a negative potential and a low Na concentration
of the cell compartment.

The effect of amphotericin B on Na and K transport provides
a means for the determination of the location of the Na:K pump.
In addition we have measured the concentration of K in the corneal
epithelium in control conditions and after amphotericin B treatment. For this the two eyeballs from frogs were isolated. One
eyeball was partially immersed in Ringer so that the cornea and
part of the sclera were in contact with the solution. The paired
eyeball was similarly immersed in Ringer with amphotericin B.
After half an hour the epithelium was peeled off, digested in
0.1 N nitric acid, and the K concentration determined by flame
photometry. Potassium concentration was 80 mM in control epithelia
and 16 mM in amphotericin B-treated epithelia. These results con-

Oscar A. Candia

clusively indicate that the Na:K pump of the corneal epithelium is
located in the basolateral membrane, and that amphotericin B increases transepithelial Na and K transport by increasing the permeability of the apical side. The following diagrams, shown in
Fig. 4, can be proposed for the Na:K transport system.

Under control conditions the tear side is rather impermeable
to Na and K. Sodium cannot reach the Na pool but it can cross the
corneal epithelium via the paracellular pathway. Under control
conditions, there is a recirculation of Na and K across the basolateral membranes. This maintains a Na gradient between the
stromal side and the cellular compartment. After treatment with
amphotericin B, the permeability of the tear side surface is increased and Na and K can now move transcellularly. As will be
shown later, mannitol fluxes are increased by amphotericin B.

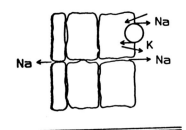

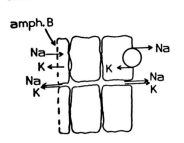

FIG. 4. Diagrams of the movement of Na and K across the
corneal epithelium. Top panel: The apical membrane is relatively
impermeable to Na and K. A Na:K pump in the basolateral membrane
maintains an ionic gradient. Na and K recirculate across the membrane. Lower panel: Amphotericin B increases Na and K permeability of the apical membrane. The Na:K pump produces a net K transport towards the tear side and a net Na transport towards the
stromal side. The permeability of the paracellular pathway is
also increased.

This suggests that the paracellular permeability is also increased
by amphotericin B.   Thus, a fraction of the transcellular movements of Na and K after the effect of the antibiotic may be paracellular.   Also, as a result of the increase in permeability of
the tear side barrier, cellular K concentration decreases.   Also,
it can be predicted, although it has not been measured, that cellular Na concentration should increase and that the electrical
negativity of the cell compartment should decrease.   Thus, the Na
gradient across the basolateral membrane is also decreased.

Recently Klyce and Wong (18) showed, in the rabbit cornea by
using PD microelectrodes, that Cl should move against an electrochemical gradient across the basolateral membrane.   They also predicted a cellular Cl concentration of 41.5 mM.   Previously, Otori
(19) has measured an epithelial Cl concentration of 23 mM.

Recently, Zadunaisky *et al.* (20) measured with microelectrodes
in the frog cornea cellular Cl activity and intracellular PD.
Their results, 29 mM Cl and −48 mV, suggests the location of the
Cl pump in the basolateral membrane.

The fact that Cl moves across the basolateral membrane against
its electrochemical potential is consistent with a Na-Cl cotransport
system but it is not definite proof of it.   It is possible that two
separate Na and Cl pumps exist in the same membrane.

V.   ENERGETIC CONSIDERATIONS FOR A COTRANSPORT SYSTEM

Figure 2, panel b, shows a tentative model of a Na-Cl cotransport system for the corneal epithelium that is similar to that
proposed for other epithelial membranes with an electrogenic chloride secretion (6).

The Na:K pump maintains a Na gradient across the basolateral
membrane.   Sodium moves down its electrochemical gradient and carries Cl with it.   The decrease in free energy per unit time,
$\Delta G/\Delta t$, of Na must be larger than the increase of free energy of Cl.

$$J_{Na} \times [RT \ \ln(Na_m/Na_c) + F\Delta\psi] > J_{Cl} \times [RT \ \ln(Cl_c/Cl_m) + F\Delta\psi]$$

The measured or estimated values for the different parameters are shown in Table VII. After treatment with amphotericin B, the Na pump is capable of a net Na transport of 1 μEq/hr/cm^2. Assuming a similar rate in control conditions (in the absence of a Net Na transport) Na recirculation into the cell would equal the pump flux and would have a value of 1 μEq/hr/cm^2. The cellular Na concentration is simply estimated from values in other epithelia.

From these values,

$$\Delta G/\Delta t_{(Na)} = 8.49 \times 10^{-3} \text{Joul/hr} > \Delta G/\Delta t_{(Cl)} = 1.13 \times 10^{-3} \text{Joul/hr}$$

Thus, even if the efficiency of the energy conversion was less than 1, the process is energetically feasible.

VI. EXPERIMENTAL TESTS FOR COTRANSPORT

As pointed out by Stein (21) a number of properties are expected from a cotransport system. These properties can be tested experimentally for a confirmation of the model. Some of them are

(a) A Na flux driven by its electrochemical gradient will drive a Cl flux up to a point at which the electrochemical gradient for Cl equals that of Na (assuming that one Na carries one Cl).

(b) If the electrochemical gradient for Cl is larger than that for Na, the Cl flux will drive a Na flux.

TABLE VII. *Values of Parameters Utilized in Calculating the Change in Free Energy*

J_{Na}	: 1 μEq/hr	J_{Cl}	: 0.5 μEq/hr
$[Na]_m$	: 104 mM	$[Cl]_m$	: 75 mM
$[Na]_c$	: 20 mM	$[Cl]_c$	: 29 mM
		$\Delta\psi$	: -47 mV

(c) If the Na gradient is reversed, the Na flux and the driven Cl flux will be reversed; and likewise for the Cl flux and the driven Na flux.

Ouabain and Na-free stromal side solution should stop net Cl flux by canceling the Na gradient. This has actually been observed. It is also expected that ouabain should stop the Na transport first and only after the cellular Na level has reached a level similar to that in the medium, will Cl transport stop. This is also the case as it can be seen in Fig. 3. The effect of ouabain on Na transport is immediate, whereas Cl transport is slowly inhibited. If an inhibitor of Na transport could be found that did not inhibit Cl transport, that would be a good argument against a cotransport of Na-Cl. This has not been the case: Na-free medium, K-free medium, ouabain, and tryptamine (22) that inhibit Na transport also inhibit Cl transport. Nevertheless specific inhibitors of Cl transport in the cornea have been found such as bumetanide (23), furosemide (24), and piretanide and MK-196 (25).

After amphotericin B treatment, the Na backflux (stroma to tear) increases more in NaCl Ringer than in Na_2SO_4 Ringer (see Tables IV and V). Since the paracellular pathway permeability increases with amphotericin B (as shown by mannitol fluxes, unpublished observations), it can be speculated that in Na_2SO_4 Ringer the increase in Na backflux is through the paracellular pathway. The additional increase in Na backflux observed with Cl can be interpreted as a transcellular Na flux coupled to Cl. After ouabain, when transepithelial unidirectional Na fluxes were the same in opposite directions, those in NaCl solution were twice as large as those in Na_2SO_4 Ringer. These results are summarized in Fig. 5. Even after ouabain incubation, when the pump is inhibited and the Na gradient is cancelled, the Na-Cl cotransport system will exchange Na and Cl across the basolateral membranes, thus allowing for a transcellular Na and Cl flux.

To conclusively prove the effect of Cl on the Na backflux, we did two sets of experiments (with amphotericin B treated corneas)

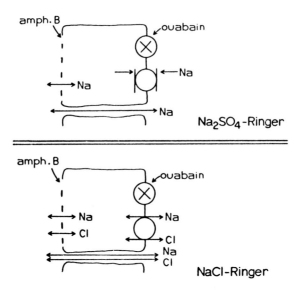

FIG. 5. *Movement of Na across the amphotericin B plus ouabain treated corneal epithelium. Top panel: In Na₂SO₄ Ringer, Na moves across the paracellular pathway. Lower panel: In NaCl Ringer, Na moves also across the transcellular pathway utilizing the Na-Cl co-transport carrier.*

in which the Na backflux was measured in the same cornea in Cl-free and Cl-rich solutions in either sequence.

    The results are shown in Table VIII. In every one of five corneas, the Na backflux increased when Cl (45 mM concentration) replaced part of the SO_4. The increase was on the average of about 0.20 μEq/hr/cm^2. Conversely, in every one of another set of five corneas, the Na backflux decreased on the average 0.30 μEq/hr/cm^2 when Cl (45 mM) in the bathing solutions was replaced by SO_4.

    It has been suggested that furosemide inhibits Cl transport by blocking the coupled entry of Na-Cl (9). We tested this hypothesis in the cornea. After amphotericin B, Cl fluxes in both directions increase, but a net in the stromal tear direction is still detectable (see reference 13). This implies that part of the SCC in amphotericin B treated corneas is Cl-originated and should be suppressed in a Cl-free medium or by furosemide or

TABLE VIII. *Unidirectional Na Backflux (Stroma to Tear) across the Short-Circuited Bullfrog Cornea in Cl-Free and 45 mM Cl Solutions*[a]

	I	II	III
	Cl-free, control	*Cl-free + amph. B*	*45 mM Cl + amph. B*
Set 1 (5)	*0.20 ± 0.02*	*0.35 ± 0.02*	*0.54 ± 0.05*[b]
	45 mM Cl, control	*45 mM Cl + amph. B*	*Cl-free + amph. B*
Set 2 (5)	*0.23 ± 0.03*	*0.70 ± 0.08*	*0.40 ± 0.05*[c]

[a]*Values in the same horizontal line are means ± SE in $\mu Eq/hr/cm^2$ from the same set of corneas.*
[b]*Significantly larger than mean in column II. P < 0.01 for paired differences.*
[c]*Significantly smaller than mean in column II. P < 0.01 for paired differences.*

bumetanide. As shown in Fig. 6, replacement of Cl by SO_4 reduced the SCC, but furosemide (or bumetanide) failed to modify the SCC.

However, when unidirectional Cl fluxes were measured, a clear inhibitory effect by furosemide was observed on both Cl fluxes as shown in Table IX. This suggests that amphotericin B and furosemide were acting in separate parallel sites of the apical membrane. This contention was further supported by the fact that furosemide did not depress the Na backflux whereas Cl removal did (see Tables VIII and IX).

Another experimental test was to measure the energetic requirements of the Na:K pump and the Cl transport mechanism. In a Na-driven Cl cotransport, the energetic requirements for the Cl transport should be less than for the Na:K pump. Oxygen consumption measurement in the cornea is consistent with a Na-Cl cotransport system. As shown by Reinach *et al.* (26), O_2 consumption in NaCl was inhibited 34% by either ouabain or Na removal, but only 16% to 19% by furosemide or Cl removal. Also, the inhibitory effect of ouabain in Na_2SO_4 medium was only 18%.

All previous and new experiments reviewed in this presentation

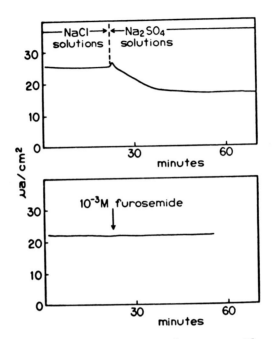

FIG. 6. In amphotericin B treated corneas, Cl removal from
the bathing solutions reduces the SCC (top panel); however, furosemide, an inhibitor of Cl fluxes, has no effect on the SCC (lower
panel).

are consistent with a model of Na-Cl cotransport located in the
basolateral membranes of the corneal epithelium. The evidence is
so overwhelming that it is difficult to defend the existence of

TABLE IX. Effect of Furosemide on Unidirectional Cl Fluxes and
Na Backflux in Amphotericin B Treated and Short-Circuited Bullfrog
Cornea[a]

	Amphotericin B Control	Amphotericin B 10^{-3} M Furosemide
Cl flux, tear to stroma (5)	0.55 ± 0.01	0.38 ± 0.03[b]
Cl flux, stroma to tear (9)	0.73 ± 0.06	0.53 ± 0.05[b]
Na flux, stroma to tear (5)	0.47 ± 0.04	0.47 ± 0.05

[a]Values are means $\pm$ SE in $\mu Eq/hr/cm^2$
[b]Significantly smaller than control. $P < 0.01$ for paired
differences.

primary "active Cl transport" in the corneal epithelium. Nevertheless, I felt that some additional experimental tests were still required to definitely assert the existence of the Na-Cl cotransport mechanism. One of them is, for example, the possibility of reversing the direction of the net Cl transport by reversing the Na gradient. Recently, I was able to reverse the normal Na gradient across the basolateral membrane (from cell:20 → bath:104 to cell:100 ← bath:25) by incubation with amphotericin B and reduction of the stromal side [Na] to 25 mM. Under this condition, measurement of the unidirectional Cl fluxes showed a net Cl flux of about 7 $\mu a/cm^2$ from the tear to stromal side bathing solution. Thus, the corneal epithelium can transport Cl in either direction depending upon the direction of the Na gradient across the basolateral membrane. These results are confirmatory of a Na-Cl cotransport mechanism and show that in the cornea, active Cl transport is not directly but rather secondarily coupled to metabolism.

SUMMARY

The active transport of Cl, Na, and K by the frog corneal epithelium is examined. The apical membrane is relatively impermeable to Na and K, and normally only a transepithelial Cl transport towards the tear side is detectable. Cl enters the basolateral membrane against an electrochemical gradient. In the presence of amphotericin B, the apical membrane permeability increases and a net transport of Na (tear to stroma) and K (stroma to tear) is also observed. Cellular K concentration is 80 mM in control conditions and 16 mM after amphotericin B incubation. Thus, the location of the Na:K pump is also in the basolateral membrane. After amphotericin B, Na backflux (stroma to tear) is larger in the presence of Cl in the medium. After ouabain (in the absence of net flux), both unidirectional Na fluxes are larger in the presence of Cl in the medium. In amphotericin B treated corneas, furosemide reduced unidirectional Cl fluxes but has no effect on Na flux or short-circuit current (SCC). However, Cl removal from the medium reduced the SCC. The experimental evidence and energetics considerations are consistent with a Na-Cl cotransport system located in the basolateral membrane. Furosemide seems to decrease Cl permeability of the apical membrane rather than to block the Na-Cl carrier.

ACKNOWLEDGMENT

I gratefully acknowledge Peter Reinach's collaboration in some of the studies described in this paper. This work was supported by Public Health Service grants EY-00160 and EY-01867.

REFERENCES

1. H. N. Nellans, R. A. Frizzell, and S. G. Schultz. Coupled sodium-chloride influx across the brush border of rabbit ileum. *Am. J. Physiol.* *225*:467-475, 1973.

2. R. A. Frizzell, M. Dugas, and S. G. Schultz. Sodium chloride transport by rabbit gallbladder: Direct evidence for a coupled NaCl influx processes. *J. Gen. Physiol.* *65*:769-795, 1975.

3. K. A. Spring and G. Kimura. Chloride reabsorption by renal proximal tubules of Necturus. *J. Membr. Biol.* *38*:233-254, 1978.

4. J. F. Quay and W. M. Armstrong. Sodium and chloride transport by isolated bullfrog small intestine. *Am. J. Physiol.* *217*:694-702, 1969.

5. J. A. Zadunaisky. Active transport of chloride in frog cornea. *Am. J. Physiol.* *211*:506-512, 1966.

6. K. J. Degnan, K. J. Karnaky, and J. A. Zadunaisky. Active chloride transport in the *in vitro* opercular skin of a teleost (*Fundulus heteroclitus*), a gill-like epithelium rich in chloride cells. *J. Physiol. (London) 271*:155-191, 1977.

7. E. Skadhauge. Coupling of transmural flows of NaCl and water in the intestine of the eel (*Anguilla anguilla*). *J. Exp. Biol. 60*:535-546, 1974.

8. C. O. Watlington, S. D. Jessee, and G. Baldwin. Ouabain, acetazolamide, and Cl-flux in isolated frog skin: Evidence for two distinct active Cl^- transport mechanisms. *Am. J. Physiol. 232(6)*:F550-F558, 1977 or *Am. J. Physiol.: Renal Fluid Electrolyte Physiol. 1(6)*:F550-F558, 1977.

9. R. A. Frizzell, M. Field, and S. G. Schultz. Sodium-coupled chloride transport by epithelial tissues. *Am. J. Physiol. 236(1)*:F1-F8, 1979 or *Am. J. Physiol.: Renal Fluid Electrolyte Physiol. 5(1)*:F1-F8, 1979.

10. T. Rosenberg. On accumulation and active transport in biological systems: I. Thermodynamic considerations. *Acta Chem. Scand. 2*:14, 1948.

11. O. Kedem. Criteria of Active Transport. *In* "Membrane Transport and Metabolism," (A. Kleinzeller and A. Kotyk, eds.), p. 87. Prague, Czechoslovak Academy of Science, 1961.

12. O. A. Candia and W. A. Askew. Active sodium transport in the isolated bullfrog cornea. *Biochim. Biophys. Acta 163*: 262-265, 1968.

13. O. A. Candia, P. J. Bentley, and P. I. Cook. Stimulation by amphotericin B of active Na transport across amphibian cornea. *Am. J. Physiol. 226(6)*:1438-1444, 1974.

14. O. A. Candia and P. Reinach. Active Na and K transport by the corneal epithelium: Coupling ratio and effect of ouabain. *The Physiologist 21(4)*:16, 1978.

15. J. A. Zadunaisky. Active transport of chloride across the cornea. *Nature 209*:1136-1137, 1966.

16. O. A. Candia. Ouabain and sodium effects on chloride fluxes

across the isolated bullfrog cornea. *Am. J. Physiol. 223(5)*: 1053-1057, 1972.

17. J. A. Zadunaisky. Sodium activation of chloride transport in the frog cornea. *Biochim. Biophys. Acta 282*:255-257, 1972.

18. S. D. Klyce and R. K. S. Wong. Site and mode of adrenaline action on chloride transport across the rabbit corneal epithelium. *J. Physiol. 266*:777-799, 1977.

19. T. Otori. Electrolyte content of the rabbit corneal stroma. *Exptl. Eye Res. 6*:356-367, 1967.

20. J. A. Zadunaisky, K. R. Spring, and T. Shindo. Intracellular chloride activity in the corneal epithelium. *Federation Proceedings 38(3)*:1059, 1979.

21. W. D. Stein. The movement of molecules across cell membranes. Academic Press, New York and London, 1967.

22. P. S. Reinach and O. A. Candia. Effects of tryptamine on active sodium and chloride transport in the isolated bullfrog cornea. *Biochim. Biophys. Acta 510*:327-338, 1978.

23. O. A. Candia and H. F. Schoen. Selective effects of bumetanide on chloride transport in bullfrog cornea. *Am. J. Physiol. 234(4)*:F297-F301, 1978 or *Am. J. Physiol.: Renal Fluid Electrolyte Physiol. 3(4)*:F297-F301, 1978.

24. O. A. Candia. Short-circuit current related to active transport of chloride in the frog cornea: Effects of furosemide and ethacrynic acid. *Biochim. Biophys. Acta 298*:1011-1014, 1973.

25. O. A. Candia, H. F. Schoen, L. Low, and S. M. Podos. Chloride transport inhibition by piretanide and MK-196 in bullfrog corneal epithelium. *Am. J. Physiol. 240*:F25, 1981.

26. P. S. Reinach, H. F. Schoen, and O. A. Candia. Effects of inhibitors of Na and Cl transport on oxygen consumption in the bullfrog cornea. *Exp. Eye Res. 24*:493-500, 1977.

CHLORIDE REABSORPTION, BICARBONATE SECRETION
AND ELECTROPHYSIOLOGICAL PARAMETERS IN THE
TURTLE BLADDER

John H. Durham

Christina Matons

William A. Brodsky

Department of Physiology and Biophysics
Mount Sinai School of Medicine
New York, New York

Gerhardt Ehrenspeck

Department of Zoology
Ohio University
Athens, Ohio

I. INTRODUCTION

Sodium Transport. When the isolated turtle bladder is bathed
by NaCl-containing media, the serosal fluid is electropositive to
the mucosal fluid. The resulting short-circuiting current, in
media devoid of exogenous HCO_3 and CO_2, was shown to be carried
almost exclusively by active Na-reabsorption (Klahr and Bricker,

1964). These findings were confirmed and extended for bladders
bathed by other media (Gonzalez *et al*, 1967a,b).

   HCO$_3$ or H Transport. Isolated bladder sacs have also been
shown to acidify the luminal fluid (Schilb and Brodsky, 1966).
When such sacs are bathed by Na-containing media, the luminal fluid
is acidified while the serosal fluid is electropositive to the mucosal solution. When bathed by Na-free solutions, the luminal
fluid is acidified while the serosal solution is electronegative to
the mucosal solution (Schilb and Brodsky, 1966). Subsequent studies
provided evidence that the acidification mechanism was that of HCO$_3$
reabsorption (Schilb and Brodsky, 1966,1972; Schilb, 1978), or
alternatively, that of H secretion (Steinmetz, 1967; Green et al,
1970; Schwartz et al, 1974). It is generally agreed that this
acidification process is electrogenic regardless of the actual
identity of the transported ionic species (Brodsky and Schilb,
1974; Steinmetz, 1974).

   With the methods used in the present study, the electrophysiological parameters associated with H secretion are the same
as those associated with HCO$_3$ reabsorption. Therefore the terms,
HCO$_3$ reabsorption and H secretion are to be taken as synonymous.

   Cl Transport. There are at present two different hypotheses
on the mode by which Cl is actively reabsorbed by the turtle bladder. One holds that an active translocation is mediated by an
electroneutral and transepithelial Cl-for HCO$_3$ exchange mechanism;
the other holds that the Cl transport is driven by an electrogenic
pump mechanism. Since each hypothesis is consistent with a specific set of data from which it had been derived, both sets will be
discussed.

   The active nature of sodium independent chloride reabsorption
by the turtle bladder was first demonstrated unequivocally in the
open-circuited sac preparation (Brodsky and Schilb, 1966). In Na-
rich media, the bladder sacs transport NaCl into an electropositive
serosal fluid; and in Na-free media, the sacs transport choline
chloride into an electronegative serosal fluid. Subsequently ob-

tained data on short-circuited turtle bladders in Na-rich media
with or without ouabain or in Na-free media, showed that the active
transport of chloride is independent of the concomitant active
transport of Na, and independent of the presence of exogenous Na
(Gonzalez et al, 1967a,b; Solinger et al, 1968). Gonzalez *et al*
(1967a) found that in short-circuited bladder preparations with Na-
free, (HCO_3 + Cl)-containing media bathing both the mucosal and
serosal surfaces, the short-circuiting current (Isc) exceeds that
of the net Cl flux (I_{net}^{Cl}):

$$I_{sc} > I_{net}^{Cl} \qquad (1)$$

Postulating that the Isc is the sum of the two parallel electrogenic ion flows of HCO_3 and Cl, Gonzalez *et al* (1967a) represented the Isc as the sum of both HCO_3 reabsorption and Cl
reabsorption:

$$I_{sc} = I_{net}^{HCO_3} + I_{net}^{Cl} \qquad (2)$$

These authors then examined the relationship between the Isc
and I_{net}^{Cl} in bladders bathed by HCO_3-poor mucosal fluids, but with
(HCO_3+CO_2)-rich serosal fluids. Under this condition, the parallel
flow of current due to HCO_3 reabsorption would be decreased, and
the Isc should decrease to approximate the value of I_{net}^{Cl}. This was
found to be the case, that is, under these conditions of HCO_3-poor
mucosal fluids, the relationships described in Eq. (2) above reduced to

$$I_{sc} \simeq I_{net}^{Cl} \qquad (3)$$

On this basis, Gonzalez *et al* (1967a) postulated that Cl transport by the bladder occurred via an electrogenic process. However
this equality is not proof of the electrogenic nature of Cl trans-

port, because it was found in the presence of transepithelial
chemical gradients for some of the ions present, most notably that
of HCO_3.

In the presence of a similar transepithelial HCO_3 gradient,
Leslie *et al* (1973) and subsequently Himmelstein *et al* (1974)
found that a net reabsorptive increment in the transepithelial Cl
flux:

   (i)   occurred only in the presence of serosal HCO_3,

   (ii) occurred concomitantly with the appearance of a nearly
         equal amount of alkali in the mucosal fluid,

         and

   (iii) was energy dependent

These authors therefore concluded that net (m to s) transepithelial Cl transport occurs via an electroneutral, energy dependent
Cl for HCO_3 exchange mechanism. It follows from this conclusion
that no electrical changes would occur following Cl removal from or
addition to the bathing fluids.

On the basis of the aforementioned studies, neither an electro
genic nor an electroneutral characteristic can be assigned with
certainty to the active Cl reabsorption in the turtle bladder. Thi
uncertainty led to the present experiments on the response of the
short-circuited turtle bladder to symmetrical changes in Cl concentration in the absence of transepithelial electrochemical gradients

II. METHODS

Turtle bladders (*Pseudemys scripta*) were excised, mounted, and
short circuit current (Isc), potential difference (PD) and resistance (R) were measured as described previously (Gonzalez *et al*,
1967a). All bladders were bathed by media containing 1-2 x $10^{-4}M$
ouabain in the serosal solution. The bathing media had the following composition (in m*M* units). *Na-free, (HCO_3+Cl) containing*

medium: Choline HCO_3, 20; choline Cl, 25; K_2SO_4, 2; K_2HPO_4, 0.61; KH_2PO_4, 0.14; $CaSO_4$, 2; $MgSO_4$, 0.8; glucose, 11; and sucrose in amounts to make the final osmolality 220 mOsM/Kg. *Na-free, Cl-free, HCO_3-containing medium:* as above, except that choline Cl was replaced with choline SO_4, 12.5. *Na-free, HCO_3-poor, Cl-containing medium* (for pH statting experiments): the same as the Na-free, (HCO_3+Cl)-containing solution except that choline HCO_3 was replaced by choline SO_4, 10; K_2HPO_4, 0.26; KH_2PO_4, 0.04. *Na-free, Cl-free, HCO_3-poor medium* (for pH statting experiments): the same as the latter solution except that choline Cl was replaced by choline SO_4, 12.5 m*M*. *Na-containing, Cl-free, HCO_3-poor medium:* Na_2SO_4, 50 (the solution used for the serosal bathing fluid contained 2 mM $CaSO_4$); glucose, 11; and sucrose in amounts to make the final osmolality 220 mOsM/kg.

All HCO_3-containing solutions were continuously bubbled with CO_2 + O_2-containing gases, all HCO_3-poor solutions with 100% O_2. The pH of all media was 7.3 except for that of the pH statting media which was 7.6.

The methods of changing bathing solutions were of 2 types: (a) For experiments depicted in Table I and Fig. 4, half the volume (5 ml) of the solution bathing the tissue was removed and replaced with 5 ml of the desired solution. This half-volume replacement was repeated 10 times. (b) For the experiments shown in Figs. 2, 3 and Table II, solution changes were made by passing the appropriate solution from a reservoir into the chamber containing 10 ml of bathing fluid, the level which was maintained constant by aspiration. The PD-detecting-electrodes were connected to the bathing solution via a readily renewable liquid-to-liquid junction (bathing solution: saturated KCl to avoid diffusion potentials in the PD measuring system occasioned by changes in solution composition (See Fig. 1). Cl determinations were done throughout the experiment to assure that the removal or addition of chloride was successfully completed. Cl fluxes were determined as described previously (Brodsky et al, 1979).

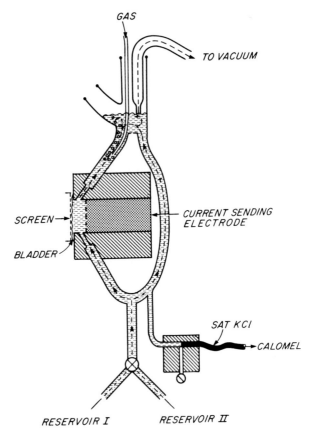

FIG. 1. Diagram of the system used for solution changes and PD measurements in those experiments depicted in Figs. 2 and 3 and Table II. Only the mucosal half-chamber is shown. To change the composition of the bathing media, aspiration was begun, and the appropriate solution from one of the reservoirs was admitted to the bathing system. During this process, the liquid-liquid (Bathing media:saturated KCl) junction was constantly renewed.

III. RESULTS AND DISCUSSION

We have investigated the effects of the removal and addition of Cl to the bathing solutions on the electrical parameters of the bladder, while maintaining the steady state transepithelial electrochemical gradients of all ions at zero. It was found that the removal and addition of Cl to the bathing media resulted in

Table I. Effect of addition of Cl to the
mucosal solution on the PD and I_{sc} of 7 bladders
in HCO_3-rich media.

Cl concentration (mV)	PD (mV)	I_{sc} (ua)
~0	26.8 ± 5.12	15.4 ± 2.24
25	45.9 ± 6.81	40.6 ± 10.2

Bladders (n=7) were bathed initially on both surfaces by Na-free (choline), HCO_3-rich, Cl-free
(SO_4) media. After a steady state had been
attained, Cl (25 mM) was substituted for SO_4 in
both solutions. Isc values are for 1.5 cm^2 of
exposed bladder surface.

decreases and increases, respectively in both the Isc and PD
(Table I). These results are in accord with the previous hypothe-
sis that Cl transport occurs via an electrogenic process
(Gonzalez, 1967a).

The next approach was to investigate the magnitude of the Cl-
induced increment in Isc which should be equal to that of the net
Cl flux, if the previous hypothesis of Gonzalez et al (1967a) were
correct (Equation 2 above).

This predicted equality was found in some of the experiments;
data from one are shown in Fig. 2. However, inequalities were
also found in a significant number of these experiments; data from
one of the most commonly observed inequality patterns are shown in
Fig. 3. The results of all 27 aforementioned experiments

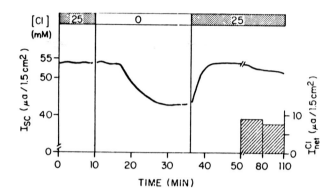

*FIG. 2. The effect of Cl removal from and addition to both
the serosal and mucosal bathing media on the Isc of a turtle
bladder initially bathed on both surfaces by Na-free, (HCO₃+Cl)-
containing media. Serosal fluid contained 10^{-4} M ouabain. Open
circuit PD (not shown) is that of serosal solution electronegative
to the mucosal solution. The Isc and PD (not shown) decreased
(after a 5 min lag period) following the removal of Cl (sulfate
substitution) from both bathing media, and increased (with little
lag time) following the addition of Cl to both bathing media. Net
Cl flux was determined after the return to the control steady
state was achieved.*

(Table II), considered en masse, indicate that there was a slight
but significant trend of proportionality between Cl-induced Isc
and the net Cl flux.

The hypothesis holding that Cl transport occurs via a transepithelial, electroneutral Cl-for-HCO₃ exchange would lead to the
prediction that there should be no Cl-induced increment in Isc.
This prediction was found to hold true in 4 of the 27 experiments.
Thus, without invoking additional assumptions, such a process is
capable of accounting for a small number of the present experiments.

The hypothesis that Cl transport occurs via an electrogenic
process which is independent of all other transport processes
would lead to the prediction that there would be a finite Cl-
induced increment in Isc, equal in magnitude to the net Cl flux.

This equality was found in 11 of the 27 experiments. Therefore, without assumptions, this hypothesis is capable of accounting for no more than 40% of the experiments.

Since neither of these predictions could be verified in all bladders, it became necessary to consider the possibilities that (i) factors other than chloride contribute to the observed changes in Isc or, (ii) one or both of the hypotheses require additional or modifying assumptions.

It is possible that the effects of Cl removal and addition on the Isc are not due to Cl per se, but to the subsequent effects of respective increases and decreases in the SO_4 concentrations of the bathing fluids. In this context, it has been demonstrated that the decreased free Ca^{2+} levels occasioned by increased SO_4^{2-} concentration can affect cellular function (Hill and Howarth, 1957). The possibility thus exists that the decreases in free Ca^{2+} concentration following Cl removal might decrease the basal Isc (HCO_3 reabsorptive or H secretory current), thereby leading to the mistaken conclusion that Cl was responsible for the decrement in Isc. Whether this phenomenon is indeed what gives rise to the results seen in the present study requires further investigation. However, such an effect due to Ca^{2+} alone would not be expected to result in a proportionality between the Cl-induced Isc and the net Cl flux (Table II).

One of the assumptions concerning the hypothesis of an electrogenic Cl transport is its independence of other transport processes. If instead of being independent, the electrogenic translocation of Cl affects the transport of another electrogenic transport process, and vice versa, then a direct proportionality or even a significant correlation between the Cl-induced Isc and net Cl flux would not necessarily be expected. The most likely candidates for the other transport process are those of HCO_3 reabsorption or HCO_3 secretion. Some of the evidence which is needed to support this latter possibility would be (i) the demonstration of a Cl-induced alkali secretion concomitant with a change in Isc, and (ii) a Cl-

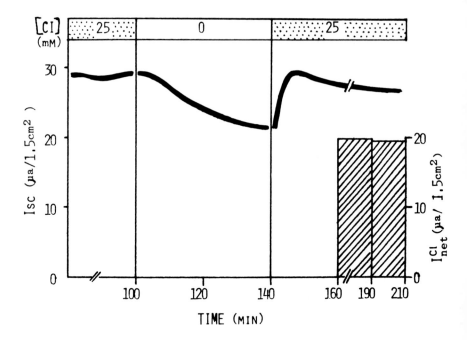

TIME (MIN)

FIG. 3. *The effect of Cl removal from and addition to both the serosal and mucosal bathing media on the Isc of a turtle bladder. Protocol as described for Fig. 2.*

Table II. *Mean steady state values of Cl-induced increment in short-circuiting current (Δ Isc) vs. those of net chloride flux (I_{net}^{Cl}) in 27 bladders. Protocol was the same as that described under Fig. 2.*

ΔI_{sc} (ua)	I_{net}^{Cl} (ua)	Correlation coefficient[1]
11.2 + 11.8	16.5 + 14	0.471
		0.05 > P > 0.01

All values are for 1.5 cm^2 of exposed bladder surface.
[1]*Product Moment Correlation Coefficient*

independent electrogenic HCO_3 secretion.

That a HCO_3 secretion is coupled to Cl reabsorption has been shown by Leslie *et al* (1973). We then looked at the effect of Cl addition to HCO_3-poor mucosal fluids (in the presence of HCO_3-rich serosal fluids) to determine its effect on Isc and alkali flow (Fig. 4). The effect was found to be an increase in Isc (as had been seen in the presence of symmetrical HCO_3 solutions) as well as an increase in alkali appearance into the mucosal fluid.

On one hand, consistent with the findings of Leslie *et al* (1973), the mucosal addition of Cl initiates a flow of alkali into the mucosal fluid in the presence of HCO_3 gradient. On the other hand, consistent with some of the above findings, the addition of

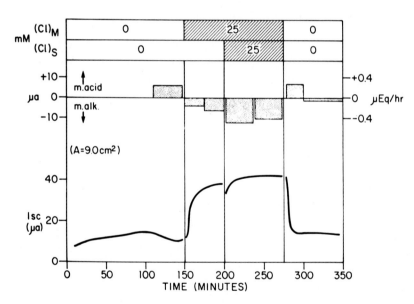

FIG. 4. The effect of the addition of Cl on the Isc and mucosal acidification rate of bladders bathed initially by Na-free, Cl-free media. The mucosal solution was a HCO_3-poor choline-SO_4 solution and was gassed with 100% O_2. The serosal solution was choline solution containing 20 mM HCO_3 and gassed with a 2% CO_2/98% O_2 mixture. Mucosal acidification (m. acid) and mucosal alkalinization (m. alk.) were measured titrimetrically, maintaining the pH at 7.6. Mucosal Cl concentration, $(Cl)_m$ Serosal Cl concentration, $(Cl)_s$. All values are for 9.0 cm^2 of bladder area. Sign convention is that as described in Fig. 2.

Cl is followed by an increase in Isc. Again, as before, neither
of the two hypotheses (electrogenic or electroneutral Cl transport)
alone can unequivocally account for these results. If however,
there was evidence that the HCO_3 secretion could be stimulated in
the absence of Cl and that this secretion carries current, it would
be possible to account for the apparent disparities between Cl-
induced increments of Isc and the net chloride fluxes.

Evidence for an Electrogenic Cl-Independent HCO_3 Secretion

    One line of evidence (Brodsky *et al*, 1979) for an electrogenic
Cl-independent secretion of HCO_3 is the observed effect(s) of
serosally-added SITS (4 acetamido, 4'isothiocyano, 2,2' disulfonic
stilbene) on bladders bathed on both surfaces by identical choline
media of two types: (i) HCO_3-rich, Cl-rich and   (ii) HCO_3-rich,
Cl-free. In the first medium the Isc and PD are decreased by SITS
from the usual control orientation under these conditions (mucosal
side positive) to zero levels while the net chloride reabsorption
remains finite. This electrically-silent post-SITS state requires
the coexistence of a charge-balancing ion flow, probably a HCO_3
secretion on the basis of data which rule out the transepithelial
reabsorption of cations or anions other than HCO_3 (Brodsky *et al*,
1979). The charge-balancing ion flow could be part of an
electroneutral exchange with the Cl reabsorption or could be due to
an electrogenic (current-carrying) HCO_3 secretion. The latter
hypothesis was verified when the experiment with SITS was carried
out in the second medium (a bathing system devoid of Cl). Under
these Cl-free, HCO_3-rich conditions, the Isc and PD following SITS
are not merely reduced to zero, but reversed so that the mucosal
side becomes electronegative to the serosal side (Table III). This
net flow of anion (probably HCO_3) from serosa to mucosa in the
absence of all transepithelial gradients can be represented as an
active, electrogenic secretion of bicarbonate ions.

TABLE III. The SITS-induced reversal in Isc and PD of bladders (n = 7) bathed on both sides by identical Na-free, Cl-free, HCO_3-rich media. Serosal solutions contained 10^{-4} M ouabain.

Condition	I_{sc}(ua)	PD(mV)
Control	14.6 $\pm$ 4.3	26.6 $\pm$ 6.7
SITS	-3.5 $\pm$ 0.4	-7.3 $\pm$ 0.3

Minus signs denote electronegativity of mucosal solution with respect to serosal solution. Isc values are for 1.5 cm^2 of exposed bladder surface (derived from Brodsky et al, 1979).

A second line of evidence also supports the idea of an electrogenic secretion of HCO_3. Donor turtles which had been adapted (for 2 weeks) to an ambient temperature of 30°C, were fed within 4-5 days prior to removal of the bladder. With these bladders, it has been shown (Ehrenspeck, 1981) that the serosal addition of IBMX (isobutyl-methyl-xanthine) also induces a reversal in the direction of Isc and PD from serosa negative to mucosa negative (Table IV). This IBMX-induced reversal, found in ouabain-treated bladders, bathed by Cl-free, Na-free media as well as by Cl free, Na-rich media, is found to be elicited only when HCO_3 and CO_2 are present in the serosal solution, regardless of whether the mucosal fluid contains HCO_3 and CO_2. The serosal addition of cyclic AMP under these IBMX-treated conditions further increased the magnitude of the "reversed" Isc so that the mucosa became more electronegative. On the other hand, the serosal addition of acetazolamide decreased this reversed Isc to near zero levels.

The aforementioned effects of IBMX and SITS are most simply explained by invoking an electrogenic secretion HCO_3 ions.

TABLE IV.   The effect of IBMX (5 x 10^{-5}M,
serosal fluid) in 4 bladders bathed by Na-SO_4-
HCO_3-solutions devoid of other ions except 2mM
Ca^{++} in the serosal fluid. 2 x 10^{-4}M ouabain
was present in the serosal solution throughout
the experiment.   The same effects are seen in
Na-free (choline) solutions.

Conditions	I_{sc} (ua)	PD(mV)
Control	18 ± 6	14 ± 6
IBMX	-13 ± 8	-20 ± 7

Footnote: minus signs denote electronegativity
of mucosal solution with respect to serosal
solution.   All changes were significant.   Isc
values are for 1.5 cm^2 of exposed bladder
surface.

Thus, there is evidence to support the conclusion that the
bladder is capable of an active, electrogenic, Cl-independent
secretion of HCO_3. When such a process is placed in parallel
with an electrogenic pathway for Cl reabsorption and another
parallel path for electrogenic HCO_3 reabsorption (or H secretion),
the bladder can be represented by a two-membrane, distributed-
parameter, equivalent electrical network which will account for:
(i) a Cl-induced increase in Isc which may or may not be equal to
the net Cl flux;   (ii) a Cl-induced increase in HCO_3 secretion via
electrical coupling between these two flows; and (iii) a Cl-
dependent electrogenic HCO_3 secretion. Such a model has been
presented previously (Brodsky et al, 1979). However, this model
cannot account for the serosal HCO_3 requirement of the Cl reabsorption.
    What can be said at this time on the basis of data from this
and other laboratories is that the turtle bladder possesses (i) an

active electrogenic mechanism for HCO$_3$ reabsorption (or proton secretion); (ii) an active, electrogenic mechanism for HCO$_3$ secretion (or proton reabsorption); and (iii) an active mechanism for Cl reabsorption, the exact nature of which remains to be determined.

ACKNOWLEDGEMENT

Supported in part by NIH grant # AM-16928-04A2 and NSF grant # PCM-7923511.

REFERENCES

1.   W. A. Brodsky and T. P. Schilb.   Ionic mechanism for sodium
     and chloride transport across turtle bladders.   *Am. J.*
     Physiol. 210: 987-986, 1966.
2.   W. A. Brodsky and T. P. Schilb.   The means of distinguishing
     between hydrogen secretion and bicarbonate reabsorption:
     Theory and applications to the reptilian bladder and mamma-
     lian kidney.   "Current Topics in Membranes and Transport,"
     5: 161-224. (F. Bonner and A. Kleinzeller, eds.) Academic
     Press, N.Y., 1974.
3.   W. A. Brodsky, J. H. Durham, and G. Ehrenspeck.   Disulfonic
     stilbene-induced effects on and the electrogenic nature of
     HCO$_3$ reabsorption, HCO$_3$ secretion and Cl reabsorption in
     turtle urinary bladders.   *J. Physiol. (London) 287:* 559-573,
     1979.
4.   G. Ehrenspeck.   IBMX reverses the direction of the HCO$_3$/CO$_2$-
     dependent PD and Isc in turtle bladder.   *Fed. Proc. (Abstr.)*
     40: 357, 1981.
5.   C. F. Gonzalez, Y. E. Shamoo, and W. A. Brodsky.   Electrical
     nature of active chloride transport across short-circuited

turtle bladder. *Am. J. Physiol. 212:* 641-650, 1967a.

6.   C. F. Gonzalez, Y. E. Shamoo, H. R. Wyssbrod, R. E. Solinger, and W. A. Brodsky. Electrical nature of sodium transport across the isolated turtle bladder. *Am. J. Physiol. 213:* 333-340, 1967b.

7.   A. V. Hill and J. V. Howarth. The effect of potassium on the resting metabolism of the frog's sartorius. *Proc. Roy. Soc. B. 147:* 21-43, 1957.

8.   H. H. Green, P. R. Steinmetz, and H. S. Frazier. Evidence for proton transport by turtle bladder in the presence of ambient bicarbonate. *Am. J. Physiol. 218:* 815-850, 1970.

9.   S. Himmelstein, J. A. Oliver, and P. R. Steinmetz. Energy dependence of urinary bicarbonate secretion in turtle bladder. *J. Clin. Invest. 44:* 1003-1008, 1975.

10. S. Klahr and N. S. Bricker. Na transport by isolated turtle bladder during anaerobiosis and exposure to KCN. *Am. J. Physiol. 206:* 1333-1339, 1964.

11. B. R. Leslie, J. H. Schwartz, and P. R. Steinmetz. Coupling between Cl^- absorption and HCO_3-secretion in turtle bladder. *Am. J. Physiol. 225:* 610-617, 1973.

12. T. P. Schilb and W. A. Brodsky. Acidification of mucosal fluid by transport of bicarbonate ion in turtle bladders. *Am. J. Physiol. 210:* 997-1008, 1966.

13. T. P. Schilb and W. A. Brodsky. CO_2 gradients and acidification by transport of HCO_3 in turtle bladders. *Am. J. Physiol. 22:* 272-281, 1972.

14. T. P. Shilb. Bicarbonate ion transport: A mechanism for the acidification of turtle urine in the turtle. *Science 200* (4339): 208, 1978.

15. J. H. Schwartz, J. T. Fine, G. Vaughn, and P. R. Steinmetz. Distribution of metabolic CO_2 and the transported ion species in acidification by turtle bladder. *Am. J. Physiol. 226:* 283-284, 1974.

16. R. E. Solinger, C. F. Gonzalez, Y. E. Shamoo, H. R. Wyssbrod,

and W. A. Brodsky. Effect of ouabain on ion transport mechanisms in the isolated turtle bladder. *Am. J. Physiol. 215:*
249-261, 1968.

17. P. R. Steinmetz. Characteristics of hydrogen ion transport in
urinary bladder of water turtle. *J. Clin. Invest. 46:* 1631-
1640, 1967.

18. P. R. Steinmetz. Cellular mechanism of urinary acidification.
Physiol. Rev. 54: 890-956, 1974.

ENERGETICS OF ACTIVE CHLORIDE TRANSPORT IN SHARK RECTAL GLAND*

Franklin H. Epstein
Patricio Silva

Department of Medicine and Thorndike Laboratory
Harvard Medical School and Beth Israel Hospital
Boston, Massachusetts
and
Mount Desert Island Biological Laboratory
Salsbury Cove, Maine

Thirty years ago, when James Gamble summarized the facts of water and electrolytes for an admiring generation of medical students, the transport of chloride was considered to be entirely passive. Excluded from cells, chloride was regarded as a kind of filler, its movement dependent entirely upon the varied whims of sodium, potassium, and bicarbonate. Gamble epitomized this notion of chloride by calling it "a mendicant ion."

It is becoming apparent that far from being a mendicant ion, chloride is in fact actively transported by a number of animal tissues, including the central nervous system, and especially by secretory epithelial organs. Study of the rectal gland of the

* Aided by a grant (AM-18078) from the U.S. Public Health Service and grants BG-5781 and PCM-77-01146 from the National Science Foundation.

shark has helped to establish the mechanism of this active transport.

I. THE OSMOTIC DILEMMA OF THE SHARK

From an evolutionary standpoint, the shark is said to have
developed in fresh water. But for millions of years sharks have
lived in the ocean, subject to special environmental stresses that
menace the composition of their body fluids (see Fig. 1). The
shark swims in a sea containing approximately 500 mEq/L of sodium
chloride, or 1000 mosm/L. The serum sodium of the shark is higher
than that of man, approximately 260-290 mEq/L, but sodium and its
associated anions amount to only half of the external osmotic pre-
sure of the ocean. To counterbalance the hypertonicity of the sea,
which if unopposed would rapidly dehydrate the fish, elasmobranchs

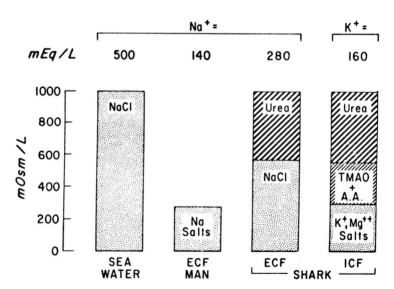

FIG. 1. Osmolar concentrations of seawater, extracellular
fluid (ECF) of man and shark, and intracellular fluid (ICP) of
shark muscle. A sizeable portion of intracellular osmolarity in
shark muscle is made up of trimethylamine oxide (TMAO) and amino
acids (AA).

manufacture urea, which circulates in high concentration in the blood and also permeates all cells. The concentration of mineral salts plus urea precisely balances the osmotic concentration of seawater, preventing the depletion of body water. Nevertheless, sharks must combat a tendency to become hypernatremic even beyond their accustomed plasma concentrations of sodium, which in comparison with bony vertebrates appears very high (260-290 mEq/L). Sodium chloride tends to diffuse inward through the gills and is absorbed when the shark swallows food. The rectal gland is the mechanism that elasmobranchs have devised for getting rid of increments of salt.

II. RECTAL GLAND OF ELASMOBRANCHS

The rectal gland of the spiny dogfish shark, *Squalus acanthias* (Fig. 2), is a relatively simple structure that looks somewhat like the appendix. It is composed of tightly packed tubules that look superficially like sections of the renal cortex. The movement of fluid and electrolytes is entirely in the secretory direction, from the blood to the lumen. The rectal gland has a single artery, vein, and central duct. This makes it easy for the rectal gland to be isolated, removed from the shark, and perfused in the laboratory with a shark Ringers solution resembling an ultrafiltrate of shark plasma. The gland is a hardy poikilothermic organ that works well at the temperature of seawater ($\sim 12^0$-20^0C) (2,3).

The rectal gland is necessary to the economy of the shark. When it is removed (Fig. 3), there is a progressive rise in serum sodium and chloride (4). The composition of rectal gland secretion consists almost entirely of hypertonic sodium chloride, at approximately the concentration of seawater, or 1/2 - 2 times the concentration of salt in the shark plasma. There are miniscule amounts of urea in the rectal gland secretion, which is isotonic (in osmotic terms) with plasma. An important feature of rectal gland secre-

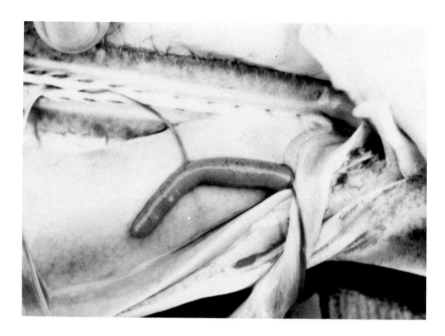

FIG. 2. Photograph of the rectal gland of Squalus acanthias.
The dorsal aorta is above, and the tail of the shark to the right.

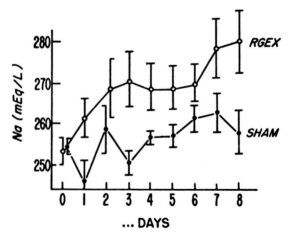

FIG. 3. Plasma sodium values of live Squalus acanthias after
excision of the rectal gland or after a sham operation.

tion observed in the artificially perfused gland is that the duct
is always electrically negative to the perfusate. Thus, chloride
appears to be moving against both a chemical and an electrical
gradient into the duct of the gland. Sodium, on the other hand,
moves down an electrical gradient although against a chemical
gradient (3).

III. CONTROL OF RECTAL GLAND SECRETION

Rectal gland secretion is under the control of cyclic AMP (5).
The addition of either cyclic AMP or theophylline to solutions perfusing the gland results in an enormous increase in secretory
activity. When secretion is stimulated in this way, the electrical
potential difference between duct and perfusate invariably becomes
more negative and the concentration of sodium chloride in the
secretion sometimes increases (3). Thus, cyclic AMP stimulates the
movement of chloride against an even higher electrochemical gradient
than existed previously, underlining the active nature of chloride
transport in rectal gland secretion.

The ease of rectal gland study, and the ability to turn on
massive chloride secretion at will, have stimulated intensive study
of the characteristics of the gland in an attempt to delineate the
mechanisms of chloride secretion.

IV. KEY PHYSIOLOGICAL FEATURES OF RECTAL GLAND SECRETION

Some key experimental features of rectal gland secretion that
must be incorporated into any general hypothesis of its mechanism
are as follows:

(1) The gland, as already indicated, is turned on by cyclic
AMP and theophylline.

(2) Chloride is actively secreted against an electrical and

chemical gradients, from blood to duct.

(3) Secretion is inhibited by ouabain and by omitting potassium from the perfusing solutions. Therefore it is likely to depend on Na-K-ATPase (3).

(4) Radioautographic studies clearly indicate that Na-K-ATPase lines the basolateral border of rectal gland cells in extensive basolateral infoldings, positioned to pump sodium out of the cell into the blood rather than into the duct (6).

(5) Secretion of chloride depends on the presence of sodium in the perfusate (3, 7).

(6) Secretion of sodium depends on the presence of chloride in the perfusate (3).

(7) Secretion is inhibited by furosemide and thiocynate, agents which inhibit the active transport of chloride in other tissues like the cornea, the gill, and the mammalian kidney (3).

(8) Finally, the intracellular concentration of chloride deduced from chemical measurements, as well as the activity of chloride in cellular cytoplasm determined directly with the chloride electrode, is considerably higher (70-80 mEq/L) than that predicted from the intracellular electrical potential (-60 to -80 mV) (3,7). Therefore chloride must be transported uphill into the cell across the basolateral cell border. It should be noted that an intracellular chloride concentration exceeding that predicted from the Nernst equilibrium is a feature of other secretory tissues as well. Examples of these include gastric mucosa, kidney, mammary gland, intestinal mucosa, and the avian salt gland.

V.   A GENERAL HYPOTHESIS FOR ACTIVE CHLORIDE TRANSPORT

These features suggest a general hypothesis for the active transport of chloride that is schematized in Fig. 4. Chloride is transported across the basolateral cell borders into the interior of rectal gland cells by a process of cotransport with sodium. The

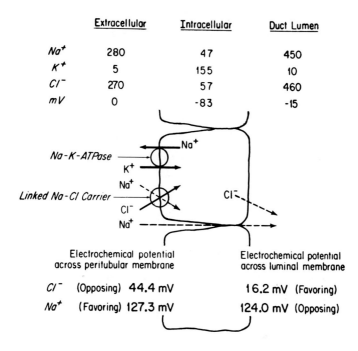

	Extracellular	Intracellular	Duct Lumen
Na⁺	280	47	450
K⁺	5	155	10
Cl⁻	270	57	460
mV	0	-83	-15

Na-K-ATPase

Linked Na-Cl Carrier

Electrochemical potential across peritubular membrane

Electrochemical potential across luminal membrane

| *Cl⁻* | (Opposing) **44.4 mV** | **16.2 mV** (Favoring) |
| *Na⁺* | (Favoring) **127.3 mV** | **124.0 mV** (Opposing) |

FIG. 4. *Schematic model for movement of chloride across rectal gland epithelium. Passive ion movements are shown by dotted lines; active transport by solid arrows. A nuetral sodium chloride carrier located in basolateral cell membrane effects active movement of chloride into cell, coupled to downhill movement of sodium. Low intracellular sodium concentration and downhill electrochemical gradient for sodium is maintained by activity of Na-K-ATPase. Chloride diffuses passively from cell into tubular lumen down an electrical gradient. Sodium moves down its electrochemical gradient into tubules through paracellular pathways, though an Na-K-ATPase pump on luminal cell border is not excluded. Lower 2 columns represent the electrochemical potentials (EC) for chloride and sodium across peritubular and luminal membranes, respectively. Calculations are based on Nernst equation where EC = PD + chemical potential, and chemical potential = (RT/zF) ln(C/C). Values for PD and electrolyte concentrations (mM) in extracellular, intracellular, and ductal fluid are shown in upper columns.*

mechanism for cotransport is entirely analogous to the mechanism well established for the cotransport of glucose and amino acids with sodium across cell membranes. The energy for chloride move-

ment derives from the passive inward movement of sodium along its
electrochemical gradient. This in turn is maintained by the action
of Na-K-ATPase. The Na-K-ATPase pump maintains a low concentration
of intracellular chloride, and at the same time is responsible for
the negative intracellular electrical potential, through the maintenance of a high intracellular concentration of potassium.
Chloride passes across the luminal border of the cell in passive
fashion, down its electrical gradient which exceeds the opposing
chemical gradient. The most probable route for the outward passage
of sodium is paracellularly, through the intracellular space, im-
pelled by an electrical gradient in the secretory direction.

The virtue of this model is that active chloride movement is
linked to the action of Na-K-ATPase even though Na-K-ATPase "faces"
in a direction opposite to that in which secretion occurs. The
model thus resolves a logical dilemma. In several secretory organs
(human sweat gland, intestinal mucosa, salivary gland), Na-K-ATPase
lines the basolateral side of the cell poised to pump sodium in a
direction opposite to that in which secretion actually occurs. In
all of these tissues chloride appears to be actively transported.

VI. DIRECT EVIDENCE FOR SODIUM AND CHLORIDE COTRANSPORT IN MEMBRANE VESICLES

Direct evidence for the cotransport of sodium with chloride
in isolated membranes has recently been obtained in membrane
vesicles derived from homogenates of shark rectal gland, as illustrated in Fig. 5 (8). Membrane vesicles were prepared from basolateral membranes (rich in Na-K-ATPase) of rectal glands of the
spiny dogfish *(Squalus acanthias)*. The uptake of radioactive
sodium into these vesicles was greatly enhanced by the presence of
chloride and proceeded much more slowly when nitrate or gluconate
was the accompanying ion. This was also the case when electrical
forces across the vesicle membrane were minimized by treating the

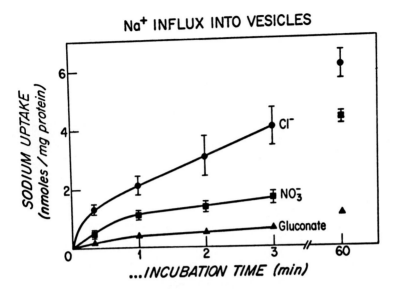

FIG. 5.   Uptake of ^{22}Na into plasma membrane vesicles of shark rectal gland.   Sodium uptake is accelerated in the presence of chloride, but retarded by nitrate or gluconate.

membrane with valinomycin.   The enhanced sodium uptake due to chloride was completely inhibited by preincubating the vesicles with furosemide.   These experiments provide the first direct evidence of the cotransport of sodium and chloride in isolated membranes and therefore provide strong support for the general hypothesis of active chloride transport outlined in Fig. 4.

VII.   CENTRAL ROLE OF Na-K-ATPase IN TRANSMEMBRANE TRANSPORT

The central role of Na-K-ATPase in providing energy for transmembrane transport of chloride via sodium coupling (9) is emphasized by these experiments.   Na-K-ATPase serves as the primary transducer mechanism to transform the chemical energy of ATP into the potential energy represented by the chemical gradient for sodium across plasma membranes.   Existence of a number of cotransporter carriers provides

the means by which this energy can be transformed into the active
transport of any variety of substances against their electrochemical gradients into the cell. The direction of transport across an
epithelial membrane can be determined simply by the placement of
the sodium cotransport attachments and the arrangement of suitable
permeabilities on either side of the cell. Transport can be regulated by regulating the velocity of the sodium ATPase pump, the
number and location of cotransporter carriers, and also by changing
the permeabilities of the plasma membrane lining the cell surfaces.
In the case of the rectal gland, it seems likely that the action
site of cyclic AMP is on the permeability to chloride of the luminal membrane of the rectal gland cell. This hypothesis can be
tested since it would predict that after stimulation of the rectal
gland the intracellular concentration of chloride would fall and the
negative intracellular potential difference would increase.

VIII.   RELATIONSHIP BETWEEN OXYGEN CONSUMPTION AND CHLORIDE SECRETION

The relationship between chloride secretion and oxygen uptake
by the isolated perfused rectal gland, stimulated with cyclic AMP
and theophylline, is illustrated in Fig. 6. The average Cl/O_2
ratio in these experiments was 30 $\mu Eq/\mu mol$. Thus, the oxygen cost
of transporting one mole of chloride is ~ 0.03 mol of O_2 (12).
The chemical gradient against which the rectal gland transports
chloride can be varied by experimental manipulation. Since fluid
secreted by the rectal gland is isotonic with the perfusing solution and little urea is found in the secretion, a change in the
concentration of urea in the perfusate results in a change in the
concentration of electrolytes in the secretion fluid. In the absence of urea there is no chemical gradient for chloride across the
gland. As urea concentration is increased in the perfusate, chloride concentration in the secretion fluid rises. Fig. 7 indicates

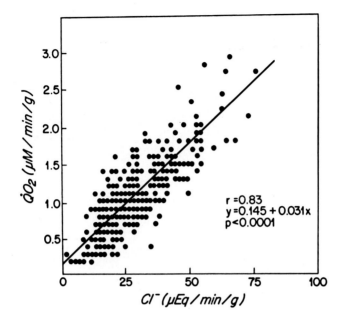

FIG. 6. *Relation between oxygen consumption and chloride secretion in stimulated perfused rectal glands (12).*

	DUCT FLUID [Cl-] mEq/L	Cl- SECRETION µ^M/min/g	Cl/QO$_2$ RATIO
No Urea	290	42±17	28±8
350 mM	450	29±9	26±7
550 mM	530	28±9	25±6
700 mM	570	23±6	31±10

FIG. 7. *Energy cost of secretion of chloride against different chemical gradients elicited by perfusing the rectal gland with varying concentrations of urea (12).*

that when the chemical gradient opposing chloride secretion is progressively increased in this way, the oxygen cost of transporting one mole of chloride remains unchanged, as indicated by the constant ratio of Cl/O_2.

The 30:1 ratio of chloride secreted to oxygen consumed by the rectal gland resembles the ratio of. sodium reabsorption to oxygen consumption observed in the mammalian kidney, but is higher than the ratio of sodium transport to oxygen consumption (14:1 - 19:1) found in frog skin and toad bladder. It is interesting that the chemical gradient opposing the secretion of chloride can vary widely without substantially changing the oxygen cost of transport per mole of chloride secreted. Analogous results have been reported in other systems (12). For example, the ratio of sodium transported inwardly across the frog skin to oxygen consumed by the skin is constant regardless of the gradient against which sodium moves. In the toad bladder as well, the number of moles of sodium transported from mucosa to serosa per mole of CO_2 produced is unchanged when the sodium concentration gradient across the bladder is altered (10). Similarly, secretion of hydrogen ions by the turtle bladder bears a constant relation to the production of CO_2, whether or not H^+ is transported against a pH gradient (11). Finally, the extrusion of Na^+ by human red-cell ghosts is accomplished at a constant metabolic cost despite changes in the gradient opposing sodium transport (13). In all these situations, the energy derived from metabolism per mole of ion translocated is presumably considerably larger than that required to overcome the electrochemical gradient opposing ion movement. It would be of interest to see whether the constant energy cost of ion movement in the face of changing chemical gradients is accomplished in the rectal gland by changing the passive movement (e.g, back-leak) of ions, by altering the electrical gradient across cell membranes, or by changing the quantity of energy released as heat per mole of ion moved.

Whether the cotransport of chloride with sodium is electrically neutral or involves the transfer of more than one Cl^- per Na^+ has not been determined. It is instructive to calculate the energy cost of active chloride transport according to this hypothesis and to compare it with the oxygen consumption actually observed. Such calculations, it should be understood, depend on a number of assumptions, including a fixed ratio of Na ions translocated to ATP hydrolyzed by Na-K-ATPase in intact cells, and a constant ratio of oxygen consumption to the production of high energy phosphate bands in ATP in living cells, extrapolated from the optimum P/O ratio of isolated mitochondria.

If 3 mol of Na^+ are transported per mol of ATP hydrolyzed by basolateral membranes of the rectal gland, and the P/O ratio of rectal gland mitochondria is 3/1, the ratio of Na^+ transported to O_2 consumed should be 18/1. If 1 mol of Cl^- were transferred for every mol of Na^+ transported, the ratio of Cl^- to QO_2 would also be 18/1, clearly lower than the figure of 30/1 that was observed. Derivation of energy for transport from nonoxidative metabolism (glycolysis) cannot account for this discrepancy because the production of lactic acid by the gland was minimal. Passive transport of chloride by means not requiring oxidative energy, e.g. solvent drag, could occur to a minor degree but seems unlikely to be significant because of the low concentration of urea in the duct secretion, averaging 10-15% of its concentration in the perfusate. If simple hydrostatic filtration or solvent drag played a major role in the formation of secretory fluid by the stimulated gland, a higher concentration of urea in duct fluid would be expected. The high ratio of Cl^- to O_2 suggests that more than one chloride ion is transported for each sodium extruded from the cell by Na-K-ATPase, raising the possibility in turn that the inward cotransport of Na^+ and Cl^- across basolateral membranes of the rectal gland may be electrogenic. This hypothesis can also be tested using vesicles derived from the plasma membranes of rectal gland cells.

IX.   ACTIVATION OF Na-K-ATPase IN THE STIMULATED RECTAL GLAND

There is strong evidence that Na-K-ATPase is activated when
the rectal gland is stimulated by cyclic AMP. Ouabain-inhibitable
oxygen consumption rises 6-7 fold when the gland is stimulated.
It is presumably the hydrolysis of ATP to ADP and Pi by Na-K-ATPase
that generates the chemical pressure that accelarates oxidative
phosphorylation and stimulates respiration. Changes in the content
of intracellular sodium and potassium that occur after cyclic AMP
stimulation also suggest that the enzyme is activated. Intracellular sodium falls and intracellular potassium rises in a way
suggesting primary stimulation of Na-K-ATPase rather than increased
activation of the enzyme by intracellular sodium. It seems possible that cyclic AMP stimulation results in some way in the rapid
activation of enzyme sites *in vivo*. Thus far it has not been
possible to demonstrate an increase in sodium-potassium activity
in rectal gland homogenates *in vitro* after stimulation. Eveloff
et al. were unable to show an increased rate of binding of ouabain
to slices of the rectal gland of *Squalus acanthias* after stimulation with cyclic AMP and theophylline (6). However, Shuttleworth
and Thompson recently reported an increase in ouabain binding sites
in slices of the rectal gland of the European dogfish, *Scyliorhinus
canicola*, after exposure to similar concentrations of cyclic AMP
and theophylline (13). It seems possible that subtle changes in
membrane conformation accompanying an increase in ion transport
may activate Na-K-ATPase *in vivo* without necessarily increasing the
maximum velocity of the enzyme reaction as measured in tissue homogenates incubated *in vitro*.

X.   SUMMARY

The rectal gland of the spiny dogfish, *Squalus acanthias*,
provides an easily studied model of active chloride transport

powered indirectly by Na-K-ATPase. Cotransport of sodium with chloride can be demonstrated in membrane vesicles isolated from basolateral membranes of the gland. Chloride secretion is stimulated by adenylate cyclase and cyclic AMP. In the stimulated state, oxygen uptake by the isolated perfused rectal gland is directly related to the rate of chloride secretion. A stoichiometric relationship exists between chloride transport and oxygen consumption, with a chloride to oxygen ratio of $\sim$ 30:1, resembling that reported for sodium in mammalian kidneys. The ratio remains constant under varying degrees and modes of stimulation, and does not change when the gland is induced to secrete chloride against varying chemical gradients by altering the concentration of urea in the perfusate. Na-K-ATPase is activated during cAMP-induced secretion but the mechanism of activation is not yet fully understood.

REFERENCES

1. J. W. Burger. Further studies on the function of the rectal gland in the spiny dogfish. *Physiol. Zool. 38:* 191-196, 1965.

2. J. P. Hayslett, D. A. Schon, M. Epstein, and C. A. M. Hogben. *In vitro* perfusion of the dogfish rectal gland. *Am. J. Physiol. 226:* 1188-1192, 1974.

3. P. Silva, J. Stoff, M. Field, L. Fine, J. N. Forrest, and F. H. Epstein. Mechanism of active chloride secretion by shark rectal gland: Role of Na-K-ATPase in chloride transport. *Am. J. Physiol. 233:* F298-F306, 1977.

4. J. N. Forrest, P. Silva, A. Epstein, and F. H. Epstein. Effect of rectal gland extirpation on plasma sodium in the spiny dogfish. *Bull. Mount Desert Island Bio. Lab. 13:* 41-42, 1973.

5. J. S. Stoff, P. Silva, M. Field, J. Forrest, A. Stevens, and F. H. Epstein. Cyclic AMP regulation of active chloride transport in the rectal gland of marine elasmobranchs. *J. Exp. Zool. 199:* 443-448, 1977.

6. J. Eveloff, K. J. Karnaky, Jr., P. Silva, F. H. Epstein, and W. B. Kinter. Elasmobranch rectal gland cell. Autoradiographic localization of [^{3}H] ouabain-sensitive Na-K-ATPase in dogfish, *Squalus acanthias*, rectal gland. *J. Membr. Biol.* 83: 16-32, 1979.

7. M. E. Duffey, P. Silva, R. A. Frizzell, F. H. Epstein, and S. G Schultz. Intracellular electrical potentials and chloride activities in the perfused rectal gland of *Squalus acanthias:* A report of preliminary data. *Bull. Mount Desert Island Biol. Lab. 18:* 73-71, 1979.

8. J. Eveloff, R. Kinne, E. Kinne-Saffran, H. Murer, J. Stoff, P. Silva, W. B. Kinter, and F. H. Epstein. Mechanism of active chloride transport: coupled Na/Cl transport by plasma membrane vesicles. *Trans. Assoc. Am. Phys. 41:* 433-443, 1978.

9. R. A. Frizzell, M. Field, and S. G. Schultz. Sodium-coupled chloride transport by epithelial tissues. *Am. J. Physiol. 236:* F1-F8, 1979.

10. R. Beauwens and Q. Al-Awqati. Further studies on coupling between sodium transport and respiration in toad urinary bladder. *Am. J. Physiol. 231:* 222-227, 1976.

11. R. Beauwens and Q. Al-Awqati. Active H$^+$ transport in the turtle bladder. Coupling of transport to glucose oxidation. *J. Gen. Physiol. 68:* 421-439, 1976.

12. P. Silva, J. S. Stoff, R. J. Solomon, R. Rosa, A. Stevens, and J. Epstein. Oxygen cost of chloride transport in perfused rectal gland of *Squalus acanthias*. *J. Membr. Biol. 53:* 215-221, 1980.

13. R. Whittam and M. F. Ager. The connection between active cation transport and metabolism in erythrocytes. *Biochem. J. 97:* 214-227, 1965.

14. T. J. Shuttleworth and J. L. Thompson. Cyclic AMP and ouabain-binding solution sites in the rectal gland of the dogfish *Scyliorhinus canicola*. *J. Exp. Zool. 206:* 297-302, 1978.

HORMONAL CONTROL OF CHLORIDE SECRETION

IN THE RECTAL GLAND OF *SQUALUS ACANTHIAS**

Patricio Silva[+]

Jeffrey S. Stoff

Franklin H. Epstein

Department of Medicine and Thorndike Laboratory
of Harvard Medical School at Beth Israel Hospital
and Mount Desert Island Biological Laboratory

I. INTRODUCTION

Because of the large difference in salt concentration between
their body fluids and that of sea water, fish swimming in the
ocean face the peculiar situation of being in danger of dehydration while surrounded by water. The composition of sea water
plasma, urine, and other fluids in the dogfish is shown in Table I.
There is a large difference in the concentration of salt in the
plasma of the dogfish and that of the ocean, but this difference
is not as large as that seen in other fish, such as the teleosts,

*Supported by grants AM 18078 from the USPHS and BG-5781 and
PCM-77-01146 from the National Science Foundation.
[+]Dr. P. Silva is an Established Investigator of the American
Heart Association.*

277

TABLE I. Composition of body fluids rectal gland secretion and
urine in the shark. All values are expressed in mM/L.

	Plasma ECF	Rectal gland secretion	Urine	Sea water
Na	270	450	337	450
K	5	10	2	9
Cl	280	460	203	490
Ca	2.6	1.0	4-10	10
Mg	3.7	1.0	25-50	51
Urea	350	15-20	50-100	0
Osmolality	1000	1000	780	1000

where it is even more striking. The dogfish is different from
other fish in another respect. Although the concentration of
salt in its plasma is lower than that of the sea, the osmolality
of its body fluids is similar to that of the ocean. The dogfish,
and in general all elasmobranchs, have increased the concentration of urea in their body fluids to reach the osmolality of the
ocean. By equilibrating their internal osmolality with that of
the sea where they live, the dogfish avoids the risk of rapid
dehydration as sea water rushes past the animal's gills. Yet the
fish faces still another problem. Every time the animal eats,
salt and water enter the gastrointestinal tract. Although theoretically salt and water can also move in through the gills, this
route does not seem to be important in the dogfish (1). To maintain salt balance the salt taken in with the food has to be eliminated preferentially over the water.

II. HISTORICAL BACKGROUND

In 1931, Homer Smith observed that in the dogfish the ratio
of Mg:Cl in the urine was greater than that of ingested seawater,
and concluded from this fact that renal excretion of salt alone
would be incapable of maintaining salt balance and, therefore, sig-

nificant extrarenal transport of chloride (and presumably sodium) must take place (2). Thirty years later Burger and Hess identified the rectal gland of the dogfish as a major organ for the excretion of salt (3). In those and later experiments Burger (4,5) observed that the secretion of the rectal gland contained salt at a concentration equal or slightly higher than that of the ocean and significantly larger than plasma. He also noted that the rate of secretion of the rectal gland varied enormously with time as shown in Fig. 1, which has been recalculated from some of his published data. He made the additional observation that intravascular infusions of hypertonic fluid were followed by increased salt secretion by the rectal gland (Fig. 1). From these experiments he concluded that the rectal gland was a principal route for the excretion of salt in the dogfish and that the gland was under

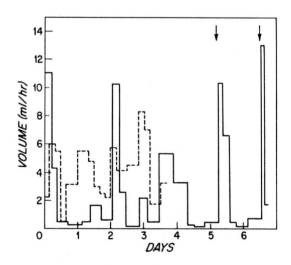

FIG. 1. Rate of rectal gland fluid secretion in two dogfish. The rectal gland ducts were surgically cannulated and the fish allowed to swim freely in a tank while collections were continued via a polyethylene catheter. The collections represent all rectal gland fluid secreted at variable time intervals and are expressed in terms of ml/hr for each time interval. Collections in one fish are depicted by a solid line, those in the other by dashed lines. At the points marked by the arrows 10 ml of 8% sodium chloride was injected. Note the marked irregularity in the rate of rectal gland secretion as well as the striking response to injections of hypertonic sodium chloride.

neurohormonal control. He then proceeded to examine the effect on
rectal gland secretion of several agents after intravascular injection. Secretion was not stimulated by either acetylcholine, meta-
choline, carbachol, pilocarpine, eserine, or eserine followed by
acetylcholine, epinephrine, or vasopressin.

III. RECTAL GLAND PERFUSION

Examination of the mechanisms by which the rectal gland secretes salt has been facilitated by the ease with which this organ
can be perfused *in vitro*. Several investigators have used this
technique in the past to examine different aspects of the mechanism of salt secretion by the gland (6, 7, 8 - 10, 11). Although
the technique has proven useful, initial experiments with the perfused rectal gland were disappointing because the ratio of volume
secretion by the gland always declines with time (9, 10, 11).
Figure 2 shows graphically this decline. Electrolyte concentra-

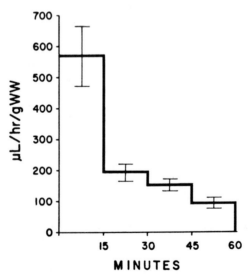

FIG. 2. *Rate of fluid secretion versus time in isolated perfused rectal glands. The rate of fluid secretion declines with
time in rectal glands removed from the dogfish and perfused in
vitro to values lower than 20% of the initial rate. Values are
mean ± SE. (Reproduced with permission from Am. J. Physiol. 233:
F298-F306, 1977.)*

tion in the secreted fluid does not change as the rate of volume secretion falls and, therefore, all parameters decline in parallel: volume and electrolyte secretion.   This fall in rectal gland function when the gland was perfused *in vitro* with an artificial medium suggested to one of the authors (J.S.S.), as it had previously suggested to Burger, that the gland was under the tonic control of a neurohormonal agent.   Once the gland was removed from the animal the putative neurohormonal agent could not reach it and its function slowly declined with time.   Since many neurohormonal agents exert their effects on electrolyte transport work through the adenylate cyclase-cyclic AMP system, the logical next step was to examine directly the effect of theophylline and cyclic AMP added to the perfusion solution.   The effect of these agents is shown in Fig. 3.   There was a rapid increase in the rate of secretion with theophylline or dibutyryl cyclic AMP.   The effect was apparent within 2 min and lasted 20 - 30 min.   The response to theophylline was dose-dependent and increased in magnitude without a plateau as the concentration of this drug was augmented in the perfusion medium from 0.01 to 5 mM (Fig. 4).   It was of interest that the concentrations of electrolytes in the rectal gland fluid remained constant and rose only when the highest (5 mM) concentration of theophylline was used.   Stimulation of rectal gland secretion with theophylline and dibutyryl cyclic-AMP resulted in a rise in the transglandular potential difference from 6.8 ± 0.9 to 15.0 ± 1.6.   Thus, after stimulation, secretion of chloride takes place against a higher electrical gradient.

Stimulation of the gland with theophylline resulted in an increase in intracellular cyclic AMP.   There was a progressive increase in the cyclic AMP content of the rectal gland as the concentration of theophylline was increased.   The basal content was 3.9 ± 0.55 pmol/mg protein and rose to 6.01 ± 0.65 pmol/mg protein (p < 0.05) at 0.01 mM theophylline and to a 15.4 ± 1.25 pmol/mg protein (p < 0.001) at a final concentration of theophylline of 5.0 mM.

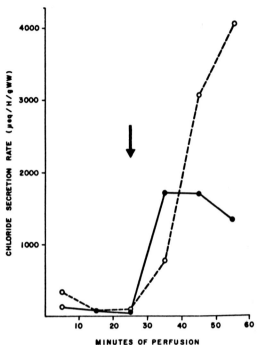

FIG. 3. *Effect of theophylline and dibutyryl-cyclic AMP on
the rate of chloride secretion in two isolated perfused rectal
glands. Two representative experiments in isolated perfused rectal glands are shown here. The solid lines and close circles show
the rate of chloride secretion in an isolated perfused rectal
gland in which theophylline 5.0 mM was added at the point marked
by the arrow. There was a twenty fold increase in the rate of
secretion. The dashed line and open circles represents another
experiment in which dibutyryl cyclic AMP 0.1 mM was added at the
point marked by the arrow. There was a greater than forty fold
increase in the secretory rate. (Reproduced with permission from
J. Exp. Zool. 199:443-448, 1977.)*

IV. HORMONAL STIMULATION OF RECTAL GLAND SECRETION

Since cyclic AMP is the intracellular mediator for many hormones, a search for the hormone(s) capable of stimulating the rectal gland was conducted. These studies were done using the isolated perfused rectal gland. The hormones were added to the perfusate at pharmacological concentrations after two or more control
periods that were considered baseline. After the hormones were

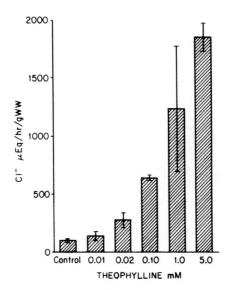

FIG. 4. Effect of theophylline on the rate of chloride secretion in the isolated perfused rectal gland. This figure shows the dose response of the rate of chloride secretion to increasing concentrations of theophylline in the perfusate of isolated rectal glands. Bars represent means ± SE. (Reproduced with permission from J. Exp. Zool. 199:443-448, 1977.)

added, one or more collection periods were allowed to elapse followed by an intraarterial bolus of theophylline. If a response was seen with theophylline but not with the hormone it was concluded that the hormone was not effective in stimulating secretion by the gland. An alternative protocol was also used when testing some hormones. Glands were perfused continuously with a small concentration of theophylline in the perfusate (an amount of theophylline that was not sufficient to increase rectal gland secretion) and the hormone to be tested was then added to the perfusate. Table II shows a list of the hormones and other agents tested in this way. Antidiuretic hormone (Pitressin [R]) and some substituted oxytocin derivatives (valitocin, vasotocin, and aspartocin) normally found in the posterior pituitary of the dogfish (kindly supplied by Dr. W. Sawyer) were tested because they could have a role in osmoregulation in the shark. Encouraged by a report that unmyelinated nerve endings with dense core vesicles are

TABLE II. Hormones studied in the isolated perfused rectal gland.

Hormone	Dose
Vasopressin	80 mU/ml
Aspartocin	1 µg/ml
Valitocin	1 µg/ml
Vasotocin	1 µg/ml
Norepinephrine	$6 \times 10^{-6}M$
Epinephrine	$3 \times 10^{-6}M$
Isoproterenol	$2.84 \times 10^{-5}M$
Serotonin	$10^{-4} - 10^{-6}M$
Substance P	1 µg/ml
Calcitonin (salmon)	1 µg/ml
Pentagastrin	$10^{-4} - 10^{-6}M$
Glucagon	1 µg/ml
Secretin	75 - 180 mU/ml
Vasoactive intestinal peptide (VIP)	$10^{-9} - 10^{-6}M$

found in this gland (12), several adrenergic agents were used. Other agents tried included Substance P, cyclic GMP, and the calcium ionophore A23187. None of these compounds stimulated the rate of rectal gland fluid or electrolyte secretion.

Since the rectal gland is an appendage of the hind gut of the shark, it was reasoned that hormones that stimulate intestinal secretion in other species might have an effect on the rectal gland. The hormones tested included pentagastrin and a structurally related group of three peptide hormones: glucagon, secretin, and vasoactive intestinal peptide (VIP). Of these only (VIP) stimulated secretion by the rectal gland (13). The effect of a bolus of $10^{-6}M$ VIP on the rate of volume and chloride secretion by the rectal gland is shown in Fig. 5. There was a rise of approximately 500% in both parameters. The response was immediate, peaking within 10 min and declining back to basal at the end of 30 min. The response to the hormone was dose-dependent over the range of concentration studied and could be elicited at a concentration as low as $10^{-8}M$. The combination of subthreshold concentrations of theophylline (0.025 mM) and VIP $10^{-8}M$ produced a synergistic response. In addition to a stimulation of volume and

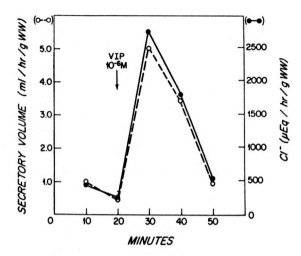

Fig. 5. Effect of VIP on the rate of fluid and chloride secretion in an isolated perfused rectal gland. This graph shows the average of two representative experiments. The open symbols show the rate of fluid secretion and the closed symbols that of chloride. VIP in the amount shown was injected intraarterially as a bolus over 1 min. A control bolus infusion of vehicle alone was given 5 min prior to the hormone. (Reproduced with permission from Am. J. Physiol. 237:F138-F144, 1979).

chloride secretion, VIP produced a doubling in transglandular PD and a rise in glandular oxygen consumption that paralleled the increase in chloride secretion. All these observations provide proof that VIP stimulates active chloride transport by the gland.

V.   VIP IN THE DOGFISH

Measurements of VIP in the plasma of the dogfish show that this hormone is present in the plasma of free-swimming fish. The levels of VIP measured in 20 fish in three consecutive summers by a radioimmunoassay averaged 407 ± 85 pg/ml (radioimmunoassays for VIP were kindly performed by Dr. Joseph Fischer). These values are considerably higher than those found in man and approximate values found in patients with the watery diarrhea syndrome (14).

VIP is present in the tissues of the dogfish in concentrations that
are generally higher than those found in mammalian tissue (15).

Since VIP is present in the dogfish and is capable of stimulating secretion by the isolated rectal gland perfused *in vitro*,
and Burger had observed that intravascular infusions of hypertonic
saline stimulated secretion by the rectal gland (4), preliminary
experiments were performed to investigate the afferent mechanisms
that might control the endogenous release of VIP. The effect of
intravascular as well as intragastric infusions of hypertonic
saline on circulating levels of VIP were studied in two sets of
four fish. After intravascular infusions of 50 ml of 1 M NaCl
there was a rapid rise in both plasma osmolality from 1254 to 1381
mOsm/kg, and sodium concentration from 245 to 350 mEq/L. Despite
the elevations in these parameters, plasma VIP actually decreased.
When 100 ml of 1 M NaCl were administered intragastrically, the
change in plasma osmolality was smaller and occurred later.
Plasma VIP levels, on the other hand, rose immediately after the
gastrointestinal infusion peaking 30 min later at 170% of the
basal levels and returning to normal after 60 min. Although this
rise in plasma VIP was not statistically significant due to large
dispersion of the values, it is encouraging.

VI. EFFECT OF VIP ON CYCLIC AMP

The effect of VIP to stimulate rectal gland secretion appears
to be mediated by cyclic AMP. Figure 6 shows the effect of VIP on
cyclic AMP content of rectal gland slices. VIP produced a dose-
dependent rise in intracellular AMP with a significant rise observed at even $10^{-8} M$ concentrations. Theophylline potentiates
the effect of VIP, resulting in a synergistic increase in intracellular cyclic AMP. Secretin structurally related to VIP does
not increase tissue levels of cyclic AMP either alone or in the
presence of theophylline.

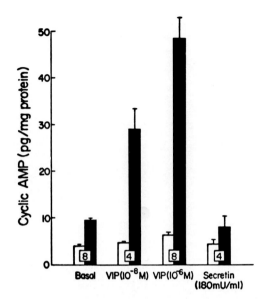

FIG. 6. Effect of VIP, VIP plus theophylline and secretin on cyclic AMP content of rectal gland slices. Clear bars represent intracellular cyclic AMP levels in the basal state or after VIP or secretin alone. Dark bars represent theophylline alone or in combination with VIP and secretin. Bars are mean ± SE. The number of experiments is inserted at the base of the bars. VIP at either $10^{-8}M$ or $10^{-6}M$ produced a significant increase in intracellular cyclic AMP. The addition of theophylline to the incubation medium evoked a synergistic response with VIP. Secretin did not increase intracellular cyclic AMP when added alone and when used in combination with theophylline the effect was not greater than with theophylline alone. (Am. J. Physiol. 237:F138-F144, 1979.)

VII.   INHIBITION OF VIP-INDUCED STIMULATION

Recent studies have shown that somatostatin, a tetradecapeptide which was first isolated from the hypothalamus and found to inhibit growth hormone release (16), inhibits VIP induced intestinal secretion (17). It is of interest that the inhibitory effect of somatostatin appears to be restricted to those agents that exert their action through the adenylate cyclase-cyclic AMP system since in the colon, for example, it inhibits only partially the effect of cyclic AMP (17), and it has no effect on the cholecystokinin-stimulated pancreatic secretion which is thought to be mediated by cyclic GMP (16).

In the rectal gland somatostatin by itself has no effect on chloride secretion; but it completely blocks the VIP induced stimulation of the gland as can be seen in Fig. 7. The effect of somatostatin on VIP stimulation of chloride secretion is reversible. Somatostatin had no effect on theophylline-dibutyryl-cyclic AMP-induced stimulation of rectal gland secretion. Somatostatin has been identified in tissues of several species, including Squalus acanthias, in sites similar to those of VIP (15, 18 - 23), which suggests a possible functional relation between the two hormones.

The stimulation of chloride secretion by VIP and dibutyryl cyclic AMP indicate that as in many other tissues electrolyte transport in the rectal gland is under hormonal control. The effect of hormones and cyclic AMP on electrolyte transport is well documented in a number of different epithelia (24, 25, 26). Although definitive proof is lacking, it seems possible that secretion by the rectal gland in the shark is stimulated by the release

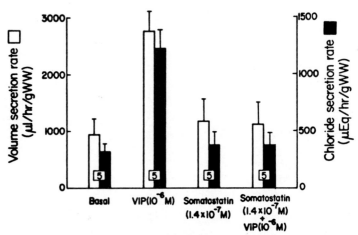

FIG. 7. *Effect of somatostatin on VIP-induced stimulation of rectal gland secretion. Clear bars represent rate of fluid secretion, dark bars show rate of chloride secretion. Bars are mean ± SE, number of experiments is indicated at the base of the bars. VIP at $10^{-6}M$ increases the rate of fluid and chloride secretion. Somatostatin $1.4 \times 10^{-7}M$ does not modify the basal rate but completely prevents the stimulation normally seen with VIP. (Am. J. Physiol. 237:F138-F144, 1979.)*

into the circulation of VIP from its stores in the intestine. The latter release would be the result of changes in intestinal tonicity or volume brought about by feeding. Thus, the intake of salt by the fish would be followed by stimulation of a mechanism for salt excretion.

The way in which cyclic AMP stimulates chloride secretion is not yet clear. There are three potential sites for regulation. The entry of chloride into the cell could be facilitated by cyclic AMP. There is no evidence for this mechanism in other tissues and the fact that intracellular chloride concentration falls rather than rises following stimulation of the gland with theophylline and dibutyryl cyclic AMP makes this mechanism unlikely. Since the secretion of chloride by the rectal gland depends on Na-K-ATPase activity, it can be proposed that cyclic AMP stimulates chloride transport by activating Na-K-ATPase. It should be pointed out that experiments from our laboratory have failed to show activation of the enzyme by cyclic AMP added to homogenates of rectal gland *in vitro* (27, 28), as well as after stimulation of the perfused rectal gland (29). Also, the binding of ^{3}H-ouabain, itself an indirect measure of the number of available Na-K-ATPase sites in the membrane, did not change with the addition of cyclic AMP in experiments performed in rectal gland slices (27) and in the perfused rectal gland (28). Others, however, working with slices of rectal gland of another shark species, *Scyliorrhinus canicola,* have shown stimulation of ^{3}H-ouabain binding after incubation with dibutyryl cyclic AMP (29). Related experiments in isolated smooth muscle cells of the bullfrog stomach have shown that protein kinase is capable of stimulating Na-K-ATPase *in vitro* (30). This may indicate that cyclic AMP has the potential capability of stimulating Na-K-ATPase. Another way in which cyclic AMP can stimulate chloride secretion by the rectal gland is by increasing the conductance for chloride at the luminal cell membrane. The available electrical and chemical evidence suggests that this might be the case. Chemical measurements of intracellular chloride show that the concentration of chloride in the cellular water falls after the gland is stimulated with cyclic AMP from 78 ± 4 to 44 ± 4 mEq/L of tissue water. Preliminary studies in rectal gland slices by William Armstrong and associates (personal communication) show that intracellular potential drops after stimulation of chloride transport with cyclic AMP from 60 to 50 mV. These observations suggest that the electrochemical driving force for the extrusion of chloride from the cell into the lumen is decreasing as the rate of chloride secretion is increasing. This necessarily means that an associated increase in the permeability of the luminal cell membrane to chloride must take place.

In summary, the rectal gland of the shark secretes salt thus helping to maintain electrolyte homeostasis. The gland appears to be under hormonal control. It seems possible that during feeding, ingestion of sea water and/or food may release VIP from its stores in intestinal tissue into the plasma. High circulating levels of the hormone would produce an increase in chloride secretory rate in the rectal gland thus contributing to electrolyte homeostasis.

VIII. SUMMARY

The rectal gland of the spiny dogfish *Squalus acanthias* actively secretes chloride. The secretion of chloride by this gland is regulated by the intracellular cyclic AMP. Intracellular production of cyclic AMP is stimulated specifically by only one of many hormones tested, vasoactive intestinal peptide (VIP). VIP is also capable of stimulating chloride secretion by the gland. VIP is present in high concentrations in the plasma of the fish and is also present in other tissues. The effect of VIP is inhibited by somatostatin another hormone that is present in the tissues of the shark in sites similar to those of VIP. It seems probable that salt homeostatis is regulated in the shark by the endogenous release of VIP. The effect of VIP and or cyclic AMP

at the cellular level appears to be to increase the chloride
conductance of the luminal membrane of the rectal gland cell.

REFERENCES

1. J. W. Burger and D. C. Tosteson. Sodium influx and efflux
    in the spiny dogfish *Squalus acanthias*. *Comp. Biochem.
    Physiol.* *19*:649-653, 1966.
2. H. W. Smith. The absorption and excretion of water and salts
    by the elasmobranch fishes. *Am. J. Physiol.* *98*:296-310, 1931.
3. J. W. Burger and W. N. Hess. Function of the rectal gland in
    the spiny dogfish. *Science 131*:670-671, 1960.
4. J. W. Burger. Further studies on the function of the rectal
    gland in the spiny dogfish. *Physiol. Zool.* *35*:205-217, 1962.
5. J. W. Burger. Roles of the rectal gland and the kidneys in
    salt and water excretion in the spiny dogfish. *Physiol.
    Zool.* *38*:191-196, 1965.
6. J. P. Hayslett, D. A. Schon, M. Epstein, and C. A. M. Hogben.
    In vitro perfusion of the dogfish rectal gland. *A. J. Physiol.*
    226:1188-1192, 1974.
7. R. F. Palmer. *In vitro* perfusion of the isolated rectal gland
    of *Squalus acanthias*. *Clin. Res. 14*:77, 1966.
8. N. J. Siegel, P. Silva, F. H. Epstein, T. M. Maren, and J. P.
    Hayslett. Functional correlates of the dogfish rectal gland
    during *in vitro* perfusion. *Comp. Biochem. Physiol. 51A*:593-
    597, 1975.
9. N. J. Siegel, D. A. Schon, and J. P. Hayslett. Evidence for
    active chloride transport in dogfish rectal gland. *Am. J.
    Physiol.* *230*:1250-1254, 1976.
10. P. Silva, J. S. Stoff, M. Field, L. Fine, J. Forrest, and
    F. H. Epstein. Mechanism of active chloride secretion by
    shark rectal gland. Role of Na-K-ATPase in chloride trans-
    port. *Am. J. Physiol.* *233*:F298-F306, 1977.

11. J. S. Stoff, P. Silva, M. Field, J. Forrest, A. Stevens, and F. H. Epstein. Cyclic AMP regulation of active chloride transport in the rectal gland of marine elasmobrachs. *J. Exp. Zool. 199*:443-448, 1977.

12. W. D. Doyle. Histology of the rectal gland of *Squalus acanthias. Bull. MDIBL 17*:34-35, 1977.

13. J. S. Stoff, R. Rosa, R. Hallac, P. Silva, and F. H. Epstein. Hormonal regulation of active chloride transport in the dog-fish rectal gland. *Am. J. Physiol. 237*:F138-F144, 1979.

14. S. R. Bloom, J. M. Polak, and A. G. E. Pearse. Vasoactive intestinal peptide and watery-diarrhea syndrome. *Lancet. 2*: 14-16, 1973.

15. J. S. Stoff, R. Hallac, R. Rosa, P. Silva, J. Fischer, and F. H. Epstein. The role of vasoactive intestinal peptide (VIP) in the regulation of active chloride secretion in the rectal gland of *Squalus acanthias. Bull. MDIBL 17*:66-69, 1977.

16. P. Brazean, W. Vale, and R. Burgus. Hypothalamic peptide that inhibits the secretion of immunoreactive pituitary growth hormone. *Science 179*:77-79, 1973.

17. R. F. Carter, K. N. Bitar, A. M. Zfas, and G. M. Makhlouf. Inhibition of VIP-stimulated intestinal secretion and cyclic AMP production by somatostatin in the rat. *Gastroenterology 74*:726-730, 1978.

18. S. Falkmer, R. P. Elde, C. Hellerström, and B. Petersson. Phylogenetic aspects of somatostatin in the gastroenteropan-creatic (GEP) endocrine system. *Metabolism 27* (Suppl. 1): 1193-1196, 1978.

19. T. S. Gaginella, H. S. Mekhjian, and T. M. O'Dorisio. Vaso-active intestinal peptide: Quantification by radioimmuno-assay in isolated cells, mucosa, and muscle in the hamster intestine. *Gastroenterology 74*:718-721, 1978.

20. S. J. Konturek, J. Tasler, W. Obtubowicz, D. H. Coy, and A. V. Schally. Effect of growth hormone-releasing inhibiting

hormone on hormones stimulating exocrine pancreatic secretion. *J. Clin. Invest. 58*:1-6, 1976.

21. A. G. E. Pearse. Peptides in brain and intestine. *Nature 262*:92-94, 1976.

22. J. M. Polak, A. G. Pearse, L. Grimelius, and S. R. Bloom. Growth-hormone release-inhibiting hormone in gastrointestinal and pancreatic D cells. *Lancet. 1*:1220-1222, 1975.

23. C. Rufener, M. P. Dubois, F. Malaisse-Lagae, and L. Orci. Immunofluorescent reactivity to antisomatostatin in the gastrointestinal mucosa of the dog. *Diabetologia 11*:321-324, 1975.

24. M. J. Berridge, B. D. Lindley, and W. T. Prince. Membrane permeability changes during stimulation of isolated salivary glands of Calliphora by 5-hydroxyptamine. *J. Physiol. (London) 244*:549-567, 1975.

25. R. A. Frizzell, M. J. Koch, and S. G. Schultz. Ion transport by rabbit colon. I. Active and passive components. *J. Membrane Biol. 27*:297-316, 1976.

26. S. D. Klyce and R. K. S. Wong. Site and mode of adrenaline action on chloride transport across the rabbit corneal epithelium. *J. Physiol. (London) 266*:777-799, 1977.

27. J. Eveloff, K. J. Karnaky, Jr., P. Silva, F. H. Epstein, and W. B. Kinter. Elasmobranch rectal gland cell. Autoradiographic localization of (^{3}H) ouabain-sensitive Na, K-ATPase in rectal gland of dogfish, *Squalus acanthias. Cell. Biol. 83*:16-32, 1979.

28. P. Silva, T. Baranano, J. Epstein, J. Stoff, and F. H. Epstein. Na-K-ATPase and rectal gland secretion in *Squalus acanthias. Bull. MDIBL. 18*:16-19, 1978.

29. T. J. Shuttleworth and J. L. Thompson. Cyclic AMP and ouabain-binding sites in the rectal gland of the dogfish *Scyliorrhinus canicula. J. Exper. Zool. 206*:297-302, 1978.

30. C. R. Scheid, T. W. Honeyman, and F. S. Fay. Mechanism of β-adrenergic relaxation of smooth muscle. *Nature 277*:32-36, 1979.

THE OPERCULAR EPITHELIUM: AN EXPERIMENTAL MODEL FOR TELEOST
GILL OSMOREGULATION AND CHLORIDE SECRETION

Kevin J. Degnan[1]

José A. Zadunaisky

The Department of Physiology and Biophysics
and the Department of Ophthalmology
New York University Medical Center, New York

The Mt. Desert Island Biological Laboratory
Salsbury Cove, Maine

I.   INTRODUCTION

Since the pioneer investigations of Smith (1) and Krogh (2),
extrarenal mechanisms have been known to play a significant osmo-
regulatory role in teleosts. Various experimental preparations,
such as the heart-gill (3), the isolated gill (4), the externally
perfused gill (5), and the isolated head (6) have demonstrated
that the teleost gill epithelium is the primary, and almost ex-
clusive site of this extrarenal osmoregulation. In particular,
the gill chloride cells, initially described by Keys and Willmer
(7), have been singled out as the cell type housing these ion
transporting mechanisms [for a review, see Karnaky (8)]. While
these experimental preparations have provided a wealth of valuable
information on ion transport mechanisms in general and teleost os-

[1]Present Address: Physiology Section, Division of Biology
and Medicine, Brown University, Providence, Rhode Island 02912.

295

moregulation in particular, they have not been able to provide the
information obtained by the application of the short-circuit current
technique of Ussing and Zerahn (9). This technique is the method
of choice in the study of ion transport across epithelia and has
been extensively employed over the past 30 years on a variety of
epithelia, such as the frog skin and toad bladder. One of the requirements for the application of this technique is the isolation
of the tissue as a relatively flat sheet of cells. For most epithelia, this presents no problem. However, the intricate and complex nature of the gill epithelium precludes the application of this
technique for the study of branchial ion transport processes.

The recent development of the isolated opercular skin preparation of the teleost, *Fundulus heteroclitus* (10), has provided an alternative approach to the study of teleost gill osmoregulation and
chloride cell function by enabling the use of the short-circuit current method in these studies. The rationale behind proposing this
skin as an experimental model for the gill is that in the teleost,
F. heteroclitus, this epithelium contains an abundance of chloride
cells (11) identical in ultrastructure to the gill chloride cells
(12). This epithelium is readily dissected off the opercular bone
and mounted in a lucite chamber. Its ion transport functions and
electrical properties are then studied under rigidly defined conditions, including the short-circuit current technique.

Since the isolated opercular epithelium preparation is a recent development, much of the initial investigations with this
tissue have focused on elucidating and characterizing its basic
properties from an ion transport point of view. The data presented
in this review are from chamber studies with opercular epithelia
from seawater-adapted *F. heteroclitus*. This epithelium has a number of properties similar to the gill epithelium (13), as well as
to a variety of both chloride absorbing- and secreting-epithelia
(see Frizzell *et al.*, 14). The data obtained to date strengthen
the proposal of this preparation, serving as a model for teleost
gill osmoregulation and chloride cell functions. In addition, the

data indicate that this is a most attractive preparation with which to pursue the details of active chloride secretion.

II.   THE ELECTRICAL PROPERTIES OF THE OPERCULAR EPITHELIUM

When isolated, mounted in a lucite chamber, and bathed on both sides with Ringer solution, the short-circuit current (Isc) across the operculum ranges between 50 - 350 $\mu A \cdot cm^{-2}$ with a mean of 137.9 ± 4.1 $\mu A \cdot cm^{-2}$ (n=280); the transepithelial potential difference ($\Delta \Psi$) ranges between 10 - 40 mV (serosal side positive) with a mean of 18.0 ± 0.4 mV (n=280); and the total tissue ionic conductance (G_t) ranges between 2.0 - 20.3 $mS \cdot cm^{-2}$ with a mean of 6.3 ± 0.2 $mS \cdot cm^{-2}$ (n=280).

The Isc is comparatively high with respect to other Cl^- transporting epithelia such as the frog cornea (15) and skin (16), and varies considerably from fish to fish. A possible explanation for this variability could be variations in the chloride cell density of the operculum of different fish. Karnaky et $al.$ (17) have demonstrated a linear correlation between the Isc and chloride cell number by using both the opercular epithelium and the epithelium lining the anterior roof of the mouth of $F.$ $heteroclitus.$ This latter tissue contains relatively few chloride cells, but with both preparations, these cells are readily visualized at relatively low magnifications (50-100X) with the fluorescent dye, DASPMI (18). A similar correlation has been observed between the chloride cell number and the Isc in the skin of $Gillichthys$ $mira$$bilis,$ with a reported current of 6.5 × 10^{-4} μA per chloride cell (9).

The potential difference across the isolated opercular epithelium, when bathed bilaterally with Ringer, has the same orientation (serosa +) and approximate magnitude as that measured across the gill epithelium in seawater (for a review of gill potentials, see Evans (20)). When bathed externally with seawater,

the potential across the isolated operculum (31.9 ± 0.7 mV; n=80)
is generally higher than that with Ringer outside. It is close to
the Na^+ equilibrium potential ($E_{Na}+$) of 26.8 mV (serosa +), and is
somewhat higher than that measured across the gill epithelium with
seawater outside.

The transepithelial resistance across the opercular epithelium
ranges between 50 - 400 $\Omega\cdot cm^{-2}$ (20 - 2 $mS\cdot cm^{-2}$) with a mean of
158.3 ± 4.7 $\Omega\cdot cm^{-2}$, which places this tissue in the category of
"leaky" epithelia. Approximately 50% of this conductance is a
partial Na^+ conductance ($G_{Na}+$), and the conductive pathway for this
cation appears to be a paracellular shunt (21). The remainder of
the G_t appears to be a combination of partial Cl^-, HCO_3^-, and K^+
conductances whose magnitudes and pathways have yet to be determined. The wide range of conductances observed with this preparation can also be explained by variations in the chloride cell
number of different epithelia. There is evidence, although indirect, that the conductive Na^+ shunt pathway is the paracellular
route between adjacent chloride cells (21 - 23). Consequently,
both the Isc and $G_{Na}+$ (which comprises the major fraction of the
G_t), would be a direct function of the number of chloride cells
and the resulting chloride cell-chloride cell junctions. This
suggestion is supported by the observation that the current and
conductance pathways across the opercular membrane of *Sarotherodon
mossambicus* are localized around the chloride cells in this tissue
(24).

III. PHARMACOLOGICAL CHARACTERISTICS OF THE OPERCULAR EPITHELIUM

A list of the effects of a variety of pharmacological agents
on the Isc across the opercular epithelium are presented in Table
I. Compounds which stimulate the Isc are the β-adrenergic activators such as isoproterenol, and the phosphodiesterase inhibitors
such as aminophylline and theophylline. It is known that phos-

TABLE I. Pharmacological Characteristics of the Opercular Epithelium[a]

Stimulants	Inhibitors	No effect
Aminophylline (10^{-4} M)	Acetylcholine (10^{-5} M)*	A23187 (10^{-4} M)
β-Adrenergics (10^{-6} M)*	α-Adrenergics (10^{-6} M)*	Acetazolamide (10^{-3} M)
isoproterenol	arterenol	Amiloride (10^{-4} M)
dB-cyclic AMP (10^{-4} M)	epinephrine	Amphotericin B (10^{-4} M)
Theophylline (10^{-4} M)	Bumetanide (10^{-4} M)	Cortisol (10^{-5} M)
	Furosemide (10^{-3} M)	dB-cyclic GMP (10^{-4} M)
	Ouabain (10^{-6} M)*	Methazolamide (10^{-3} M)
	SITS (10^{-3} M)	Prolactin (10^{-4} M)
	TAP^{+} (10^{-2} M)	
	Thiocyanate (10^{-3} M)	

[a] The effect of a variety of pharmacological agents on the Isc across the opercular epithelium. Concentrations which give a significant response, or no response, are given in parentheses. Compounds marked with an asterisk are effective from the serosal side only. All other compounds were tested by addition to both sides of the epithelium.

phodiesterase inhibition produces an elevation in the cell 3',5' cyclic AMP levels (25), and it presently appears that activation of the β-receptors in the operculum leads to an elevation in the tissue cyclic AMP levels. Preliminary data (unreported) have demonstrated a 6-fold increase in cyclic AMP levels in opercular epithelia incubated in media containing 10^{-5} M isoproterenol. In gill epithelium, Cuthbert & Pic (26) demonstrated an increase in cyclic AMP levels with epinephrine, which could be abolished by propranolol, a β-blocking agent.

Among the inhibitors of the Isc across the operculum are the compounds known to inhibit active Cl^- transport, such as furosemide, SITS, thiocyanate, and ouabain. TAP^+, a compound which blocks paracellular cation shunts (27), also inhibits the Cl^- secretion across the operculum in addition to its blocking effect on the Na^+ shunt pathway (21). Epinephrine, and α-adrenergic activators such as arterenol, are fast acting inhibitors of the Isc. Preliminary findings (unreported) have indicated that α-adrenergic activation has no significant effect on cyclic AMP levels in the operculum, suggesting another mechanism, not involving cyclic AMP, for this effect. The possible involvement of Ca^{+2} in this response is questionable, since the Ca^{+2} ionophore, A23187, has no effect on the Isc. In the gill, activation of the α-receptors (6) has no significant effect on cyclic AMP levels, but does inhibit Cl^- secretion (28 - 31).

Among the compounds which have no effect on the Isc are amiloride and amphotericin B, two compounds known for their influences on transcellular Na^+ transport (32, 33), and the carbonic anhydrase inhibitors acetazolamide and methazolamide, known for their affect on Cl^-/HCO_3^- exchanges (34). These observations indicate that neither Na^+ nor Cl^-/HCO_3^- exchange contribute to the generation of the Isc across the opercular epithelium. The hormones cortisol and prolactin were tested because of their involvement in the adaptation of euryhaline fish to different environmental salinities (35, 36). No effect of these hormones was evident up to 3 hrs after introduction to the serosal bathing solution.

IV. IONIC FLUXES ACROSS THE OPERCULAR EPITHELIUM

Table II summarizes the results of a number of paired Na^+ and Cl^- isotope flux experiments across isolated opercular epithelia. When bathed on both sides with Ringer and short-circuited, the mean Cl^- efflux (serosa to mucosa $J_{sm}^{Cl^-}$) is 3.8 times greater than the mean Cl^- influx (mucosa to serosa $J_{ms}^{Cl^-}$), resulting in a significant net Cl^- secretion equivalent to a current of 128.3 μA cm^{-2}. This Cl^- current is not significantly different from the mean measured Isc of 130.1 μA cm^{-2}. On the other hand, the unidirectional Na^+ fluxes are not significantly different from each other, resulting in no significant net Na^+ flux under these conditions. These results demonstrate that the entire measured Isc is a Cl^- secretory current under these *in vitro* conditions.

The conditions where the isolated opercular epithelia are bathed externally with seawater and internally with Ringer, and the spontaneous transepithelial potential is allowed to remain, simulate the conditions of both the opercular and gill epithelium in the intact fish. From thermodynamic considerations, this situation favors the net influx of Cl^- down its electrochemical gradient, while the electrical and chemical gradients for Na^+ movements are operating in opposite directions. According to Ussing's (37) criteria, if these ions behaved passively under these conditions, the influx/efflux ratio for Cl^- and Na^+ would be approximately 11.0 and 1.0 respectively. The results of isotope flux experiments across paired opercular epithelia under these conditions are summarized in Table III. No significant net flux of either Na^+ or Cl^- was observed. Comparison of the predicted and observed flux ratios, however, suggests that the Na^+ movements are passive and that Cl^- moves across the epithelium by mechanisms other than simple passive diffusion. 10^{-6} M ouabain on the serosal side pro-

TABLE II. *Short-Circuited Na^+ and Cl^- Fluxes across Isolated Opercular Epithelia from Seawater-Adapted Fundulus heteroclitus.*[a]

	J_{sm}	J_{ms}	P	J_{net}	I_{net}	I_{sc}	P
	$\mu Eq \cdot cm^{-2} \cdot hr^{-1}$				$\mu A \cdot cm^{-2}$		
Cl^- fluxes (30)	6.49 ± 0.70	1.70 ± 0.35	<0.01	4.79 ± 0.52	128.3 ± 13.9	130.1 ± 9.1	<0.60
Na^+ fluxes (25)	1.95 ± 0.21	2.00 ± 0.18	<0.70	-0.05 ± 0.06	-1.2 ± 1.6	100.9 ± 6.3	<0.001

[a] *Data expressed as mean ± s.e.m. with the number of experiments in parentheses. P values are for the difference in the unidirectional fluxes and the difference in the ionic current and the Isc. Fluxes were determined across paired epithelia from the same fish.*

TABLE III. Open-Circuit Na^+ and Cl^- Fluxes and Flux Ratios across Isolated Opercular Epithelia from Seawater-Adapted Fundulus heteroclitus[a]

	J_{sm} ($\mu Eq \cdot cm^{-2} \cdot hr^{-1}$)	J_{ms} ($\mu Eq \cdot cm^{-2} \cdot hr^{-1}$)	P	J_{net} ($\mu Eq \cdot cm^{-2} \cdot hr^{-1}$)	J_{ms}/J_{sm} Predicted	Observed	P
Na^+ fluxes							
control (15)	3.02 ± 0.52	3.08 ± 0.38	>0.80	-0.06 ± 0.31	0.94 ± 0.08	1.14 ± 0.12	>0.20
(ΔΨ)	(30.3 ± 1.5)	(28.2 ± 1.2)	>0.50				
Ouabain, 10^{-6}	2.01 ± 0.36	3.32 ± 0.86	<0.01	-1.32 ± 0.41	1.70 ± 0.14	1.72 ± 0.18	>0.70
(ΔΨ)	(15.6 ± 2.2)	(16.2 ± 2.5)	>0.30				
Cl^- fluxes							
control (10)	2.53 ± 0.49	2.52 ± 0.31	>0.90	0.01 ± 0.49	11.4 ± 0.9	1.38 ± 0.27	<0.001
(ΔΨ)	(28.8 ± 2.3)	(29.4 ± 2.5)	>0.60				
isoproterenol, 10^{-5}	3.14 ± 0.49	3.09 ± 0.42	>0.90	0.07 ± 0.61	12.7 ± 0.9	1.24 ± 0.25	<0.001
(ΔΨ)	(31.7 ± 2.0)	(33.2 ± 2.7)	>0.30				
arterenol, 10^{-5}	1.05 ± 0.20	2.14 ± 0.29	<0.01	-1.09 ± 0.37	7.8 ± 0.7	2.95 ± 0.78	>0.025
(ΔΨ)	(16.7 ± 2.6)	(17.4 ± 2.7)	>0.20				

[a] Data expressed as mean ± s.e.m. with the number of experiments and the mean paired potentials in parentheses. The P values are for the difference in the unidirectional fluxes and the flux ratios.

304 Kevin J. Degnan and José A. Zadunaisky

duces a significant reduction in both the Na^+ efflux and potential difference in these preparations (38). However, the agreement between the predicted and observed Na^+ flux ratios after ouabain inhibition, indicates that this effect of ouabain on the Na^+ efflux is indirect and secondary to the ouabain-induced depolarization of the epithelium. Isoproterenol stimulates and arterenol inhibits the Cl^- efflux in these preparations (38). However, these effects occur opppsitely to those predicted by the change in the potential difference produced by these adrenergic activators. The changes in the Cl^- influx correspond to the potential changes.

These open-circuited flux data are consistent with the short-circuited flux data. Under both conditions, Cl^- is actively secreted while the unidirectional Na^+ fluxes and the Cl^- influx appear to be passive diffusional fluxes. Under open-circuited conditions, this Cl^- secretory mechanism operates against a considerable electrochemical gradient, and appears to just balance the passive inward "leak" of Cl^-. In the marine teleost gill, large unidirectional fluxes of both Na^+ and Cl^- are observed, resulting in a small net efflux of both ions (see data compiled by Maetz & Bornancin, 39). The fact that no significant net ionic flux across the operculum was demonstrable, may have resulted from an insufficient number of experiments necessary to demonstrate a small net flux, or possibly from the absence of certain neurohumoral influences which are present in the intact fish. Other than this, the operculum appears to behave like the gill from the point of view of ionic movements. Branchial Na^+ and Cl^- fluxes are believed to occur principally by exchange diffusion mechanisms (40). The equality of the unidirectional Na^+ and Cl^- fluxes across the operculum under open-circuited conditions could also be explained by exchange diffusion mechanisms. However the uncoupling of the fluxes by the action of ouabain and arterenol suggests that exchange diffusion contributes little, if any, to the transepithelial movements of these ions in the isolated opercular epithelium preparation.

V. IONIC CONTRIBUTIONS TO THE ELECTRICAL PROPERTIES OF THE OPER-
    CULAR EPITHELIUM

The transepithelial potential difference across the isolated
opercular epithelium is directly proportional to the log of the
seawater ion concentration, with a slope of 39.5 mV/10-fold percent concentration change (41). The potential reduces to 0 at approximately 17% seawater (80-90 mM NaCl), and reverses its orientation to serosal side negative at concentrations below 17% seawater. Specific seawater ion substitutions (Table IV) indicate
that Na^+ was the predominant ion involved in the generation of the
potential across the operculum in seawater. This observation,
coupled with the facts that Na^+ behaves passively and the measured
potential approximates the E_{Na^+}, suggests that the potential is
primarily a Na^+-diffusion potential. Similar conclusions have
been reached for the gill epithelium (42, 43). The interesting
feature of the seawater ion substitutions is the depolarization of
the operculum in response to seawater Cl^- substitution. This effect is opposite to that expected if Cl^- diffusion down its elec-

TABLE IV. *The Effect of Seawater Ion Substitutions on the Transepithelial Potential Difference across Isolated Opercular Epithelia
from Seawater-Adapted Fundulus heteroclitus.*[a]

Seawater	$\Delta\Psi$ (mV)	P	Slope	Intercept	r
Na^+-rich (6)	32.1 ± 0.7				
Na^+-free	-3.7 ± 0.3	<0.001	26.6	-38.8	0.925
Cl^--rich (6)	31.9 ± 1.8				
Cl^--free	21.8 ± 2.5	<0.005	7.2	11.4	0.932
K^+-rich (8)	33.7 ± 1.4				
K^+-free	29.4 ± 1.5	<0.005	4.8	27.1	0.806

[a]*Data expressed as mean ± s.e.m. with the number of experiments in parentheses. Regression data was calculated by least
square analyses of the points above 50 mM. The normal seawater ion
concentrations were Na^+ = 480 mM, Cl^- = 533 mM, K^+ = 10 mM. Na^+
was replaced with equimolar amounts of choline or Tris, Cl^- with
methylsulphate, and K^+ with Na^+.*

trochemical gradient acted to partially shunt the Na^+-diffusion potential. This observation can be explained by an external Cl^--dependent, electrogenic Cl^- secretory mechanism, whose transport potential is additive to the Na^+ diffusion potential. A number of lines of evidence support this explanation (31, 38, 41).

When bathed bilaterally with Ringer, the Na^+-diffusion potential across the opercular epithelium is abolished, and the spontaneous measured transepithelial potential may then represent partially or totally a Cl^--transport potential. The substitution of Cl^- in the Ringer bathing the mucosal side of the operculum results in a 53% and 44% reduction in the Isc and potential respectively (41), which demonstrates an externally Cl^--dependent secretory mechanism and potential. In seawater, this effect of external Cl^- on the Cl^- secretion and its associated transport potential apparently override any shunting effects of Cl^- on the Na^+-diffusion potential.

The Isc across the opercular epithelium exhibits a significant bilateral Cl^- dependency, a mucosal Na^+ sensitivity, and a serosal K^+ and HCO_3^- sensitivity (41, 44). Table V lists the mean percentage decreases in the Isc across the operculum in response to bilateral and unilateral ion substitutions. The lack of a significant decrease in the Isc in response to serosal Na^+ substitution is not consistent with the proposed model for Cl^- secretion where the movement of Cl^- into the cell is tightly coupled to the movement of Na^+ down its electrochemical gradient (45). Subsequent investigations into this aspect of Cl^- secretion have demonstrated that the Cl^- secretion is considerably dependent on serosal Na^+ and that the lack of any significant effect of serosal Na^+ substitution on the Isc results from Na^+ diffusion from the mucosal side (21, 46). The relatively high Na^+ permeability of 9.3×10^{-6} cm $\times sec^{-1}$, which is independent of serosal Na^+, apparently is responsible for a large diffusion artifact which acts to increase the Isc under these serosal Na^+ substituted conditions. The serosal K^+-dependence of the Isc is in keeping with the model for Cl^- se-

TABLE V.   The Inhibitory Effect of Bilateral and Unilateral Ion Substitutions on the Isc across Isolated, Short-Circuited Opercular Epithelia from Seawater-Adapted Fundulus heteroclitus.[a]

| Ion Substituted | Mean percent decrease in the Isc | | |
	Mucosa and Serosa	Mucosa only	Serosa only
Na^+	97.9 (14)[*]	85.0 (14)[*]	17.3 (19)
Cl^-	96.6 (9)[*]	53.4 (9)[*]	90.6 (6)[*]
K^+	70.0 (7)[*]	4.4 (12)	81.0 (18)[*]
HCO_3^-	--	13.7 (8)	51.6 (8)[*]

[a]*Number of experiments is given in parentheses. Na^+ was substituted with equimolar amounts of choline or Tris, Cl^- with methylsulphate, K^+ with Na^+, and HCO_3 with phosphates. Percentages marked with asterisks indicate statistical significance (P < 0.01).*

cretion and the basal-lateral localization of Na^+- , K^+-ATPase in chloride cells (47, 48). This enzyme is ultimately responsible for maintaining the Cl^- entry step by maintaining the electrochemical gradient for Na^+ entry. Consequently, the inactivation of this enzyme accompanying serosal K^+ substitution should reduce the Cl^- entry and the Isc. The serosal HCO_3^- (or CO_2) dependency of the Cl^- secretion suggests a possible Cl^-/HCO_3^- exchange as a cellular entry step for Cl^- (44). The lack of any effect of carbonic anhydrase inhibitors suggests other possibilities for this HCO_3^- (or CO_2) dependence. This may be an intracellular pH effect or a metabolic effect (49). The recent demonstration of the presence of a HCO_3^--ATPase in uncontaminated microsomal fractions, including those from gill (50, 51), makes the possible involvement of this enzyme in Cl^- transport more plausible than previously believed.

The Na^+ and Cl^- dependency of the Isc across the operculum prompted further investigations into the nature of these influences for two reasons: the known Na^+ dependency of Cl^- transport in a variety of both absorptive and secretory epithelia (14) and the known 'trans' sensitivity of branchial Na^+ and Cl^- secretion to the external Na^+ and Cl^- concentration (39). Table VI summarizes some preliminary results of the effects of unilateral Na^+ and Cl^-

TABLE VI. The Effect of Unilateral Na^+ and Cl^- Substitutions on the Cl^- Efflux and Ionic Conductance of Isolated, Short-Circuited Opercular Epithelia from Seawater-Adapted Fundulus heteroclitus.[a]

Mucosal concentration	$J_{sm}^{Cl^-}$ ($\mu Eq\ cm^{-2}\ hr^{-1}$)	G_t ($mS\ cm^{-2}$)	Serosal concentration	$J_{sm}^{Cl^-}$ ($\mu Eq\ cm^{-2}\ hr^{-1}$)	G_t ($mS\ cm^{-2}$)
Na^+ = 150 mM	6.10 ± 0.31	8.5 ± 0.7	Na^+ = 150 mM	4.92 ± 0.49	8.8 ± 1.9
75 mM	4.50 ± 0.16	6.0 ± 0.4	75 mM	4.47 ± 0.55	5.5 ± 1.4
38 mM	3.44 ± 0.07	4.0 ± 0.9	38 mM	2.34 ± 0.42	4.3 ± 0.9
0 mM	1.61 ± 0.32	1.9 ± 0.2	0 mM	1.24 ± 0.22	3.9 ± 0.4
Cl^- = 142 mM	8.63 ± 1.68	7.8 ± 1.1	Cl^- = 142 mM	4.49 ± 0.33	5.9 ± 1.3
71 mM	6.16 ± 1.50	7.3 ± 1.3	71 mM	2.32 ± 0.52	5.1 ± 0.8
35 mM	3.80 ± 1.05	5.2 ± 1.4	35 mM	1.32 ± 0.14	4.2 ± 0.6
0 mM	2.51 ± 0.73	3.9 ± 1.2	<5 mM	0.12 ± 0.02	2.9 ± 0.6

[a]Data expressed as mean ± s.e.m. N = 4 for each substitution series. Na^+ was substituted with equimolar amounts of choline and Cl^- was substituted with equimolar amounts of methylsulphate. Preparations were short-circuited throughout the substitution series. During substitutions the concentration of all other ions remained constant.

substitutions on the $J_{sm}^{Cl^-}$ and G_t of short-circuited opercular epithelia. The gradual reduction in either the mucosal Na^+ or Cl^-
concentration results in a corresponding reduction in the $J_{sm}^{Cl^-}$,
which declines to approximately 25% of the control level at zero
mucosal Na^+ and Cl^- concentrations. With complete mucosal Na^+
substitution, the G_t was reduced to 22% of the control level,
while with mucosal Cl^- substitution, the G_t was reduced to 50% of
the control level. Serosal Na^+ substitution had similar effects,
reducing the $J_{sm}^{Cl^-}$ and G_t to 25% and 44% of their control levels
relatively in the absence of serosal Na^+. At low serosal Cl^- concentrations (<5 mM), the $J_{sm}^{Cl^-}$ and G_t were reduced to 3% and 40%
of their control levels respectively. These results clearly
demonstrate the mucosal Na^+ and Cl^- dependency of Cl^- secretion
across this epithelium. In addition, they demonstrate the serosal
Na^+ dependency of the Cl^- secretion but do not provide sufficient
evidence to distinguish between a Na^+-dependent and a Na^+-coupled
secretory mechanism. The observed linear relationship between
the serosal Cl^- concentration and the $J_{sm}^{Cl^-}$ confirms previous observations (13, 21), suggesting that this mechanism is operating
below its maximum rate under these conditions.

The G_t of the opercular epithelium is predominantly a partial Na^+ conductance (21). The range of this conductance is between 41% - 69%, with a mean of 54%. The G_{Na^+} varies directly
with the G_t, while the passive Cl^- and urea permeabilities demonstrate no correlation to the G_t. The Na^+-conductive pathway in
this tissue has characteristics of a single rate-limiting barrier
(52), is blocked by TAP^+ (27), and is unaffected by amiloride or
amphotericin B. These findings are suggestive of a low-resistance
paracellular Na^+ shunt pathway across the operculum. In the gill
(23) and opercular epithelium (47) a low-resistance extracellular
pathway is believed to occur between adjacent chloride cells, with
the shallow chloride cell-chloride cell "tight" junctions forming
the single rate-limiting barrier. In the operculum, this pathway
is dependent on mucosal Na^+, independent of serosal Na^+, and ap-

pears to be separate from the passive transepithelial Cl^- and urea pathways (21).

VI. NATURE OF THE IONIC MOVEMENTS ACROSS THE OPERCULAR EPITHELIUM

Exchange diffusion is believed to account for most of the transepithelial Na^+ and Cl^- movements across the gill epithelium (40). This has been based on the observed 'trans' effect of seawater (mucosal) Na^+ and Cl^- substitutions on the branchial Na^+ and Cl^- efflux rates, which in many instances exhibit typical saturation kinetic-like responses. Similar studies on the branchial Na^+ and Cl^- influxes in response to plasma (serosal) Na^+ and Cl^- substitutions can not easily be performed with isolated gill preparations because of the numerous technical problems. The isolated operculum epithelium preparation, on the other hand, allows easy access to both the mucosal and serosal compartments, and the accurate determination of the unidirectional fluxes unaffected by transepithelial potential changes in response to ionic gradients (i.e., short-circuited). Table VII summarizes the effects of 'trans' Na^+ and Cl^- substitutions on the unidirectional Na^+ and Cl^- fluxes across short-circuited opercular epithelia. Both mucosal Na^+ and Cl^- substitutions produce significant reductions in their respective efflux rates. The Na^+ efflux is reduced 60% in the absence of mucosal Na^+, and the Cl^- efflux is reduced 71% in the absence of mucosal Cl^-. This 'trans' sensitivity is suggestive of exchange diffusion, and in the case of mucosal Cl^- substitution, typical saturation kinetics are observed for the Cl^- efflux (46). However, corresponding serosal Na^+ and Cl^- substitutions have no effect on their respective influxes, indicating the absence of any Na^+/Na^+ or Cl^-/Cl^- transepithelial exchanges across this tissue.

These effects of mucosal ion substitutions can be explained by conductance changes in both the paracellular and transcellular ionic pathways. The paracellular Na^+ shunt pathway is known to be

TABLE VII. *The Effect of 'Trans' Ionic Substitution on the Uni-*
directional Short-Circuit Fluxes across Isolated Opercular Epi-
thelia from Seawater-Adapted Fundulus heteroclitus.[a]

	J_{sm}	J_{ms}
	$\mu Eq \cdot cm^{-2} \cdot hr^{-1}$	
Control J^{Na^+}	3.24 ± 0.37	3.94 ± 0.51
'Trans' $Na^+ = 0$	1.30 ± 0.21	3.31 ± 0.33
P (n)[b]	<0.001 (5)	>0.20 (5)
Control J^{Cl^-}	8.63 ± 1.68	0.38 ± 0.09
'Trans' $Cl^- = 0$	2.51 ± 0.73	0.37 ± 0.09
P (n)	<0.01 (4)	>0.50 (6)

[a]*Data expressed as mean ± s.e.m.*
[b]*Number of experiments given in parentheses.*

dependent on the mucosal Na^+ concentration (21). However, the
magnitude of the G_t reduction in response to mucosal Na^+ substitution is greater than the G_{Na^+}, suggesting that mucosal Na^+ is
also involved in maintaining the conductance of other ions. The
observed dependence of the Cl^- secretion on mucosal Na^+ (Table VI),
suggests that mucosal Na^+ may be involved in maintaining the transcellular Cl^- conductance pathway. Whether mucosal Cl^- affects the
paracellular Na^+ pathway has yet to be determined. The magnitude
of the reduction in the G_t in response to mucosal Cl^- substitution
(approximately 50%), and the large reduction in the $J_{sm}^{Cl^-}$, suggests
that the Na^+ conductive pathway is little affected by mucosal Cl^-
substitution, and that most of the reduction in the G_t results
from a reduction in the transcellular Cl^- pathway. It appears
then that mucosal Na^+ is an important determinant of the paracellular Na^+ conductance, while mucosal NaCl is an important determinant of the transcellular Cl^- conductance.

Although presently speculative, such conductance changes as
described can explain the observation that isolated opercular epi-

thelia from freshwater-adapted and seawater-adapted *F. heteroclitus* have similar transepithelial resistances and Cl^- secretion rates when bathed on both sides with Ringer (13). The G_t and $J_{sm}^{Cl^-}$ may be reduced when epithelia from seawater-adapted fish are bathed externally with Ringer, and increased when epithelia from freshwater-adapted fish are bathed externally with Ringer ($\sim$150 mM NaCl). Such a direct regulatory influence of the external salinity, which occurs in isolated preparations devoid of neurohumoral influences, can also explain the rapid reduction in the branchial Na^+ and Cl^- efflux rates when fish are transferred from seawater to freshwater (40, 53) before the neurohumoral influences become active (36, 54).

SUMMARY

The opercular skin of the teleost, *Fundulus heteroclitus,* is a chloride cell-rich epithelium that can be studied *in vitro* with methods not applicable to other teleost gill preparations. When isolated and mounted as a flat membrane between two well stirred fluid compartments, this tissue exhibits many physiological and pharmacological similarities to the marine teleost gill and other Cl^- transporting epithelia. This tissue actively secretes Cl^- at a rate which is directly proportional to the number of chloride cells. By all criteria, Na^+ behaves passively and traverses the epithelium via a paracellular shunt, which morphologically may be the extracellular pathway between adjacent chloride cells.

No transepithelial ionic exchanges are operating in the opercular epithelium. However, both the Na^+ and Cl^- efflux rates are considerably dependent on the 'trans' (musocal) Na^+ and Cl^- concentrations. In addition, the Cl^- secretion is considerably dependent on the 'cis' (serosal) Na^+, K^+, and HCO_3^- (CO_2) concentrations. Activation of β-adrenergic receptors in this tissue leads to an elevation in the cyclic AMP levels and a stimulation in the Cl^- secretion rate. Activation of the α-adrenergic receptors has no effect on the cyclic AMP levels and inhibits the Cl$^-$ secretion rate.

The similarities between the isolated opercular epithelium preparation and the marine teleost gill epithelium, as well as the similarities to most Cl$^-$ absorbing and secreting epithelia, make this preparation a most attractive model with which to pursue the details of teleost osmoregulation and Cl$^-$ secretion.

ACKNOWLEDGMENTS

This work was supported by National Institutes of Health grants GM 25002 and EY 01340, NSF grant DEB-7826821, and NIH grant SO7-RR-05764 to the Mt. Desert Island Biological Laboratory.

REFERENCES

1. H. W. Smith. The absorption and excretion of water and salts by marine teleosts. *Am. J. Physiol. 93*:485-505, 1930.
2. A. Krogh. Osmotic regulation in fresh water fishes by active absorption of chloride ions. *Z. Vergl. Physiol. 24*:656-666, 1937.
3. A. B. Keys. The heart-gill preparation of the eel and its perfusion for the study of a natural membrane *in situ*. *Z. Verol. Physiol. 15*:352-363, 1931.
4. M. Kamiya. Changes in ion and water transport in isolated gills of the cultured eel during the course of salt adaptation. *Annotnes Zool. Jap. 40*:123-129, 1967.
5. T. H. Kerstetter, L. B. Kirschner, and D. D. Rafuse. On the mechanisms of sodium ion transport by the irrigated gills of rainbow trout (*Salmo gairdneri*). *J. Gen. Physiol. 56*:342-359, 1970.

6. P. Payan. A study of the Na^+/NH_4^+ exchange across the gill of the perfused head of the trout (*Salmo gairdneri*). *J. Comp. Physiol.* *124*:181-188, 1978.

7. A. B. Keys and E. N. Willmer. "Chloride-secreting cells" in the gills of fishes with special reference to the common eel. *J. Physiol.* *76*:368-378, 1932.

8. K. J. Karnaky, Jr. Ion-secreting epithelia: Chloride cells in the head region of *Fundulus heteroclitus*. *Am. J. Physiol.* *238*:R185-R198, 1980.

9. H. H. Ussing and K. Zarahn. Active transport of sodium as the source of electric current in the short-circuited frog skin. *Acta Physiol. Scand.* *23*:110-127, 1951.

10. K. J. Karnaky, Jr., K. J. Degnan, and J. A. Zadunaisky. Chloride transport across isolated opercular epithelium of killifish: a membrane rich in chloride cells. *Science 195*: 203-205, 1977.

11. J. Burns and D. E. Copeland. Chloride excretion in the head region of *Fundulus heteroclitus*. *Biol. Bull. Mar. Biol. Lab. Woods Hole 99*:381-385, 1950.

12. K. J. Karnaky, Jr. and W. B. Kinter. Killifish opercular skin: A flat epithelium with a high density of chloride cells. *J. Exp. Zool.* *199*:355-364, 1977.

13. K. J. Degnan, K. J. Karnaky, Jr., and J. A. Zadunaisky. Active chloride transport in the *in vitro* opercular skin of a teleost (*Fundulus heteroclitus*), a gill-like epithelium rich in chloride cells. *J. Physiol.* *271*:155-191, 1977.

14. R. A. Frizzell, M. Field, and S. G. Schultz. Sodium-coupled chloride transport by epithelial tissues. *Am. J. Physiol.* *236*:F1-F8, 1979.

15. J. A. Zadunaisky. Active transport of chloride in frog cornea. *Am. J. Physiol.* *211*:506-512, 1966.

16. J. A. Zadunaisky, O. A. Candia, and D. J. Chiarandini. The origin of the short-circuit current in the isolated skin of the South American frog, *Leptodactylus ocellatus*. *J. Gen.*

Physiol. 47:393-402, 1963.

17. K. J. Karnaky, Jr., K. J. Degnan, and J. A. Zadunaisky. Correlation of chloride cell number and short-circuit current in chloride-secreting epithelia of *Fundulus heteroclitus. Bull. Mt. Desert Isl. Biol. Lab. 19*:109-111, 1979.

18. J. Bereiter-Hahn. Dimethylaminostyrylmethylpyridiniumiodine (DASPMI) as a fluorescent probe for mitochondria *in situ. Biochim. Biophys. Acta 423*:1-14, 1976.

19. W. S. Marshall and R. S. Nishioka. Relation of mitochondria-rich chloride cells to active chloride transport in the skin of a marine teleost. *J. Exp. Zool. 214*:147-156, 1980.

20. D. H. Evans. Kinetic studies of ion transport by fish gill epithelium. *Am. J. Physiol. 238*:R224-R230, 1980.

21. K. J. Degnan and J. A. Zadunaisky. Passive sodium movements across the opercular epithelium: The paracellular shunt pathway and ionic conductance. *J. Membrane Biol. 55*:175-185, 1980.

22. S. A. Ernst, W. B. Dodson, and K. J. Karnaky, Jr. Structural diversity of zonulae occludentes in seawater-adapted killifish opercular epithelium. *J. Cell Biol. 79*:242a, 1978.

23. C. Sardet, M. Pisam, and J. Maetz. The surface epithelium of teleostean fish gills. Cellular and junctional adaptations of the chloride cell in relation to salt adaptation. *J. Cell Biol. 80*:96-117, 1979.

24. J. Scheffey, J. K. Foskett, and T. E. Machen. Identification of the transporting cell type, the chloride cell, in a heterogenous epithelium, the teleost opercular membrane. *Biophys. J. 33*:220a (abstract), 1981.

25. J. S. Handler, R. W. Butcher, E. Sutherland, and J. Orloff. The effect of vasopressin and of theophylline on the concentration of adenosine 3',5'-phosphate in the urinary bladder of the toad. *J. Biol. Chem. 240*:4524-4526, 1968.

26. A. W. Cuthbert and P. Pic. Adrenoceptors and adenyl cyclase in gills. *Br. J. Pharm. 49*:134-137, 1973.

27. J. Moreno. Blockage of gallbladder tight junction cation-selective channels by 2,4,6-triaminopyrimidium (TAP). *J. Gen. Physiol.* *66*:97-116, 1975.

28. J.-P. Girard. Salt secretion by the perfused head of trout adapted to seawater and its inhibition by adrenaline. *J. Comp. Physiol.* *111*:77-91, 1976.

29. A. B. Keys and J. B. Bateman. Branchial responses to adrenaline and to pitressin in the eel. *Biol. Bull. Mar. Biol. Lab. Woods Hole* *63*:327-336, 1932.

30. P. Pic, N. Mayer-Gostan, and J. Maetz. Branchial effects of epinephrine in the seawater-adapted mullet. II. Na^+ and Cl^- extrusion. *Am. J. Physiol.* *228*:441-447, 1975.

31. T. J. Shuttleworth. The effect of adrenaline on potentials in the isolated gills of the flounder. *(Platichthys flesus* L.). *J. Comp. Physiol.* *124*:129-136, 1978.

32. T. E. Andreoli, V. W. Dennis, and A. M. Weigl. The effect of amphotericin B on water and non electrolyte permeability of thin lipid membranes. *J. Gen. Physiol.* *53*:133-156, 1969.

33. P. J. Bentley. Amiloride: A potent inhibitor of sodium transport across the toad bladder. *J. Physiol.* *195*:317-330, 1968.

34. T. Maren. Carbonic anhydrase: Chemistry, physiology, and inhibition. *Physiol. Rev.* *47*:595-781, 1967.

35. N. Mayer, J. Maetz, D. K. O. Chan, M. Forster, and I. Chester-Jones. Cortisol, a sodium excreting factor in the eel *(Anguilla anguilla L.)* adapted to sea water. *Nature* *214*:1118-1120, 1967.

36. G. E. Pickford and J. G. Phillips. Prolactin, a factor in promoting survival of hypophysectomized killifish in freshwater. *Science* *130*:454-455, 1959.

37. H. H. Ussing. The alkali metal ions in isolated systems and tissues. *In* "Handbuch der Experimentallen Pharmakologie," (O. Eichler and A. Farah, eds.), Vol. 13, Part 1, pp. 1-195. Springer Verlag, Berlin, 1960.

38. K. J. Degnan and J. A. Open-circuit sodium and chloride
     fluxes across isolated opercular epithelia from the teleost
     Fundulus heteroclitus. *J. Physiol.* *294*:484-495- 1979.

39. J. Maetz and M. Bornancin. Biochemical and biophysical as-
     pects of salt excretion by chloride cells in teleosts.
     Fortschr. Zool. *23*:322-362, 1975.

40. R. Motais, F. Garcia-Romeu, and J. Maetz. Exchange diffusion
     effect and euryhalinity in teleosts. *J. Gen. Physiol.* *50*:391-
     422, 1966.

41. K. J. Degnan and J. A. Zadunaisky. Ionic contributions to the
     potential and current across the opercular epithelium. *Am. J.*
     Physiol. *238*:R231-R239, 1980b.

42. C. R. House and J. Maetz. On the electrical gradient across
     the gill of the seawater-adapted eel. *Comp. Biochem. Physiol.*
     A47:917-924, 1974.

43. L. B. Kirschner, L. Greenwald, and M. Sanders. On the
     mechanism of sodium extrusion across the irrigated gill of
     sea water-adapted rainbow trout (*Salmo gairdneri*). *J. Gen.*
     Physiol. *64*:148-165, 1974.

44. K. J. Degnan, J. A. Zadunaisky, and N. Mayer-Gostan. The bi-
     carbonate sensitivity of chloride secretion across the oper-
     cular epithelium. *Bull. Mt. Desert Isl. Biol. Lab.* *20*:114-
     116, 1980.

45. P. Silva, J. Stoff, M. Field, L. Fine, J. N. Forrest, and
     F. H. Epstein. Mechanism of active chloride secretion by
     shark rectal gland: role of Na-K-ATPase in chloride trans-
     port. *Am. J. Physiol.* *233*:F298-F306, 1977.

46. K. J. Degnan and J. A. Zadunaisky. The sodium and chloride
     dependence of active Cl$^-$ secretion across the opercular epi-
     thelium. *Fed. Proc.* *40*:370 (abstract), 1981.

47. S. A. Ernst, W. C. Dodson, and K. J. Karnaky, Jr. Structural
     diversity of occluding junctions in the low-resistance chlor-
     ide-secreting opercular epithelium of seawater-adapted killi-
     fish (*Fundulus heteroclitus*). *J. Cell Biol.* *87*:488-497, 1980.

48. K. J. Karnaky, Jr., L. B. Kinter, W. B. Kinter, and C. E. Stirling. Teleost chloride cell. II. Autoradiographic localization of gill Na,K-ATPase in killifish *Fundulus heteroclitus* adapted to low and high salinity environments. *J. Cell Biol.* *70*:157-177, 1976.

49. D. W. Martin and B. Murphy. Carbamyl phosphate and glutamine stimulation of the gallbladder salt pump. *J. Membrane Biol.* *18*:231-242, 1974.

50. M. Bornancin, G. de Renzis, and R. Noan. Cl^--HCO_3^--ATPase in gills of the rainbow trout: evidence for its microsomal localization. *Am. J. Physiol.* *238*:R251-R259, 1980.

51. E. Kinne-Saffran and R. Kinne. Further evidence for the existence of an intrinsic bicarbonate-stimulated Mg^{2+}-ATPase in brush border membranes isolated from rat kidney cortex. *J. Membrane Biol.* *49*:235-251, 1979.

52. L. J. Mandel and P. F. Curran. Response of the frog skin to steady-state voltage clamping. I. The shunt pathway. *J. Gen. Physiol.* *59*:503-518, 1972.

53. F. H. Epstein, J. Maetz, and G. de Renzis. Active transport of chloride by the teleost gill: Inhibition by thiocyanate. *Am. J. Physiol.* *224*:1295-1299, 1973.

54. N. Mayer-Gostan and T. Hirano. The effects of transecting the IXth and Xth cranial nerves on hydromineral balance in the eel *Anguilla anguilla*. *J. Exp. Biol.* *64*:461-475.

CHLORIDE TRANSPORT IN FROG SKIN

Poul Kristensen

Institute of Biological Chemistry A
August Krogh Institute
Copenhagen, Denmark

I. INTRODUCTION

In order to get a full understanding of epithelial function,
it is necessary to have deep insight into anion as well as cation
transport mechanisms. While sodium transport has been studied
very extensively in model epithelia like frog skin and toad urinary
bladder for many years, until recently our information concerning
chloride transport has been scarce and contradictory.

This paper will attempt, on the basis of a general two-membrane hypothesis (1), to provide the reader with some ideas concerning the localization of the pathways through which chloride
traverses the skin of frogs (*Rana temporaria, esculenta,* and
pipiens), and by what mechanisms this transport occurs.

The first problem in pathway localization is to decide
whether chloride crosses the skin via a transcellular or a para-

cellular pathway. The next problem arises when there is more than
one possible transcellular pathway (as in the case of frog skin),
where molecules may pass through the most abundant cell type of
the *Stratum granulosum*, through the mitochondria rich cells, or
through the mucous glands.

In the case of the study of transport mechanism, the aim will
be to distinguish between free diffusion, exchange diffusion, and
active transport.

II. A TRANSCELLULAR PATHWAY

The first attempt to investigate anion transport in frog skin
was made by Ussing (2) in his famous flux ratio paper. The experiments were made by studying unidirectional fluxes of ^{131}I in
skins with different spontaneous transepithelial potential difference. In some cases flux ratios were observed to be in accordance with the flux ratio equation (which describes the variation
of flux ratio with transmembrane potential difference). But in
many cases the experimental flux ratios were considerably lower
than what had been anticipated. This led Ussing to the conclusion
that iodide may cross the skin partly by free diffusion and partly
by exchange diffusion.

In a following paper (3), chloride transport was investigated with $^{36}Cl^-$ and chemical determination of chloride. Calculated and observed flux ratios often agreed, which indicates completely passive behavior of chloride. (It may be important to
note that these experiments were performed with 10 times diluted
chloride Ringer on the outside.)

While these two papers (2, 3) are important with respect to
transport mechanism, they do not give evidence permitting localization of chloride transport pathways. An important step concerning this problem occurred (4) when it was observed that removal of sodium from the outside solution (substitution with K^+),

or the addition of amiloride (a specific inhibitor of the transport of sodium across the outside barrier of the frog skin), significantly reduced the efflux as well as the influx of chloride.
The chloride transport's dependence on Na^+ in the outside solution
had been noted before, but had not been explained (5). We suggested (4) that inhibition of the passive diffusion of sodium
across the outer barrier may result in a hyperpolarization of the
intracellular potential, and that this alone can represent an increase in the energy barrier for chloride movement. We had previously developed a steady-state mathematical model for a simple two-
membrane hypothesis involving a sodium and chloride permeable
outward-facing membrane and a potassium and chloride permeable
inward-facing membrane (6). When supplied with information concerning sodium pump characteristics, membrane permeabilities,
bath compositions, and clamping potential (transepithelial), this
model will calculate clamping current, intracellular potential,
intracellular ion concentrations, net sodium transport, and chloride fluxes. Figure 1 shows how chloride efflux is expected to
vary with outside sodium concentration under short circuit conditions (clamping potential = 0). The relationship has been calculated for two outward facing membrane chloride permeabilities at
constant sodium permeability. Since sodium permeability is a
function of outside sodium concentration (7), the relationship has
been calculated for such a situation. In accordance with our tentative suggestion, the calculations show, under all circumstances
that although the chloride permeabilities of the two membranes in
series were kept constant, inhibition of sodium permeability led
to a reduction of transepithelial chloride permeability. Calculations on even simpler systems also suggest a similar mechanism
for the action of amiloride on chloride fluxes (8). Figure 2
shows experiments relating chloride efflux and outside sodium concentration. Since it is a well-established fact that amiloride
blocks the entry of sodium from the outside solution into the
cells, it can be concluded that 25 - 90% of the chloride fluxes

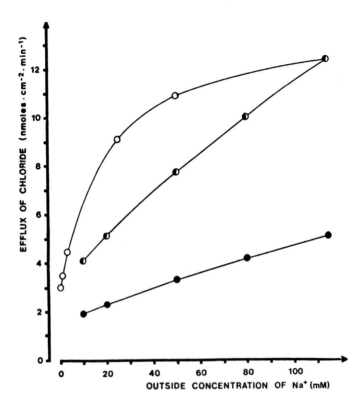

FIG. 1. Two-membrane model predictions for the relationship between chloride efflux and outside sodium concentration under short circuiting. ○: *calculations made with* $P_{Cl}^{o} = 6 \times 10^{-6}$ *cm* sec^{-1} *and a sodium concentration dependent sodium permeability according to ref. (6).* ◑: P_{Cl}^{o} *as before and constant sodium permeability.* ●: $P_{Cl}^{o} = 2 \times 10^{-6}$ *cm* sec^{-1} *and constant* P_{Na}^{o} *as before.*

are localized to transcellular pathways.

This conclusion is supported by experiments with elevated potassium concentrations in the inside bathing solution (4). In the presence of amiloride in the outside solution, the intracellular potential of the epithelial cells is probably very nearly equal

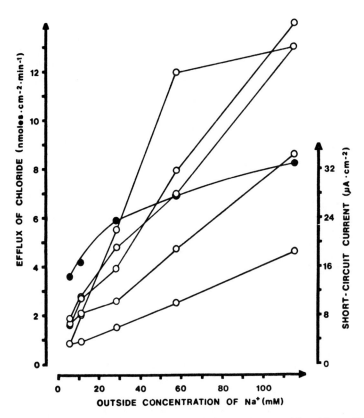

FIG. 2. *Experimental relationship between chloride efflux and outside sodium concentration, measured under short circuit conditions with* $^{36}Cl^-$. *The inside was bathed with ordinary chloride Ringer's, and the sodium concentration in the outside bathing solution was changed by substitution with potassium for sodium.* o: *chloride efflux, left ordinate.* ●: *average of recorded short circuit currents, right ordinate.*

to the potassium equilibrium potential. If inside K^+ concentration is raised (e.g., by substituting 75% of Na^+ with K^+), the membrane potential will be considerably depolarized. The effect of high K^+ is completely reversible and not harmful (9).

When inside K^+ concentration was raised to about 80 mM on

amiloride treated skins, chloride fluxes were reactivated to levels
even higher than before the addition of amiloride (4). This shows
that arguments for the intracellular potential as a mediator of the
effect of amiloride on chloride transport are correct, and that the
results are completely in agreement with computer calculated effects of both K^+ and amiloride. (Details not given here.) A similar conclusion with respect to localization has been arrived at by
Mandel (10).

III. THE EFFECT OF CHLORIDE

An interesting property of the chloride transport is that it
depends on outside sodium concentration as well as on the concentration of chloride in the outside bathing solution (4). Although
very different levels of chloride transport are observed in different animals, chloride efflux (under short-circuit conditions)
is, in all cases, roughly linearly-dependent on outside chloride
concentration (Fig. 3). This behavior (trans-effect) is remi-
niscent of exchange diffusion with a very large half-saturation
constant. Since chloride flux ratios are predictable from the
flux ratio equation, chloride transport must be conductive and
therefore cannot occur as exchange diffusion. Its conductive
properties are also revealed by the fact that chloride permeability varies inversely with the spontaneous transepithelial potential
difference (2, 3). Furthermore, inhibition of chloride transport
with $10^{-5}M$ $CuSO_4$ results in considerable increases in transepithelial potential difference in low potential skins (11, 12).
Therefore, a working hypothesis explaining the behavior of
chloride transport may involve the possibility that chloride
passes the frog skin via a transcellular pathway. The chloride
permeability of the outward-facing membrane may be visualized as
localized to channels that are open in the presence of chloride
and closed in its absence.

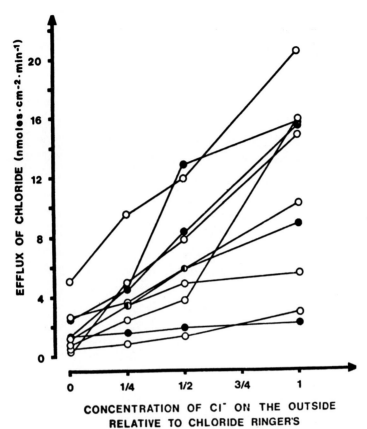

FIG. 3. *The transeffect: Variation of chloride efflux with*
outside chloride concentration in short circuited frog skin.
Changes in chloride concentration was obtained by substitution
with sulfate. O : *experiments on R. temporaria,* ● : *R. esculenta.*

Some new experiments may give interesting information regarding the mechanism of transport and transport regulation. Table I shows experiments in which chloride efflux was measured under short-circuit conditions with Ringer solutions on the outside containing anions other than chloride. Besides chloride, only bromide is capable of stimulating chloride efflux; iodide and nitrate are completely ineffective. So too is thiocyanate (see Table II).

When the outside is bathed with chloride Ringer containing 1 mM Br⁻, I⁻, or SCN⁻, the ratios between the transepithelial per-

TABLE I. *Chloride Efflux Measured with Different Ringer's Solutions as the Outside Medium.*[a]

Efflux of chloride ($nmol\ cm^{-2}min^{-1}$)			
NaCl-Ringer's outside	NaBr-Ringer's outside	NaI-Ringer's outside	$NaNO_3$-Ringer's outside
20.64	8.25	0.83	1.25
2.88	1.54	0.50	0.61
6.70	3.41	0.80	1.02
11.94	4.19	0.67	1.40
24.65	18.57	3.58	4.77
7.77	2.56	0.95	0.76

[a]*The inside solution was NaCl-Ringer's throughout. Short circuit conditions.*

TABLE II. *Chloride efflux with chloride Ringer's and thiocyanate Ringer's bathing the outside.*[a]

Chloride efflux ($nmol\ cm^{-2}min^{-1}$)		Potential diff. (mV)		S-c current ($\mu A/cm^2$)		Tissue conductance (mS/cm^2)	
Control	Exp.	Control	Exp.	Control	Exp.	Control	Exp.
14.7	1.41	25.6	52.7	35.7	24.3	1.40	0.46
16.6	0.52	17.3	78.7	27.1	25.7	1.57	0.33
11.3	0.20	22.5	89.0	27.1	27.1	1.21	0.31
9.5	0.78	47.0	106.0	52.3	52.0	1.11	0.49

[a]*Control: Chloride outside, Exp.: Thiocyanate outside, short circuit conditions.*

meabilities are ($P_{Cl^-}P_{SCN}:P_I:P_{Br}$): 1.0 : 1.7 : 0.8 : 1.3. The two types of results mean that the abilities of the ions to cause an opening of the anion tansport pathway are very different from the relative abilities with which they pass through the open pathway.

The working hypothesis may be extended therefore, to include the possibility that regulation of chloride permeability by chloride occurs at a site with characteristics different from the channel through which chloride and other anions are transported.

IV. DIFFERENT TRANSCELLULAR PATHWAYS

There can be very little doubt that the sodium- and
chloride-stimulated chloride transport, which constitutes the
major fraction of chloride transport, is localized to a transcellular pathway. There are three candidates for such a pathway:
the *granulosum* cells, the mitochondria rich cells (MR cells), and
the mucous glands of the skin. The third has in fact, been shown
to be responsible for the transfer of chloride ions under certain
conditions, but its response to the appropriate stimuli (epinephrine or high inside K^+) is of a rather short duration compared
to the time interval used for the flux measurements referred to
in this paper (13, 14).

The MR cells were suggested by Voûte and Meier (15) to play
a role in chloride transport. They showed that the chloride transport level and the number of MR cells having access to the outside solution were linearly correlated.

The normal *stratum granulosum* cells are undoubtedly responsible for sodium transport. Numerous experiments with microelectrodes seem to support this statement (16), although other reasons
could also be given. The experiments showing that chloride transport is dependent on sodium in the outside bathing medium suggest
that chloride passes the skin through cells that have a sodium
permeable outward-facing membrane, *i.e.*, through the same cells
as sodium. But it has not been shown that the MR cells do not
have a sodium permeable outward-facing membrane. So if chloride
transport does take place through these cells, these experiments
indicate that the MR cells may take part in sodium transport.

Until now it is still not possible to decide which cellular
pathway is responsible for chloride transport. Evidence does not
prove that the MR cells constitute the pathway, yet it does remain
a possibility.

V. THE PARACELLULAR PATHWAY AND ACTIVE TRANSPORT

Since we have discussed the cellular passive pathway in detail, we should also examine the significance of the paracellular
pathway, and the possible presence of active transport of chloride.

The paracellular pathway can be studied with reasonable accuracy only in skins where the above described transcellular pathway
is of minor significance. This is the case if the animals used
for experiments have been stored for more than a few weeks at temperatures about 4°C (11, 12). In skins from such animals, the dependence of chloride fluxes on transepithelial clamping voltage
can be predicted reasonably well by the constant field equation
(17, 18). (Which concludes that chloride transport occurs across
one simple barrier, namely, the tight junctions between the outermost cells of Stratum granulosum.)

Simultaneously, it was shown that an active transport of
chloride from the outside to the inside was taking place (17, 19)
in a Ringer's solution containing 1 mM Cl$^-$ in sulfate. Investigations of the nonsteady-state chloride fluxes under these circumstances immediately after addition of tracer, revealed that
active transport and passive transport of chloride in cold-adapted
frogs occur via two kinetically different pathways - tentatively
suggested to be a cellular and a paracellular pathway (20).

The active transport capacity is very small and in cold-
adapted frogs will only contribute about 10% of the total chloride
fluxes when the skin is bathed on both sides with chloride Ringer's
(in warm adapted frogs it is likely to be less than 1%). Therefore, active transport capacity does not have quantitative significance under the circumstances of the experiments concerning
passive transport mechanisms discussed in the beginning of this
paper. However, the active transport may have physiological significance for the animals in some of the conditions they encounter
in nature (21, 22).

VI. CONCLUDING REMARKS

Space does not allow for a detailed discussion of small differences in opinion concerning chloride transport. It seems evident that lack of knowledge about the effect of outside concentrations of chloride and sodium may lead to different points of view.
The varying conditions, particularly storage temperature, under
which the animals are kept also plays a role. Variations of
chloride transport with season has also been discussed (23).

The present discussion of some relevant literature and new
experiments has resulted in the following picture of chloride
transport in skins from some of the most often used frogs (*Rana
temporaria, esculenta,* and *pipiens*): The major fraction (conductance 1-2 mS/cm^2) of chloride transport occurs passively via a
transcellular pathway, and depends on the presence of sodium as
well as chloride in the outside bathing solution. In skins from
animals stored in the cold (hibernating), this pathway is reduced
to a very low level, and chloride transport is then dominated by
a paracellular pathway (which often has a very low chloride conductance - 10^{-5} - 10^{-4} mS/cm^2). Together with these two passive
mechanisms, the presence of a very small exchange-diffusion cannot
be excluded, and a small active transport is clearly demonstrable.
In other frogs the relative quantitative significance of the various pathways may be different (24, 25).

It is understandable that a system of such complexity has led
to disagreements, but hopefully the present short discussion has
explained some of their causes.

VII. SUMMARY

The mechanisms responsible for chloride transport in frog
skin, and the pathways used by this transport has been discussed.
Based on earlier flux ratio studies and recent experiments des-

cribing the dependence of chloride transport on outside sodium concentration, it is concluded that chloride is passively transported through a transcellular pathway.

It is also shown that chloride efflux is dependent on outside chloride concentration. This cannot be explained as exchange diffusion because the chloride transport pathway in frog skin is conductive.

While only chloride, and, to a smaller extent bromide can activate the anion pathway, both Cl^-, Br^-, I^-, and SCN^- can pass the transport system in its open state. This leads to the conclusion that transport and transport activation are different processes. This leads to a working hypothesis according to which anions pass a channel by free diffusion. The opening of the channels are determined via a regulatory subunit that can react with chloride.

REFERENCES

1. V. Koefoed-Johnsen and H. H. Ussing. The nature of the frog skin potential. *Acta Physiol. Scand.* *42*:298, 1958.
2. H. H. Ussing. The distinction by means of tracers between active transport and diffusion. *Acta Physiol. Scand.* *19*: 43, 1949.
3. V. Koefoed-Johnsen, H. Levi, and H. H. Ussing. The mode of passage of chloride ions through the isolated frog skin. *Acta Physiol. Scand.* *25*:150, 1952.
4. P. Kristensen. Effect of amiloride on chloride transport across amphibian epithelia. *J. Membrane Biol.* *40S*:167, 1978.
5. R. I. Macey and S. Meyers. Dependence of chloride permeability on sodium in the isolated frog skin. *Am. J. Physiol.* *204*: 1095, 1963.
6. E. Hviid Larsen and P. Kristensen. Properties of a conductive cellular pathway in the skin of the toad (*Bufo bufo*). *Acta Physiol. Scand.* *102*:1, 1978.

7. W. Fuchs, Hviid Larsen, and B. Lindemann. Current-voltage
    curve of sodium channels and concentration dependence of
    sodium permeability in frog skin. *J. Physiol. 267*:137, 1977.

8. O. A. Candia. Reduction of chloride fluxes by amiloride
    across the short circuited frog skin. *Am. J. Physiol. 234*:
    F437, 1978.

9. H. H. Ussing, T. U. L. Biber, and N. S. Bricker. Exposure
    of the isolated frog skin to high potassium concentrations
    at the internal surface. *J. Gen. Physiol. 48*:425, 1965.

10. L. J. Mandel. Actions of external hypertonic urea, ADH, and
    theophylline on transcellular and extracellular solute per-
    meabilities in frog skin. *J. Gen. Physiol. 65*:599, 1975.

11. V. Koefoed-Johnsen, I. Lyon, and H. H. Ussing. Effect of
    Cu ion on permeability properties of isolated frog skin
    (*Rana temporaira*). *Acta Physiol. Scand. Suppl. 396*:102,
    1973.

12. V. Koefoed-Johnsen and H. H. Ussing. Transport pathways in
    frog skin and their modification by copper ions. *In* "Secre-
    tory mechanisms of exocrine glands." *Alfred Benzon Symposium
    VII*, Munksgaars, Copenhagen, 1974, p. 411.

13. V. Koefoed-Johnsen, H. H. Ussing, and K. Zerahn. The origin
    of the short-circuited current in the adrenaline stimulated
    frog skin. *Acta Physiol. Scand. 27*:38, 1952.

14. L. Share and H. H. Ussing. Effect of potassium on the move-
    ment of water across the isolated amphibian skin. *Acta Phy-
    siol. Scand. 64*:109, 1965.

15. C. L. Voûte and W. Meier. The mitochondria rich cell of frog
    skin as hormone sensitive shunt path. *J. Membrane Biol. 40S*:
    151, 1978.

16. W. Nagel. Effects of antidiuretic hormone upon electrical
    potential and resistance of apical and basolateral membranes
    of frog skin. *J. Membrane Biol. 42*:99, 1978.

17. P. Kristensen. Chloride transport across isolated frog skin.
    Acta Physiol. Scand. 84:338, 1972.

18. L. J. Mandel and P. F. Curran. Response of the frog skin to steady-state voltage clamping. I. The shunt pathway. *J. Gen. Physiol.* *59*:503, 1972.

19. D. W. Martin and P. F. Curran. Reversed potentials in isolated frog skin. II. Active transport of chloride. *J. Cell. Comp. Physiol.* *67*:367, 1966.

20. P. Kristensen. Anion transport across frog skin. *In* "Transport mechanisms in epithelia." *Alfred Benzon Symposium V.* Munksgaard, Copenhagen, 1972, p. 148.

21. C. B. Jørgensen, H. Levi, and K. Zerahn. On active uptake of sodium and chloride ions in anurans. *Acta Physiol. Scand.* *30*:178, 1954.

22. A. Krogh. Osmotic regulation in the frog (*R. esculenta*) by active absorption of chloride ions. *Scand. J. Physiol.* *76*: 60, 1937.

23. C. O. Watlington and F. Jessee. Chloride flux across frog skins of low potential difference. *Biochim. Biophys. Acta* *330*:102, 1973.

24. Ques-von Petery, C. A. Rotunno, and M. Cereijido. Studies on chloride permeability of the skin of *Leptodactylus ocellatus*: I. Na^+ and Cl^- effect on passive movements of Cl^-. *J. Membrane Biol.* *42*:317, 1978.

25. J. A. Zadunaisky, O. A. Candia, and D. J. Chiarandini. The origin of the short circuit current in the isolated skin of the South American frog *Leptodactylys ocellatus*. *J. Gen. Physiol.* *47*:393, 1963.

CHLORIDE CURRENT RECTIFICATION IN TOAD SKIN EPITHELIUM[*]

Erik Hviid Larsen

Zoophysiological Laboratory A
August Krogh Institute
University of Copenhagen
13, Universitetsparken
DK-2100 Copenhagen Ø
Denmark

I. INTRODUCTION

Epithelial cells specialized for maintaining osmotic and
ionic concentration differences between the two extracellular
compartments they separate, typically rectify ion fluxes and
water movement. This has led to the concepts of absorbing versus
secreting epithelia. In this paper I shall discuss some basic
properties of the Cl pathways in the epithelial cells of toad
skin, which is an absorbing epithelium transporting Na^+ and Cl^-
from the external environment toward the serosal side. This tissue seems to provide a useful model system for the study of mech-

[*]*Supported by the Danish Natural Science Research Council,
Grants Nos. 511-7120 and 511-10629.*

Copyright © 1982 by Academic Press, Inc.
All rights of reproduction in any form reserved.
ISBN 0-12-775280-3

anisms of epithelial Cl⁻ transport.

Our studies are based upon the transepithelial voltage-clamp method. This method has been used because (i) the distinction between simple passive transport and complex transport processes, like active transport and exchange diffusion, relies on a method which allows clamping of the transepithelial electrochemical potential difference while unidirectional isotope fluxes are measured (1,2). (ii) Combined with isotope technique, it also allows a precise measurement of the steady-state transepithelial ionic currents. By comparing the ionic currents with the steady-state clamping current, the contribution of the transepithelial ionic conductances to the total skin conductance can be analyzed. (iii) Conductances with dynamic properties (complex voltage or concentration dependence) will be disclosed and are best studied by this method.

From transepithelial voltage-clamp studies we learn about transport processes operating across the entire epithelial structure. Thereby we can make the first proposals to the nature of the mechanisms operating at the membrane levels. Such proposals will then be a useful guide for a subsequent and direct study of the individual membranes using microelectrode technique, the precision of which at present is difficult to evaluate in connection with amphibian skin (3,4). Recently published microelectrode studies of amphibian skin (5,6,7) focus on Na⁺ transport with little or no attention paid to anion pathways.

II. FLUX-RATIO ANALYSIS

If the translocation of Cl⁻ across the skin is driven exclusively by the chloride electrochemical potential difference between the bathing solutions, and if the ions do not interact with other molecules (moving or stationary) in the membrane the following relation holds (1):

$$\left(\frac{J_{Cl}^{in}}{J_{Cl}^{out}}\right) = \frac{(Cl)_o}{(Cl)_i} \exp\left[-FV/(RT)\right], \quad V = \psi_o - \psi_i \qquad (1)$$

where J_{Cl}^{in} and J_{Cl}^{out} are the inward and outward unidirectional fluxes, $(Cl)_o$ and $(Cl)_i$ the chemical activities in the outer and inner bathing solutions, V the transepithelial potential difference $\left[\text{outside } (\psi_o) - \text{inside } (\psi_i)\right]$, and F, R, and T have their usual meanings. After logarithmic transformation and introduction of the Cl-equilibrium potential (E_{Cl}) by the Nernst equation, *Eq. (1)* takes the form,

$$\log_{10}\left(\frac{J_{out}^{in}}{J_{Cl}^{out}}\right) = -(V-E_{Cl})/58 \qquad (2)$$

Here $-F(V-E_{Cl})$ is proportional to the externally applied force per mole of chloride ions. This force is positive in an inward direction.

In practice, care must be taken when imposing a large outwardly directed force on the chloride ions. Reversal of the transepithelial potential difference to values above +50 mV often leads to damage of the skin as revealed by unphysiologically high transepithelial nonspecific conductances (8). To avoid this difficulty the flux ratio in this region was studied by varying also the chemical activity of Cl$^-$ in the outer solution. Thereby we could vary $(V-E_{Cl})$ symmetrically around zero and cover the considerable range of 225 mV, corresponding to a change in the theoretical flux ratio of approximately four orders of magnitude.

The results are depicted in Fig. 1. Reasonably good agreement between the theoretical (full line) and experimental (symbols) flux ratios is obtained for an inwardly driving force (left hand side of the diagram), indicating that *a simple passive pathway* is available for a Cl-uptake driven by an electrical potential difference. A significant deviation from the theoretical line is

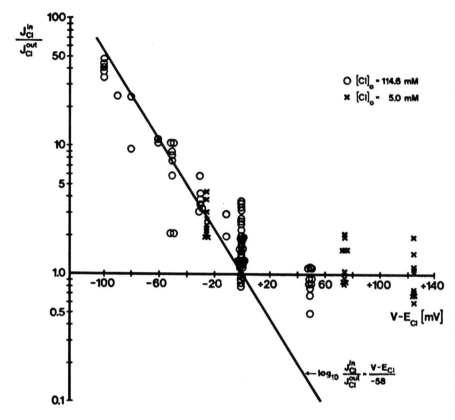

FIG. 1. *Flux-ratio analysis of Cl⁻ transport across the isolated toad skin. Each of the 101 data points is calculated from influx and outflux in paired skin halves. The different values at each (V-E$_{Cl}$) are from different toad skin preparations. Each set of paired skins was tested at two or three clamping voltages. Total number of animals is 40. Data from (9) and Hviid Larsen, H. Løkkegaard, and K. Simonsen, unpublished.*

manifest in the range of (V-E$_{Cl}$) from zero towards large positive values (right hand side of the diagram). Here, two other types of mechanisms can be identified. First, *an active transport pathway*, disclosed by the finding that the flux ratio is different from unity when the chloride ions are distributed in electro-chemical equilibrium across the skin (V-E$_{Cl}$ = 0). This active transport is inwardly directed. The interpretation of the flux ratios obtained for an outwardly directed driving force is less trivial. The flux ratios are up to three orders of magnitude larger than expected for a simple passive transport. This could be due to active transport. On the other hand, the experimental flux ratios are independent of the driving force in this region and always close to one, which means that the net charge transfer by this pathway is close to zero. This important finding holds whether the driving force is manipulated by voltage clamping or by varying the outer Cl-concentration. Therefore, the most likely interpretation is that *a 1:1 chloride exchange pathway* governs the steady-state chloride transport in this region. Additional experimental support of this hypothesis is discussed in references 9,10,11.

The existence of 1:1 exchange diffusion pathways in biological membranes was originally suggested by Ussing to account for the very fast Na^+ exchange in frog satorius muscle (12). In the meantime, Cl:Cl exchange pathways have been disclosed in a number of different cell types, e.g., gastric mucosa (13), red cells (14, 15,16), kidney cortical collecting tubule (17), frog skin (18,19, 20), and Ehrlich ascited tumor cells (21,22).

III. THE CONDUCTIVE CHLORIDE PATHWAY

A. *The Steady-State Current-Voltage relation and Its Ionic Components*

In toad skin exposed to NaCl Ringer's on both sides, the steady-state transepithelial current is a nonlinear function of clamping voltage. This I-V relation was studied in greater detail by measurements of steady-state transepithelial in- and outfluxes of Na^+ and Cl^- as function of voltage from which net fluxes and currents were calculated (9). It is seen from Fig. 2 that the

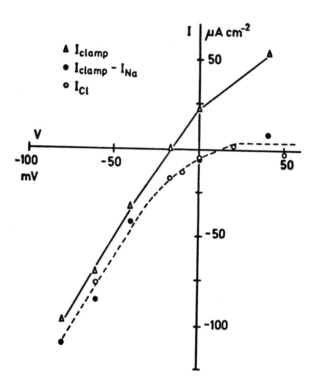

FIG. 2. *Steady-state current-voltage relation for the iso-*
lated toad skin bathed in NaCl Ringer's on both sides. Ionic
currents are calculated from simultaneous measurements of influxes
and outfluxes of Na$^+$ and Cl$^-$ in paired skins, $I_j = z_j F(J_j^{in} - J_j^{out})$,
j = Na or Cl. Instead of plotting I_{Na} directly, $I_{clamp} - I_{Na}$
is plotted in order to visualize the identity, $I_{clamp} = I_{Na} + I_{Cl}$.
Data from ref. 9.

steady-state I-V curve shows outward-going rectification, and that
the contributions of the Na$^+$ and Cl$^-$ components vary with voltage.
In the hyperpolarized skin the clamping current consists of a
large outward going Cl$^-$ component, (*i.e.*, a net inward Cl$^-$ flux).
at reversed transepithelial potentials Na$^+$ carries the clamping

current.

This agrees with the conclusion derived from the above flux ratio analysis, viz. that an inward driving force maintains a dissipative Cl^- transport and that the transport is determined by an electrically silent exchange pathway at reversed potential difference. The nonzero I-axis intercept of the I_{Cl}-V curve is statistically significant (9) and shows that the active Cl^- transport is rheogenic, as is also the case for the active Cl^- transport across the skin of *Leptodactylus ocellatus* first described by Zadunaisky and his group (23,24). Notably, the I_{Cl}-V curve also shows that the steady-state transepithelial Cl^- conductance is a function of the transepithelial voltage. Our analysis of this anion current rectification is reviewed in the remaining part of this paper.

B.   The Chloride Current Response to a Stepwise Voltage Clamp

The transition of the chloride pathway from the low to the high conducting state can be studied by recording the current response to a hyperpolarizing stepwise change in transepithelial potential (25). Examples are shown in Fig. 3. With NaCl Ringer's outside, the instantaneous current response is followed by a slow current response attaining a new steady level in the course of minutes. The time course of the slow response is not a simple exponential but shows an initial delay. In some skins the initial part of the slow current response shows a transitory decrease before activation sets in (Figs. 4, 8, and 9). Since, under these conditions, the transepithelial potential is held constant, the changes in clamping current reflect changes in a transepithelial conductance. The specificity of this conductance is well illustrated by the other current traces of Fig. 3, showing that the conductance activation does not require the presence of Na^+ in the outer solution. When external Cl^- is replaced by gluconate ions,

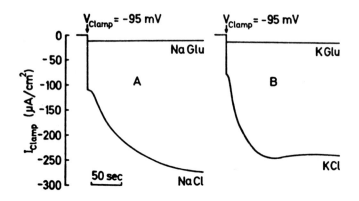

FIG. 3. Clamping current response to a hyperpolarizing
transepithelial potential change. At the time indicated by an
arrow the potential was stepped from its open circuit value to
-95 mV (outside negative). Inner bathing solution is NaCl
Ringer's. Composition of the Ringer's in the outer chamber is
indicated at the records. Data from ref. 25.

the instantaneous current response is reduced one order of magnitude and the time-dependent component disappears, indicating its
high Cl$^-$ specificity. This conclusion was verified by proving the
identity between clamping current and Cl$^-$ influx by means of Cl-36
(25).

Since the outer membrane of the transporting cells is prac-
tically impermeable to K$^+$ (26), substitution of Na$^+$ by K$^+$ in the
outer solution reduces the transcellular cation current to zero.
Thus the Cl$^-$ conductance now rules the clamping current even at
reversed transepithelial potentials. Furthermore, improvement of
the voltage clamp method has been achieved by introducing a two-
level voltage-clamp program. This allows preequilibration of the
skin at any desired holding potential and subsequent fast switch
to the clamping potential at which the kinetics of the conductance

activation are studied. Results of a typical experiment are shown in Fig. 4. The voltage-clamp program given on top brings the skin from a holding potential of +40 mV (outside positive) to a clamping potential of -70 mV, and after 200 sec back again to +40 mV. Note the small outward going on-response of the clamping current to the large voltage change of 110 mV. The current, in turn builds up slowly to a maintained level of -100 $\mu A/cm^2$. By stepping the voltage back to +40 mV, the instantaneous *inward* going current becomes large and is followed by a slow current decrease towards the initial steady-state value. The time constants of

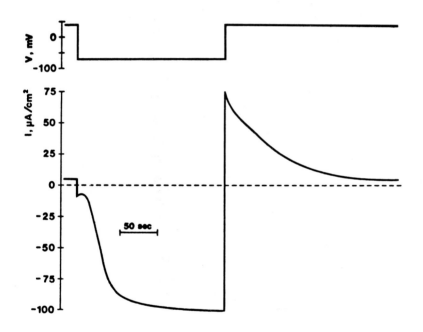

FIG. 4. *Time course of transepithelial current in response to a stepwise change in transepithelial potential from +40 mV to -70 mV. Note the small on response of the current and the small transitory decrease in current preceding the current activation. Note also the large inward going off response of the current as the potential is stepped back to +40 mV. NaCl Ringer's inside, KCl Ringer's outside. (Unpublished).*

current activation and decrease are of the same order of magnitude.
These results offer two additional, important properties of the
Cl^- conductive pathway. (i) The conductance activation is a fully
reversible process. (ii) The conductance in the activated state
does not rectify. The Cl^- current passes the external border of
the skin in outward as well as inward directions depending on the
sign of the transepithelial potential as expected for a simple
ohmic conductor. This similarity is even more striking if we
notice that the "fully activated" current-voltage relation is
linear with the same slope for outward and inward currents (25).
This is to be expected if transport of Cl^- across the outer border
of the skin is predominantly passive.

C. Localization of the Voltage-Dependent Cl^- Conductance

In the skin of room temperature adapted frogs passive Cl^-
transport takes place along a transcellular route (10,27,40). Also
for toad skin there is evidence that the Cl^- current passes through
and not between the outermost living cells of the epithelium: Firs
the Na^+ and urea permeabilities of the paracellular shunt are independent of voltage in the domain where the transepithelial Cl^- conductance shows strong voltage dependence (9). Second, the Cl^- current is depressed by phloretin, furesemide, diamox, and SCN^-, all
known to inhibit Cl^- transport systems in surface membranes of othe
cell types (11). Third, the fully activated chloride conductance i
strongly dependent on cellular energy metabolism. Cyanide (2 mM)
treatment or oxygen depletion of the bathing solutions leads to a
fast, and reversible inhibition of the chloride current and increas
in the electrical resistance of the skin (28). A contribution of
the deeply located gland cells to the Cl^- conductance is ruled out
by the finding that the Cl^- current-decrease following sudden removal of Cl^- in the outer solution proceeds with a time constant simi
lar to the Na^+ current response to removal of outside Na^+. The so-

dium selective membrane and the chloride selective membrane, therefore, seem to be located at the same distance from the outer bulk
solution. However, a thorough quantitative analysis could not be
carried through owing to the presence of an appreciable diffusion
barrier on top of the skin (25). The simple linear relation
between the transcellular Cl^- current and the outside Cl^- concentration (see below) also indicates that the rate-limiting Cl^-
conductance is located in a membrane facing the outer solution.
Finally, the Cl^- channels must be separated from the Na^+ channels
since blocking the Na^+ transport by amiloride does not affect the
Cl^- current (25).

Although the above arguments obtained from quite different
experimental approaches all point to the same anatomical localization of the voltage-dependent Cl^- channels, testing of the above
hypothesis by microelectrode studies is desirable. Such a study
may also reveal whether the Cl^- current passes through the mitochondria rich cells or through the granulosa cells, a question
which could not be settled by the experiments discussed above.

D. Time- and Voltage-Dependence of the Cl^- Conductance

The time it takes the Cl^- conductance to reach steady-state
after a hyperpolarizing voltage step depends on the voltage at
which the activation takes place. A family of conductance-time
curves is presented in Fig. 5. From a holding value of +40 mV,
the transepithelial potential was stepped at zero time to values
indicated on each curve. Note that there is a brief delay before
the conductance activation starts, and that the rate of activation
depends on the clamping potential. This dependence may conveniently be illustrated by calculating the time ($T_{\frac{1}{2}}$) it requires
to reach half maximal conductance change. The $T_{\frac{1}{2}}$ -V relation
(Fig. 5, upper diagram) is bell-shaped with a maximum near -20 mV
which is not far from the spontaneous skin potential with NaCl

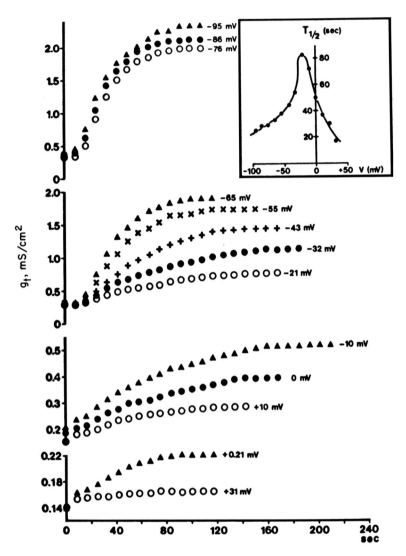

FIG. 5. *Rise in transepithelial conductance following a stepwise change in potential from* V_{hold} = +40 *mV to the clamping potentials indicated at the curves. Inserted diagram depicts the relation between clamping potential and* $T_{\frac{1}{2}}$ *-values of conductance activation. See text for definition of* $T_{\frac{1}{2}}$. *NaCl Ringer's inside, KCl Ringer's outside. Data from ref. 38.*

Ringer's outside (cf. Fig. 2).

A typical example of the steady-state transepithelial conductance (g_t^∞) as function of clamping potential is given in Fig. 6. The relation is S-shaped with its steepest part in the region where the $T_{\frac{1}{2}}$ -V curve has its maximum. In our experiments, V_{clamp} is seldom brought to values below -100 mV because g_t^∞ has usually reached the plateau above this potential. It is noteworthy that even for -100 mV > V > 150 mV, g_t^∞ does not change much. With K-gluconate on the outside, the transepithelial conductance is voltage independent. Thus, the S-shaped g_t^∞-V relation is a characteristic feature of a pathway with high chloride ion specificity, i.e., the transcellular Cl^- conductive pathway.

IV. THE MECHANISM OF CHLORIDE CURRENT RECTIFICATION

A. Requirements of a Model

Any model of the transcellular chloride pathway has to account for the observed two types of Cl^- current response to a stepwise change in the transepithelial potential. (i) The instantaneous, simple ohmic response observed immediately upon potential change, and (ii) the subsequent slow time-dependent response where the Cl^- current increases following a hyperpolarizing potential step, and decreases following potential steps of opposite sign. The slow response which is caused by conductance changes leads to a strong outward-going rectification of the steady-state transcellular Cl^- current. Also this feature as well as the sigmoidal g_t^∞ -V curve have to be explained by the model.

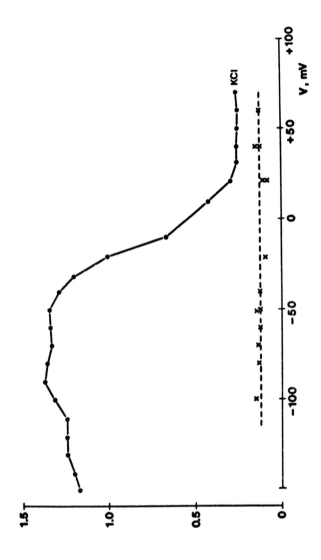

FIG. 6. Steady-state skin conductance (g_t^∞) as function of transepithelial potential (V). NaCl Ringer's inside. The two g_t^∞-V relations shown are from the same skin exposed on the outside to KCl and K-gluconate Ringer's, respectively. The g-V relations of toad skin are discussed in ref. 25.

B. *A Goldman-Type Rectification Is Insufficient*

Our first attempt at explaining rectification of the transcellular Cl^- current was to assume homogeneous, constant field
pathways in the two opposite membranes of the NaCl transporting
cells. Thus, we assumed constant permeabilities, and tentatively
ascribed the voltage-dependence of the Cl^- conductance to variations in mean ionic concentration within the surface membrane. In
order to examine this possibility it was necessary to estimate
variations of the intracellular Cl^- concentration, since it is the
voltage-dependence of this variable which is assumed to determine
the degree of steady-state outward-going rectification. This
required computer analysis of the two-membrane model. Mathematical
details of the model are given in (25). The most important results
are summarized in Fig. 7.

With proper choice of permeability constants, the model cell
displays nonlinear I-V relations which are like the observed
steady-state I-V relations (compare Fig. 7A with Fig. 2). In this
example, the intracellular steady-state Cl^- concentration increases
from 3.2 mM to 12.8 mM when stepping the potential from +40 mV to
-90 mV. If it is assumed that only the outermost cells contribute
to the transporting compartment, the predicted increase in $[Cl]_c$
requires less than 30 sec if all of the chloride ions entering
across the outer membrane during the nonsteady-state period are
trapped by the cells. However, other predictions do not stand
when compared to experimental results, and are listed below.

(a) The large outward Cl^- currents at -90 mV presuppose such
a large chloride permeability of the outer membrane (P_{Cl}^o) that
also the Cl^- currents at reversed potentials will contribute
significantly to the total clamping current. That this is not the
case in the skin has been established sufficiently well by isotope
flux measurements. To assume here a rectification of the chloride
channels of the outer membrane in excess of a Goldman-rectification

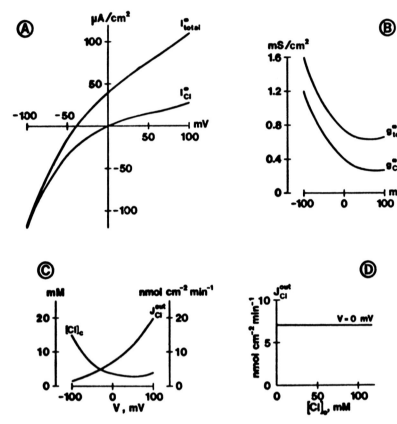

FIG. 7. Computer analysis of the two-membrane model with
constant permeability coefficients. (A) Steady-state transepithelial current (I_{total}^∞) and steady-state transcellular Cl⁻ current
(I_{Cl}^∞) as function of transepithelial potential (V). (B) Steady-
state conductance-voltage relations. g_{total}^∞: total transepithelial
conductance. g_{Cl}^∞: transcellular Cl⁻ conductance. (C) Voltage
dependence of steady-state intracellular Cl⁻ concentration ($[Cl]_c$)
and steady-state transepithelial Cl⁻ outflux (J_{Cl}^{out}). (D) Relation
between steady-state Cl⁻ outflux and outside Cl⁻ concentration
($[Cl]_o$) at short circuit conditions. Values of the independent
variables used for these computations are given in (25) with exception of the inner membrane pump constants, which in the example
shown here are $J_{Na}^{pump,max}$ = 1 nmol/cm² sec, $K_{Na,\frac{1}{2}}^{pump}$ = 10 mM, Na/K
coupling ratio of the pump = 3/2 as given in ref. 29 and 39.

is not justified by the observed linear instantaneous I-V curve in skins exposed to KCl Ringer's outside (25).

(b) The steady-state g_t-V and g_{Cl}-V relations of the model cell are not sigmoidal as they are in the skin (compare Figs. 7B and 6).

(c) The unidirectional fluxes in the model cell at V = 0 mV are very close to the Cl^- fluxes across the short-circuited skin. However, the outflux of Cl^- across the skin is independent of V between 0 mV and +50 mV (9). Clearly, this is not the case for the model cell (Fig. 7C).

(d) In the model cell, the Cl^- outflux is independent of $[Cl]_o$ at V = 0 mV (Fig. 7D). In the short-circuited skin, the Cl^- outflux is reduced to about 1/nmol (cm^2sec) following a substitution of outside Cl^- by a nonpenetrating anion (9,11).

The findings mentioned in (c) and (d) together with the results of the flux ratio analysis imply that a major component of the Cl^- outflux at V > 0 mV is due to exchange diffusion. The remaining Cl^- outflux presupposes a small P_{Cl}^o, and in this case the steady-state outward-going rectification disappears [see for example Fig. 4 in (29)].

The results of the above analyses suggested to us that a chloride permeability of the transcellular pathway increases slowly following hyperpolarizing voltage clamp, and decreases following clamping at reversed potentials. In the sections below additional experimental evidence for this hypothesis is discussed.

C. The Conductance Activation at Different Cl_o-Concentrations

It is implicit in the constant-permeability model discussed above that the conductance activations shown in Fig. 5 are caused by a slow filling up of Cl^- into a compartment just beneath the outer Cl^- selective membrane. Accordingly, the rate of conductance

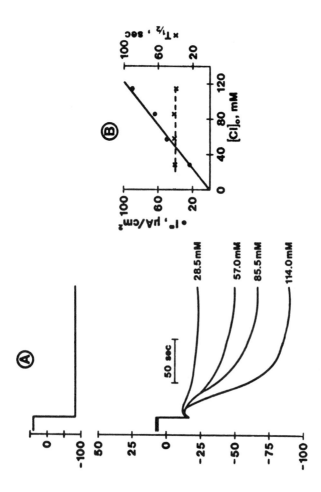

FIG. 8. (A) Transepithelial current response to change in potential from +40 mV to −80 mV. NaCl Ringer's inside, K Ringer's outside with the Cl^- concentrations indicated at the records (Cl^-/gluconate substitutions). Note the almost identical time course of currents. (B) Steady-state clamping currents and $T_{1/2}$ -values of current activation as function of outside Cl^- concentration. Note constancy of $T_{1/2}$ despite large variation in clamping current. Data from ref. 38.

increase is expected to depend on the Cl^- current through the outer membrane. As shown in Fig. 8 this is not the case. The steady-state clamping current varies linearly with $[Cl]_o$, whereas the time constant of conductance activation, expressed by its $T_{\frac{1}{2}}$-value, is independent of the Cl^- current flowing during the nonsteady-state period. The common explanation of this kind of result is that a voltage-controlled gating reaction governs the conductance time course, and that this membrane process operates independently whether or not current is flowing through the conductive paths.

D. Effects of Varying Temperature

Another method of separating experimentally a gating reaction from a transfer reaction is to investigate their temperature dependences. This powerful method was used by Hodgkin and Huxley in their analysis of clamping currents in squid axon (30) and has been used since by others (31,32).

As shown by the current records in Fig. 9, a decrease in ambient temperature from 22.5^o to 12.0^oC (i) reduces the steady-state clamping current at V = -60 mV, and (ii) increases the time it takes the conductance to attain its new steady value following a change of V from 0 mV to -60 mV. Table I summarizes the results hitherto obtained from this type of experiment. The temperature coefficient, Q_{10} of the steady-state conductance is always smaller than Q_{10} of the rate of conductance increase (indicated by the reciprocal $T_{\frac{1}{2}}$ values[1]). The Q_{10} = 1.14 of the steady-state conductance indicates a low activation energy of Cl^- translocation through the conducting paths ($E_A \simeq 10$ kJ/mol), and is in agreement with electrodiffusion. The significantly larger Q_{10} of the conductance

[1]*Since the activation time course is recorded at a potential where the Cl^- conductance becomes fully activated the reciprocal of the time constant is a good measure of the rate coefficient of conductance activation (conf. Eq. (8)).*

activation rate demonstrates an additional voltage-dependent process controlling the number of conducting paths in the membrane.

These results, therefore, give substantial support to our
hypothesis that a membrane-localized gating reaction, which is

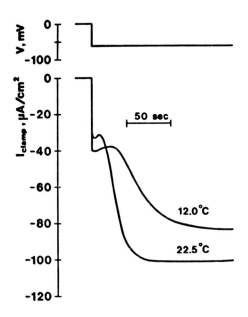

FIG. 9.   Temperature dependence of clamping current activation.   Temperature of the outer bath indicated at the records,
NaCl Ringer's inside, KCl Ringer's outside.   The transepithelial
potential (V) is given on top.   At both temperatures the Cl^- conductance becomes fully activated at the chosen clamping potential (-60
mV).   Data from ref. 38.

Table I. Temperature-Dependence of Steady-State Clamping current (I^∞_{clamp}) and Rate of Change in Clamping Current ($1/T_{1/2}$) Following a Stepwise Change in Transepithelial Potential (data from ref. 39)

V_{hold} (mV)	V_{clamp} (mV)	Temp. (°C)	I^∞_{clamp} (μA/cm²)	$T_{1/2}$ (sec)	Temp (°C)	I^∞_{clamp} (μA/cm²)	$T_{1/2}$ (sec)	$Q^{I^\infty}_{10}$	$Q_{10}(1/T_{1/2})$
+1	-60	23.0	84	84	13.7	90	132	0.93	1.63
+1	-80	23.0	154	54	11.4	110	108	1.34	1.82
+2	-80	23.0	138	57	11.6	108	114	1.24	1.84
+2	-60	22.8	85	66	12.1	82	92	1.03	1.36
+2	-60	22.5	129	48	12.5	114	91	1.13	1.90
0	-60	22.5	102	26	12.0	83	55	1.22	2.04
0	-80	22.5	128	21	12.1	123	40	1.04	1.86
0	-80	22.5	143	24	12.1	118	41	1.20	1.67
							Mean	1.14[a]	1.77[a]
							±SEM	0.05	0.07

[a] significantly different (P<0.05%).

separated from the Cl⁻ transfer reaction, controls the time-
dependent behavior of the Cl⁻ conductor.

E. *Kinetic and Steady-State Properties in Common with*
    Membrane Currents in Excitable Cells

Application of a gate type model to the chloride channels in
toad skin is based upon the above experimental verification of two
types of voltage controlled membrane reactions, viz., a translocation process and a permeability activation. The hypothesis, however, is also supported by the close resemblance to the activation
of membrane currents in nerve and muscle cells.

(1) The activation reaction is initiated by a voltage change.

(2) The mechanism is polarized, *i.e.*, only potential dis-
placements of one sign bring about activation.

(3) The system returns to its initial state when the voltage
clamp is released.

(4) The time constant of activation is a bell-shaped function
of clamping potential.

(5) The steady-state conductance is an S-shaped function of
potential, and the system is brought from a nonconductive state
to full activation by a transmembrane voltage change of less than
100 mV.

(6) In the conductive state the channels allow current to
pass in both directions.

IV. THE TRANSITION PROCESS

In gate-type models a membrane reaction of the type is
imagined,

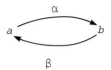

in which a corresponds to a conductive state, and b to a nonconductive state; α and β are voltage-dependent rate coefficients of the transitions between these functionally different states. We have evidence that such different states exist, but it should be emphasized that we have no reason to believe that the Cl^- channels can be found in only two configurations. More than one configuration of a and b states may exist at a given moment. Thus a description of the transitions between a and b states by means of only two rate coefficients may turn out to be insufficient.

The experimental analysis of the kinetics of the conductance activation presupposes that the membrane in which the gating reaction occurs is voltage clamped. This is a prerequisite to maintaining constant values of the rate coefficients during transition from one steady-state to another. A voltage step imposed on the skin is divided between the outward and inward facing membranes of the transporting cells, depending on their relative conductance. Although the fraction which appears across the outer membrane is expected to be large (5,6) it will not stay constant if the transcellular voltage divider ratio is time-variant. On the other hand, by inspection of the conductance activation curves [(25), Fig. 5] it is difficult to ignore the striking resemblance, for example, with the activation course of the potassium current in nerve cells. To illustrate this, let us assume that the activation reaction is characterized by a single gating variable according to the following set of Hodgkin-Huxley equations (30),

$$g = \bar{g} \, c^p \tag{3}$$

$$dc/dt = \alpha(1-c) - \beta c \tag{4}$$

in which $\bar{g}$ is a constant corresponding to the maximum conductance,

c a dimensionless variable assuming values between 0 and 1, and
the rate coefficients α and β are voltage dependent variables (constant temperature assumed). The power, p is a positive constant.
If, after a stepwise change in transepithelial potential, the outer
membrane potential stays constant at its new value, it follows that
g becomes a function of time only. The n, solution of Eqs. (3) and
(4) leads to

$$g = \{g_\infty^{\frac{1}{p}} - (g_\infty^{\frac{1}{p}} - g_o^{\frac{1}{p}})\ \exp\ (-t/\tau)\}^p \tag{5}$$

where $g = g_o$ for $t = 0$, g_∞ is the steady-state conductance at the
clamped voltage, and $\tau = 1/(\alpha + \beta)$.

Examples of fitting *equation 5* to observed conductance time
courses are shown in Fig. 10. It is evident that a single time
constant is sufficient to describe each of these conductance
activations. Probably, this means that the outer membrane resistance, even after full activation of its chloride permeability, is
much larger than the resistance of the inner membrane.

If, for the purpose of illustration, we consider the simple
case of a single energy barrier, the voltage dependence of the
rate coefficients are given by (33),

$$\alpha = \alpha_o\ \exp\ [\gamma z (V - V_{\frac{1}{2}})/(RT)] \tag{6}$$

$$\beta = \beta_o\ \exp\ [-z(1-\gamma)(V-V_{\frac{1}{2}})/(RT)] \tag{7}$$

where $V_{\frac{1}{2}}$ is the membrane voltage for $c = 0.5$, and α_o and β_o are
the rate coefficients for $V = V_{\frac{1}{2}}$ ($\alpha_o = \beta_o$). z is the charge of
the mobile membrane particles responding to a change in V, and γ
is a measure of the symmetry of the system. Thus we obtain,

$$\tau = (\alpha_o\{\exp[\gamma z(V-V_{\frac{1}{2}})/(RT)] + \exp[-z(1-\gamma)(V-V_{\frac{1}{2}})/(RT)]\})^{-1} \tag{8}$$

It is seen that $\tau \to 0$ for $V \to \pm\infty$, and that τ has its maximum
value ($\tau_{max} = 1/(2\alpha_o)$) for $V = V_{\frac{1}{2}}$. In other words, τ is a bell-
shaped function of V. The position of the $T_{\frac{1}{2}} - V$ curve relative

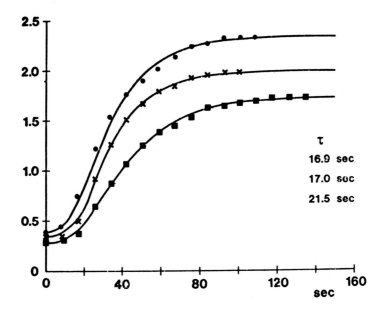

FIG. 10. *Rise in transepithelial conductance following a stepwise change in potential V_{hold} = +40 mV to the clamping potentials indicated. The symbols represent the experiment conductance values. The curves are drawn according to Eq. (5) in the text with the following parameters,* •: $p = 4$, $g_{Cl,o} = 0$ mS/cm^2, $g_{Cl,\infty} = 1.92$ mS/cm^2, $g_{leak} = 0.38$ mS/cm^2, $\tau = 16.9$ sec. x: $p = 4$, $g_{Cl,o} = 0$ mS/cm^2, $g_{Cl,\infty} = 1.65$ mS/cm^2, $g_{leak} = 0.33$ mS/cm^2, $\tau = 17.0$ sec. ■: $p = 4$, $g_{Cl,o} = 0$ mS/cm^2, $g_{Cl,\infty} = 1.44$ mS/cm^2, $g_{leak} = 0.28$ mS/cm^2, $\tau = 21.5$ sec (Unpublished).

to the g_{Cl}^{∞}-V curve as well as the shape of the $T_{\frac{1}{2}}$-V relation (compare Figs. 5 and 6) are consistent with the above predictions.

These results are encouraging for our way of thinking. It is not possible, however, to carry through the quantitative analysis of the Cl$^-$ current kinetics unless the outer membrane potential and the intracellular Cl$^-$ activity is known and clamped at preselected values during the activation. Also, it seems appropriate here to emphasize that in some skins there is a transitory decrease in current before the activation starts

(*e.g.* Fig. 9). This observation is not consistent with the kinetics of the simple gating reaction defined by the above set of
equations. The potassium current in frog node has a similar time
course when activated from a hyperpolarizing conditioning potential
(34). The cation conductance of alamethicin or monazomycin treated
black lipid membranes also shows an initial transient phase of a
sign opposite to that expected from the steady-state conductance-
voltage curve (35).

VI. CONCLUSION

Investigation of the potential dependence of unidirectional
fluxes reveals that transport of Cl^- ions across toad skin epithelium is carried by pathways of different energy requirements.
For $V_{clamp} > 0$ mV (*i.e.*, at potentials with a sign opposite to the
spontaneous potential) an electrically silent exchange diffusion
pathway and an active pathway of minor importance govern the
transport. For $V_{clamp} < 0$ mV, the Cl^- ions are carried predominantly by a simple passive pathway as revealed by the following
observations: (1) The unidirectional fluxes obey the flux-ratio
equation (Fig. 1). (2) The translocation process has a low energy
of activation (Table 1). (3) Cl^- is allowed to pass in both
directions depending on the polarity of the driving force (Fig. 4).

The passive conductive Cl^- pathway becomes fully activated by
bringing the transepithelial potential to values below -50 mV
(Fig. 6). The time constant of this activation is large, *i.e.*, in
order of 10 sec.

Recently, Finn and Rogenes concluded that the conductance
increase following hyperpolarization of toad skin and bladder
(*B. marinus*) was due to a slow activation of a paracellular Na^+
permeability (36). An increase in Na^+-efflux due to hyperpolarization is also seen in the skin of *B. bufo*, but this can easily

be accounted for by the change of the driving force *per se*, and need not be explained by a Na^+ permeability activation (9). Furthermore, unidirectional flux measurements as well as ion-substitution experiments both showed that the conductance activation in this tissue is localized to a pathway specific to Cl^- ions. (Figs. 2,3, and 6, and ref. 25).

The steady-state and time dependent properties of this passive pathway are qualitatively like those of the K^+ conductance in excitable tissues. Evidence is discussed above and in two recent communications from our laboratory (37, 38) that Cl^- permeability of an outward facing membrane depends on membrane potential in a way consistent with a potential controlled Hodgkin-Huxley gate-type mechanism.

Thus the potential has two effects on Cl^- transport across this tissue. One is to control the permeability of a passive pathway, the other is to drive Cl^- ions through this pathway.

This hypothesis accounts for steady-state as well as for the most important nonsteady-state properties of transepithelial ionic currents in voltage clamp experiments.

Since the spontaneous skin potential depends on the active Na^+ transport, our hypothesis implies that the Cl^- permeability depends indirectly on active Na^+ transport, i.e., it increases with the inward active Na^+ flux, and decreases if the active Na^+ flux is abolished.

VII. SUMMARY

Flux-ratio analysis discloses three modes of Cl^- transport across the skin of the toad *Bufo bufo* (L.): (i) active rheogenic transport in inward direction; (ii) simple passive transport; (iii) electrically silent exchange diffusion. Studies of transport via the simple passive pathway, which is an almost perfect

Cl⁻ current rectifier, is reviewed in this paper. A steady-state net flux of Cl^- is allowed to pass the skin when the driving force is inwardly directed, whereas the steady-state net flux of Cl^- is close to zero along an outwardly directed driving force. Evidence that the rectifying mechanism is situated in a membrane facing the outside bathing solution is discussed. Computer analysis of the Koefoed-Johnsen-Ussing two-membrane model indicates that the observed strong rectification cannot be obtained unless the permeability of Cl^- is a voltage-dependent variable. Therefore a Hodgkin-Huxley gate-type model is suggested and experimentally tested. The model assumes that entry of Cl^- into the epithelium is a voltage dependent process, and that an additional voltage dependent process controls the number of conductive paths in the membrane. It is shown that these two processes can be separated experimentally. Further evidence for the gate type model comes from the close resemblance between kinetics and steady-state properties of the Cl^- current in toad skin, and K^+ and Na^+ currents in nerve. A thorough description of the Cl^- channel kinetics presupposes one-membrane voltage clamp which has not yet been attempted.

REFERENCES

1. H. H. Ussing. The distinction by means of tracers between active transport and diffusion. *Acta Physiol. Scand.* *19*: 43-56, 1949.
2. H. H. Ussing. Interpretation of tracer fluxes. In "Membrane Transport in Biology," (G. Giebisch, D. C. Tosteson and H. H. Ussing, eds.), 116-140. Springer Verlag, Berlin, Heidelberg, New York, 1978.
3. B. Lindemann. Impalement artifacts in microelectrode recordings of epithelial membrane potentials. *Biophys. J. 15:* 1161-1164, 1975.

4. D. J. Nelson, J. Ehrenfeld, and B. Lindemann. Volume changes
    and potential artifacts of epithelial cells of frog skin
    following impalement with microelectrodes filled with
    3 M KCl. *J. Membr. Biol. 40S:* 91-119, 1978.

5. W. Nagel. The Dependence of the electrical potentials across
    the membranes of the frog skin upon the concentration of
    sodium in the mucosal solution. *J. Physiol. (London)*
    269: 777-796, 1977.

6. S. I. Helman and R. S. Fisher. Microelectrode studies of the
    active Na transport pathway of frog skin. *J. Gen.*
    Physiol. 69: 571-604, 1977.

7. S. I. Helman, W. Nagel, and R.S. Fisher. Ouabain on active
    transepithelial sodium transport in frog skin. Studies
    with microelectrodes. *J. Gen. Physiol. 74:* 105-127, 1979.

8. L. J. Mandel and P. F. Curran. Response of the frog skin to
    steady-state voltage clamping. I. The shunt pathway. *J.*
    Gen. Physiol. 59: 503-518, 1972.

9. K. Bruus, P. Kristensen, and E. Hviid Larsen. Pathways for
    chloride and sodium transport across toad skin. *Acta*
    Physiol. Scand. 97: 31-47, 1976.

10. P. Kristensen. Effect of amiloride on chloride transport
    across amphibian epithelia. *J. Membr. Biol. 40S:* 167-185,
    1978.

11. P. Kristensen and E. Hviid Larsen. Relation between chloride
    exchange and a conductive chloride pathway across the
    isolated skin of the toad *(Bufo bufo)*. *Acta Physiol.*
    Scand. 102: 22-34, 1978.

12. H. H. Ussing. Interpretation of the exchange of radiosodium
    in isolated muscle. *Nature (London) 160:* 262-263, 1947.

13. C. A. M. Hogben. The chloride transport system of the gastric
    mucosa. *Proc. Nat. Acad. Sci. USA 37:* 393-395, 1951.

14. M. J. Hunter. A quantitative estimate of the nonexchange
    restricted chloride permeability of red cell. *J. Physiol.*
    (London) 218: 49P, 1971.

15. U. V. Lassen. Membrane potential and membrane resistance of
    red cells. In "Oxygen Affinity of Hemoglobin and Red
    Cell Acid Base Status Alfred Benzon Symposium IV",
    Munksgaard, Copenhagen 291-304. (M. Rorth and P. Astrup,
    eds.) 1972.

16. U. V. Lassen, L. Pape, and B. Vestergaard-Bogind. Chloride
    conductance of the Amphiuma red cell membrane. *J. Membr.
    Biol. 39:* 27-48, 1978.

17. L. C. Stoner, B. Burg, and J. Orloff. Ion Transport in
    cortical collecting tubule; effect of amiloride. *Am. J.
    Physiol. 227:* 453-459, 1974.

18. P. Kristensen. Chloride transport across the isolated frog
    skin. *Acta Physiol. Scand. 84:* 338-346, 1972.

19. P. P. Idzerda and J. F. G. Slegers. Deviations from the flux
    ratio equation for chloride ions in ouabain and acetazo-
    lamide-treated frog skin. *Biochem. Biophys. Acta 419:*
    530-539, 1976.

20. T. U. L. Biber, T. C. Walker, and T. L. Mullen. Influence of
    extracellular Cl concentration on Cl transport across
    isolated skin of *Rana pipiens. J. Membr. Biol. 56:* 81-92,
    1980.

21. F. Aull. The effect of external anions on steady-state chlo-
    ride exchange across ascites tumor cells. *J. Physiol.
    (London) 221:* 755-771, 1972.

22. E. K. Hoffmann, L. O. Simonsen, and C. Sjøholm. Membrane
    potential, chloride exchange, and chloride conductance
    in Ehrlich ascites tumor cells. *J. Physiol. (London) 296:*
    61-84, 1979.

23. J. A. Zadunaisky and O. A. Candia. Active transport of sodium
    and chloride by the isolated skin of the South American
    frog *Leptodactylus ocelatus. Nature (London) 195:* 1004,
    1962.

24. J. A. Zadunaisky, O. A. Candia, and D. J. Chiarandini. The
    origin of the short-circuit current in the isolated skin

of the South American frog *Leptodactylus ocellatus*. *J. Gen. Physiol. 47:* 393-402, 1963

25. E. Hviid Larsen and P. Kristensen. Properties of a conductive cellular chloride pathway in the skin of the toad (*Bufo bufo*). *Acta Physiol. Scand. 102:* 1-21, 1978.

26. V. Koefoed-Johnsen and H. H. Ussing. The nature of the frog skin potential. *Acta Physiol. Scand. 42:* 298-308, 1958.

27. O. A. Candia. Reduction of chloride fluxes by amiloride across the short circuited frog skin. *Am. J. Physiol. 234:* F437-F445, 1978.

28. E. Hviid Larsen and P. Kristensen. Effects of anoxia and of theophylline on transcellular Cl^--transport in toad skin. *Proc. Int. Union. Physiol. Sci. XII:* 1002, 1977.

29. E. Hviid Larsen. Computed steady-state ion concentration and volume of epithelial cells. Dependence on transcellular Na^+ transport. *In* "Osmotic and Volume Regulation," Alfred Benzon Symposium XI. Munksgaard, Copenhagen, 438-456. (C. Barker Jørgensen and E. Skadhauge, eds.), 1978.

30. A. L. Hodgkin and A. F. Huxley. A quantative description of membrane current and its application to conduction and excitation in nerve. *J. Physiol. (London) 117:* 500-544, 1952.

31. B. Frankenhaeuser and L. E. Moore. The effect of temperature on sodium and potassium permeability changes in myelinated nerve fibres of *Xenopus laevis*. *J. Physiol. (London) 169:* 431-437, 1963.

32. J. W. Moore. Temperature and drug effects on squid axon membrane ion conductances. Fed. Proc. *17:* 113, 1958.

33. J. J. B. Jack, D. Noble, and R. W. Tsien. *Electric Current Flow in Excitable Cells*. Clarendon Press, Oxford, 1975.

34. Y. Palti, G. Ganot, and R. Stämpfli. Effect of conditioning potential on potassium current kinetics in the frog node. *Biophys. J. 16:* 261-273, 1976.

35. P. Mueller. Membrane excitation through voltage-induced aggregation of channel precursors. *Ann. N.Y. Acad. Sci. 264:* 247-264, 1975.

36. A. L. Finn and P. Rogenes. The effects of voltage clamping on ion transport pathways in tight epithelia. *Current Topics in Membranes and Transport. 13:* 245-255, 1980.

37. E. Hviid Larsen, B. E. Rasmussen, and N. Willumsen. Computer model of transporting epithelial cells. Analysis of current-voltage and current-time curves. *Adv. Physiol. Sci. Vol. 3. Physiology of Non-excitable Cells,* 115-127. (J. Salanki ed.), Pergamon Press, 1981.

38. E. Hviid Larsen and B. E. Rasmussen. Chloride permeability of outward facing membrane in toad skin is gated by membrane potential. VII. International Biophysics Congress, 348. Mexico City, 1981.

39. E. Hviid Larsen, W. Fuchs, and B. Lindemann. Dependence of Na-pump flux on intracellular Na-activity in frog skin epithelium (R. esculenta). Eur. J. Physiol. *S382:* R13, 1979.

40. V. Koefoed-Johnsen and H. H. Ussing. Transport pathways in frog skin and their modification by copper ions. *In* "Secretory Mechanisms of Exocrine Glands," Alfred Benzon Symposium VI. Munksgaard, Copenhagen, 411-422. (N. A. Thorn and O. H. Petersen, eds.), 1974.

REGULATION OF ACTIVE CHLORIDE TRANSPORT IN FROG SKIN

Charles O. Watlington

Department of Medicine
Division of Endocrinology
Medical College of Virginia
Virginia Commonwealth University
Richmond, Virginia

A major thrust of our research endeavor is directed to the
role of chloride transport alteration in the regulation of NaCl
metabolism. Previous studies from this laboratory have confirmed
such a regulatory system in amphibians. The frog skin model has
provided an excellent tool for evaluation of the characteristics
of active and passive chloride transport mechanisms. More recently,
we have discovered a factor in urine of chronically NaCl loaded man
which stimulates active Cl^- transport outward in isolated frog skin.
The apparent "secretory" nature of its effects may be important in
adaptation to states of altered extracellular fluid volume in man
and other animals. A major portion of this presentation will deal
with our work with this chloride transport stimulatory factor. However, we will first review our earlier experiences with active
chloride transport mechanisms and with chloride transport regulation in skin of *Rana pipiens*. The term active Cl^- transport across
the skin is used here to indicate that besides electrochemical

Copyright © 1982 by Academic Press, Inc.
All rights of reproduction in any form reserved.
ISBN 0-12-775280-3

driving forces residing in the external salt solutions, a metabolic force, generated within the skin, drives Cl⁻ in a given direction. Consequently, in the electrically short-circuited skin
exposed on both sides to identical salt solutions, one still
observes a net flux of Cl⁻ in a given direction. By contrast,
passive Cl⁻ transport indicates that the flows of Cl⁻ in both directions are solely a function of the electrochemical gradient
across the skin. Thus, no net flux of Cl⁻ is observed under the
conditions mentioned above. A change in net flux is considered a
change in active transport. A change in Cl⁻ flux in the absence
of active Cl⁻ transport change is considered an alteration in
passive "permeability."

I. ACTIVE Cl⁻ TRANSPORT MECHANISMS

     Three active Cl⁻ transport systems can be demonstrated in
isolated short-circuited skin of *Rana pipiens* (Fig. 1). The first

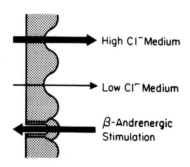

    *FIG. 1. The three active Cl⁻ transport systems demonstrable
in isolated skin of Rana pipiens.*

system is detectable in skins bathed in ordinary Ringers' solution
and consists of net Cl^- influx. The net average influx varies
from 0.2 to 0.8 $\mu Eq/cm^2 hr$ (5-30% of the active Na^+ transport) dependent on the permeability of the skin to Cl^- (1,2). The average
flux ratio J_i^{Cl}/J_o^{Cl} (influx of Cl^-/outflux of Cl^-) is approximately
1.5, although this varies greatly in small series. We have found
J_{net}^{Cl} (net flux or $\left[\left(J_i^{Cl} - J_o^{Cl}\right)\right]$ in the first system to be ouabain-
sensitive, acetazolamide-insensitive, and dependent upon Na^+ on the
cis side and K^+ on the *trans* side (2,3). It is probably the same
mechanism as the ouabain-sensitive and Na^+- dependent active Cl^-
influx originally described in 1963 by Zadunaisky and co-workers
(4) in the skin of *L. ocellatus*. A second system consists of net
Cl^- influx, which is relatively small in magnitude and demonstrable
only in skins bathed in low Cl^- sulfate Ringers. It may be mediated by Cl^-/HCO_3- exchange. It is acetazolamide-sensitive, ouabain-
insensitive, and cation-independent (3). Martin and Curran first
demonstrated this system *in vitro* (5), but it is probably the same
as the *in vivo* mechanism described by August Krogh in 1937 (6). A
third system, net Cl^- outflux, is produced by beta adrenergic
stimulation (7,8), and theophylline, presumably by activation of
mucous glands, although the evidence is largely circumstantial. It
has the same characteristics as the first except that its K^+
dependence is demonstrable only by K^+ removal from both sides (3).
This outflux system was originally described in 1953 by Koefoed-
Johnson *et al.* (9).

On the basis of our work and many others work in frog skin,
and a review of active Cl^- transport in numerous epithelia, we
suggested that at least two distinctly different mechanisms may
exist; one ouabain-sensitive and cation-dependent, the other
acetazolamide-sensitive and cation-independent (3). The models
such as proposed by Epstein *et al.* (10) for the "outward" (serosal
to mucosal) active Cl^- transport system of the shark rectal gland
(and presented elsewhere in this symposium) gives appealing fabric
to the above operational categorization of the ouabain-dependent

systems. Chloride enters the cell by a coupled NaCl mechanism, thus explaining dependence on the presence of Na^+ on the trans side. The entry is powered by a Na^+ gradient established by the pumping of Na^+ out of the cell. This basolateral Na^+K^+ - ATPase mechanism would explain K^+ dependence and ouabain sensitivity. The Cl^- then exits from the cell along established electrochemical gradients. These ouabain-sensitive cation-dependent systems seem to be directed both mucosal to serosal and vice versa, and to be in relative preponderance compared to the acetazolamide-sensitive systems (3). Frizzell, Field, and Schultz have presented an excellent review of these Na-dependent systems and the evidence for NaCl coupling in them (11).

The acetazolamide-sensitive, ouabain-insensitive and cation-independent systems are best exemplified by the mucosal-to-serosal mechanism originally described in turtle bladder by Brodsky and co-workers (12). The mechanism of transport often proposed is Cl^-/HCO_3^- exchange and the direction of Cl^- transport mucosal to serosal. It is important to point out that the proposed exchange mechanism has not been clearly established. Also, acetazolamide's effects are not necessarily mediated by carbonic-anhydrase inhibition. The drugs are generally used at concentrations of 10^{-4} -10^{-5M} to inhibit Cl^- transport, which is much greater than needed for their inhibition of the enzyme system (13).

II.   REGULATION OF Na^+ TRANSPORT BY Cl^- ADAPTATION

Our first investigations in the area of regulation of Cl^- transport mechanisms were directed toward the role of passive Cl^- permeability change in NaCl transport alteration. Unfortunately, we did not then realize that a portion of J_i^{Cl} was active. We assumed J_i^{Cl} and J_o^{Cl} were equal and that changes in J_i^{Cl} as well as J_o^{Cl}, under short-circuit conditions when the passive ions are electrically uncoupled, reflected passive permeability change.

Therefore, we used the average of approximately equal numbers of J_i^{Cl} and J_o^{Cl} as an index of passive permeability and did not distinguish between the two.

Our initial finding was that chronic preconditioning of frogs in high NaCl media not only decreased short-circuit current (which is predominantly active Na^+ transport) but also decreased chloride flux in the short-circuited state (14). We postulated that this "passive permeability" decrease represented a NaCl transport regulatory system separate from the active Na^+ transport mechanism, i.e., the Cl^- permeability decrease would inhibit Na^+ transport in the electrically coupled state by a "drag" effect. To support this point we produced a chloride flux decrease (J_i^{Cl} and/or J_o^{Cl}) without changing short-circuit current. This was done by arginine-Cl preconditioning or by NaCl preconditioning coincident with aldosterone administration (15). Despite the absence of short-circuit current change, net Na^+ influx was inhibited in the open circuit or electrically coupled state, indicating to us that selective Cl^- permeability decrease in these studies did decrease Na^+ transport by a drag effect.

Subsequently, a humoral mediator of this Cl^- flux change in chronic salt preconditioning was sought (16). To reduce the anticipated effects of Na^+ transport regulatory factors, we again used a non-Na^+ salt (arginine-Cl) for preconditioning. In these studies, J_i^{Cl} alone was used as an index of passive permeability. we were able to show that plasma from preconditioned frogs, when exposed to isolated skin of other animals (1) inhibited J_i^{Cl} in the short-circuited state, (2) did not significantly alter short-circuit current, but (3) did inhibit net Na^+ flux in the open circuit or electrically coupled state. We concluded that a humoral factor existed which inhibited Na^+ transport by Cl^- permeability change and postulated that this humoral factor was also operative in states of NaCl preconditioning. However, the J_i^{Cl} decrease could have been in the active component of influx. Therefore, we can only state that Cl^- transport mechanisms can be modulated

independently of a direct active Na^+ transport change and that the
effect on chloride flux in the short-circuited state can produce a
change in Na^+ transport when the ions are electrically coupled.
We cannot state whether the changes in Cl^- flux involve active or
passive mechanisms or both. And, we cannot state whether the inhibitory humoral factor from arginine-Cl-conditioned animals is
operative in NaCl preconditioning. Indeed, we were subsequently
unable to demonstrate a humoral Cl^- flux inhibitory factor in
plasma ultrafiltrates or NaCl preconditioned frogs or in plasma of
chronically NaCl loaded man (unpublished). Possibly, chloride flux
inhibition in NaCl and in arginine-Cl preconditioning occurs
through two different mechanisms. For instance, the arginine-Cl-
induced mechanism could be related to acid-base regulation and the
adaptation of Cl^- flux to NaCl preconditioning be related to volume
regulation. Obviously, this problem required restudy. Whatever
the future outcome of these investigations is, our evidence
indicates that a selective Cl^- transport inhibitory mechanism
exists in amphibians which can retard Na^+ transport inward and a
humoral factor is demonstrable which also has this capability.

III.   CHLORIDE TRANSPORT STIMULATORY FACTOR IN HUMAN URINE

The most promising avenue at present for demonstration of
humoral regulation of Cl^- transport is a factor in urine of chronically NaCl loaded man, which stimulates active Cl^- transport
outward (blood to pond) in isolated frog skin. These studies are
usually performed upon urine collected from normal ambulatory
young adult male and female volunteers on the fifth day of high
salt and the third day of low salt diet. Urinary Na^+ in most
cases is greater than 200 and less than 25 µEq/24 hr, respectively.
Methylene chloride-extracts of the urine are then tested in isolated frog skin placed in Ussing type chambers and studied under
short-circuit conditions. Two types of experiments are performed

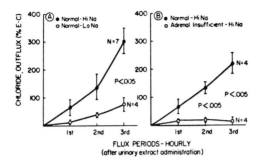

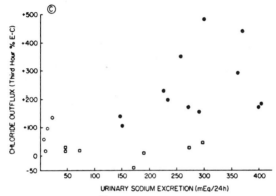

FIG. 2.   *Change in Cl⁻ outflux (%) in isolated frog skin
after administration of methylene chloride extracts of human urine
(3) (A) compares normal subjects on high and low NaCl diets,
(B) compared normal and adrenal-insufficient subjects, matched for
age and sex, on high NaCl diets, (C) summarizes third-hour flux
changes as a function of Na⁺ excretion in all subjects studied;
results from random urines on two normal subjects and urines of
three adrenal-insufficient subjects on NaCl restriction are included.   In skin pairs from the same frog, experimental skin (E)
received extracts of 1/20 of a 24 hr urine in the 8 ml inside
chamber, and control skin (C) received vehicle (0.02 ml ethanol)
alone after a 1 hr flux period.   Measurements were made in three
subsequent 1 hr flux periods.   The change, as percent of first
hour, was calculated for each skin, and the difference in experimental and control skin for each hour is expressed as % E-C.   The
average of five skin pairs was calculated for each subject.   The
mean ± S.E. of these individual values for each subject are shown
in (A) and (B).   N indicates number of subjects studied.   P values
are shown for flux periods with statistically significant differences.*

to demonstrate urinary factor activity. The first method consisted
of simultaneous ^{36}Cl outflux determination in each of two skins
from the same animal. The second method used to demonstrate the
presence of urinary factor activity consists of bidirectional ^{36}Cl
flux measurements. Details of the experimental method are included in the legend of Fig. 2.

The percentage stimulation of Cl$^-$ outflux (%E-C) by the
extracts from the normal subjects on high and low Na$^+$ diets are
shown in Fig. 2A (17). The extracts from subjects on high NaCl
intake HiNa produced a much greater increase in Cl$^-$ outflux than
did extracts from NaCl deprived subjects (LoNa). There was no
overlap in values for the two groups. In contrast, NaCl-loaded
adrenal insufficient subjects (Fig. 2B) showed no demonstrable
activity in the urine. This group included two adrenalectomized
females and two males with spontaneous idiopathic adrenal insuf-
ficiency. The control subjects are sex matched and approximately
age matched. A summary of our experience up to this point is
shown in Fig. 2C and includes a few random urines. Percent of
E-C for the third hour after extract is a positive function of
urinary Na$^+$ excretion in normal subjects, and the factor is
essentially undectable in adrenal insufficient subjects.

The increase in Cl$^-$ influx produced by urinary extracts could
be the result of a change in passive flux or the result of stimulation of an active Cl$^-$ outflux mechanism. Bidirectional flux
measurements are necessary to differentiate between the two possibilities. Fig. 3 presents the results of simultaneous determination of influx and outflux (17). The important finding resides in
the net flux values, the cross hatched areas. In both control and
Na$^+$ deprived series a gradual and modest decline in net influx
occurred. However, in the Na$^+$ loaded series net outflux developed.
This seem to begin in the second hour after extract administration
and was statistically significant by the third hour compared to
both low Na$^+$ and control series. Thus, extracts from sodium
loaded subjects produce active Cl$^-$ transport outward with a time

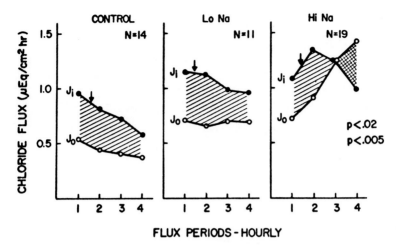

FLUX PERIODS - HOURLY

FIG. 3. Bidirectional Cl⁻ fluxes in isolated frog skin after administration of methylene chloride extracts of human urine (3). See legend of Fig. 2 for more complete experimental details. N indicates number of skin pairs studied. Each skin received the extract of 1/24 of a 24 hr urine. Control skins received diluent alone. P values compare net flux in third hr after extract in Na loaded series (HiNa) to the third hour value in LoNa and control series respectively.

course compatible with a steroid. It is noteworthy that change in short-circuit current appropriate to the Cl⁻ flux change did not occur in the high Na⁺ series. In fact, the short-circuit current was not significantly different within the three series (17). Therefore, we performed a repeat study of the effects of extracts of HiNa and LoNa urine on simultaneous Na⁺ and Cl⁻ fluxes in skin pairs.

In the double-labeling studies no change in Cl⁻ or Na⁺ was produced by LoNa extracts (1). The J^{Cl} changes subsequent to HiNa extract was almost identical to that shown in Fig. 3. The significant finding in the Na⁺ fluxes was an increase in Na outflux in the third hour after extract in the Na⁺ loaded series. Again, short-circuit current was unchanged by the extracts.

Fig. 4 summarizes the changes in net fluxes of Na⁺ and Cl for the three hours after extract, using the appropriate electrical

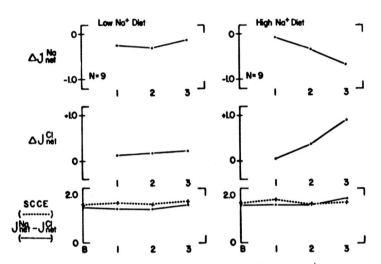

FIG. 4. Bidirectional simultaneous Cl⁻ and Na⁺ fluxes in
isolated frog skin after administration of methylene chloride extracts of human urine (18). Change in net flux (ΔJ_{net}) and short-
circuit current (SCCE) are in units of $\mu Eq \; cm^{-2} \; hr^{-1}$. See legend
of Fig. 2 for more complete experimental details. N indicates
number of skin pairs studied. Each skin received the extract of
1/24 of a 24 hr urine. Urines used were from the same three
subjects in both series.

sign, and compares the sum of net fluxes to the short-circuit current for the entire 4 hours. After extract from salt loaded
subjects, Na⁺ current fell and Cl⁻ current rose. The result of
these opposing effects was no change in short-circuit current and
maintenance of equivalence between the sum of net fluxes and short-
circuit current. This stimulation of both Na⁺ and Cl⁻ transport in
a secretory direction may represent activation of a neutral Na^{Cl}
pump or simultaneous activation of two electrogenic pumps to an
approximately equivalent degree. The site of action is unclear
and may be epidermal or in the mucous glands.

As a first purification step urine was submitted to chromatography on Sephadex LH20 (1). This system (Fig. 5) produced

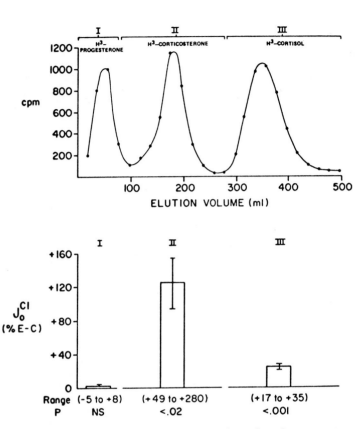

FIG. 5. *Sephadex LH20 chromatography of urinary extracts of NaCl loaded subjects and assessment of chloride outflux by three fractions (%E-C, third hour after extract) (1). Five skin pairs were studied for each five subjects and a mean value calculated for each subject as in the experiments described in Fig. 2. P values were calculated by treating average % E-C for each subject as paired data. Each skin received extract equivalent to 1/20 of a 24 hr urine. The solvent system for chromatography was benzene: methanol: cyclohexane::60:40:10 and a 20 gm, 2.5 cm column was used.*

reasonable separation of the steroids progesterone, corticosterone, and cortisol. Results of chromatography of five urines is presented here. Only Fraction II, having the elution volume of corticosterone, showed a major outflux stimulatory activity. Three of these

urines were rechromatographed and Fraction II tested on bidirectional chloride fluxes to be sure of specificity of effect. Again,
as with crude extract, reversal of chloride flux from in to out
occurred, confirming that the factor has the mobility of corticosterone.

Since these studies, we have repeated Sephadex chromatography
with another organic system which has certain practical benefits
for future purification effects. Again, activity was found in the
fraction with the mobility of corticosterone. The active fraction
from Sephadex chromatography has now been submitted to reverse
phase high pressure liquid chromatography as a second purification
step. It has been highly purified using sequentially more polar
mobile phases to achieve an estimated 100 fold increase in specific
activity and a fraction with only a few U.V. absorbing peaks.
Thus, we should soon have sufficient purity for identification by
combined gas chromatograph-mass spectroscopy, infared spectroscopy,
etc.

The chromatographic findings raise the possibility that corticosterone is responsible for the Cl^- transport stimulatory effects of urinary extracts. Therefore, we have performed a limited
and preliminary examination of the effect of corticosterone on both
J_o^{Cl} and on J_{net}^{Cl} (1). In a dose-response evaluation, stimulation of
Cl^- outflux was clearly detected at $1x10^{-5}M$ and above but not at
$3.3x10^{-6}$ and $1x10^{-6}M$. The estimated amount of free corticosterone
in the urine is too low, by two orders of magnitude, to be soley
responsible for the outflux effect produced by extracts. Also, in
bidirectional flux studies (using $1x10^{-4}$ corticosterone), both influx and outflux increased to a modest degree resulting in no major
change in net flux compared to control. The findings are in marked
contrast to the effects of urinary extracts on net flux of Cl^-.

A large portion of urinary corticosteroids are conjugated in
the form of glucuronides and mono-and disulfates. Fig. 6 demonstrates the effects of glucuronidase-sulfatase (500 units/ml) upon
extractable chloride transport stimulatory activity of the urine

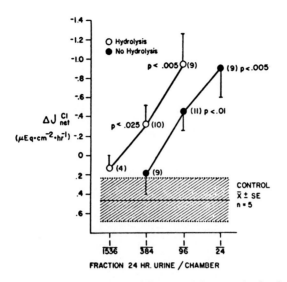

FIG. 6. *Effect of glucuronidase-sulfatase hydrolysis on chloride transport stimulatory activity of urinary extracts of NaCl loaded man (18). Change in chloride net flux (ΔJ_{net}^{Cl}) is J_{net}^{Cl} in third hr after extract minus J_{net}^{Cl} in hr before extract. J_{net}^{Cl} inward is designated (+) and outward is designated (-). The number of skin pairs used for bidirectional Cl^- flux at each dose level is shown in parentheses. P values compare skins which received extract to control skins which received dilutent alone. Urines used were from three subjects.*

of salt loaded subjects (18). In this case only bidirectional chloride flux studies were performed. In the nonhydrolyzed urine, activity is clearly present at 1/24 and 1/96 of a 24 hr urine per experimental chamber, but not at 1/384. After hydrolysis, activity could be detected at 1/384 of a 24 hr urine. Thus, hydrolysis seemed to increase activity approximately four-fold, suggesting that a considerable amount of chloride stimulatory factor appears in the urine as glucuronide and/or sulfate conjugates. The enhancement of activity should facilitate our subsequent preparative efforts.

IV. CONCLUSION

(1) The hypothesis of two distinct mechanisms of active Cl^- transport, based on operational characteristics, seems to hold for isolated frog skin of *R. pipiens* and may be generally applicable for epithelia of varied origin. The postulated model of NaCl coupled entry to transporting cells, powered by Na^+ gradients from Na -K-ATPase active Na^+ transport is an attractive explanation for the ouabain-sensitive and Na^+-K^+ dependent type of active Cl^- transport. The other type, acetazolamide-sensitive and cation-dependent, may be mediated by Cl^-- HCO_3^- exchange.

(2) Regulation of Na^+ and Cl^- transport has been examined in skins removed from animals preconditioned in various electrolyte media. The selective Cl^- transport regulatory system(s) may be related to acid-base change and or the presence of Cl^- in the conditioning medium. The evidence for a humoral factor in plasma of frogs maintained in one of the preconditioning media is preliminary.

(3) The best evidence, at present, for a hormonal factor which regulates Cl^- transport directly is from studies of lipid extracts of urine of chronically NaCl loaded man. The chloride transport stimulatory factor may be of adrenal origin and its effect on active ion transport in frog skin is antithetical to that of aldosterone. Aldosterone produces NaCl conservation in states of salt deprivation. This factor may facilitate adaptation to states of excess salt ingestion.

V. SUMMARY

Three active Cl^- transport systems are demonstrable in isolated short-circuited skins of Rana pipiens. The first, detectable in the basal state in skins bathed in ordinary Ringers' solution, consists of net Cl^- influx approximating 5-30% of active Na^+

transport inward. It is ouabain-sensitive and cation-dependent.
The second system, also net Cl^- influx, is relatively small in magnitude and demonstrable only in skins bathed in low Cl^- medium. It
is acetazolamide-sensitive and cation-independent and may be mediated by anion exchange. The third system, net Cl^- outflux induced
by beta adrenergic stimulation, has the same characteristics as the
first. Studies in frog skin and a survey of Cl^- transport in other
epithelia have led us to believe that many or most systems may
ultimately be categorized into one of these two operational categories, cation-dependent and ouabain-sensitive, or cation-independent
and acetazolamide-sensitive.

A Cl^- transport regulatory system(s), which is selectively
induced by preconditioning frogs in various electrolyte media, can
be demonstrated and shown to secondarily influence Na^+ transport.
However, the role of humoral substances as mediators of this system
is still unclear. Evidence for a hormonal factor in man, which
stimulates another Cl^- transport system in frog skin, is clearer.
Urinary extracts of man on high NaCl intake produce net Cl^- outflux
as well as Na^+ outflux stimulation in isolated short-circuited frog
skin. Little activity is found in urine of NaCl deprived normal
subjects or salt loaded adrenal insufficient subjects. The factor
may be of adrenal origin or dependent upon the presence of the
adrenal for its production. Characteristics which are compatible
with it being a steroid are its lipid solubility, its time-course
of action, and its enhanced activity by glucuronidase-sulfatase
hydrolysis. On Sephadex LH20 chromatography, it has the mobility
of cortocosterone. However, the latter is insufficient in quantity
and in qualitative effects to explain the urinary activity. This
Cl^- transport stimulatory factor may play a role in adaptation to
chronic volume expansion in man.

REFERENCES

1. C. O. Watlington and F. Jessee, Jr.   Chloride flux across
   skins of low potential difference. *Biochim. Biophys.
   Acta.* 330-102, 1973.

2. C. O. Watlington and F. Jessee, Jr.   Net Cl$^-$ flux in short-
   circuited skin of Rana pipiens: Ouabain sensitivity and
   Na$^+$-K$^+$ dependence. *Biochim. Biophys. Acta. 382:* 204-212,
   1975.

3. C. O. Watlington, S. D. Jessee, and G. Baldwin. Ouabain,
   acetazolamide, and Cl$^-$ flux in isolated frog skin:
   Evidence for two distinct active Cl$^-$ transport mechanisms
   Am. J. Physiol. 232: 550-558, 1977.

4. J. A. Zadunaisky, O. A. Candia, and D. J. Chiarandini. The
   origin of the short-circuit current in the isolated skin
   of the South American frog *Leptodactylus ocellatus. J.
   Gen. Physiol. 47:* 393-402, 1963.

5. D. W. Martin and P. F. Curran. Reversed potentials in isolated frog skin. II. Active transport of chloride. *J.
   Cell. Comp. Physiol. 67:* 367-374, 1966.

6. A. Krogh. Osmotic regulation in the frog (*R. exculenta*) by
   active absorption of Cl ions. *Skand. Arch. Physiol.
   76:* 60-74, 1937.

7. M. W. Pinschmidt, D. C. Addison, and E. G. Huf. Role of Na$^+$
   and anions in the triple response of isolated frog skin
   to norepinephrine. *Biochim. Biophys. Acta 323:* 309-325,
   1973.

8. C. O. Watlington. Effect of catecholamines and adrenergic
   blocking agents on Na transport across isolated frog skin
   Am. J. Physiol. 214: 1001-1007, 1968.

9. V. Koefoed-Johnson, H. Ussing, and K. Zerahn. The origin of
   the short-circuit current in the adrenaline-stimulated
   frog skin. *Acta Physiol. Scand. 27:* 38-48, 1953.

10. C. O. Watlington. Effect of adrenergic stimulation on ion

transport across skin of live frogs. *Comp. Biochem. Physiol. 24:* 965-974, 1968.

11. R. A. Frizelle, M. Field, and S. G. Schultz. Sodium-coupled chloride transport by epithelial tissues Am *J. Physiol.* 236F1-F8, 1979; or *Am. J. Physiol.: Renal Fluid Electrolyte Physiol. 5:* F1-F8, 1979.

12. C. F. Gonzalez, Y. E. Shamoo, and W. A. Brodsky. Electrical nature of active chloride transport across short-circuited turtle bladder. *Am. J. Physiol. 212:* 641-650, 1967.

13. P. J. Wistrand, J. A. Rawls, Jr., and T. H. Maren. Sulphonamide carbonic anhydrase inhibition and intra-ocular pressure in rabbits. *Acta Pharmacol. Toxicol. 17:* 337, 1960.

10A. P. Silva, J. Stoff, M. Field, L. Fine, J. N. Forrest, and F. H. Epstein. Mechanism of active chloride secretion by shark rectal gland: Role of Na-K-ATPase in chloride transport. *Am J. Physiol. 233:* F298-F306, 1977 or *Am. J. Physiol.: Renal Fluid Electrolyte Physiol. 2:* F298-F306, 1977.

12A. C. O. Watlington and E. G. Huf. Beta adrenergic stimulation of frog skin mucous glands; non-specific inhibition by adrenergic blocking agents. *Comp. Gen. Pharmacol. 2:* 295-305, 1971.

14. C. O. Watlington. Regulation of NaCl transport; relation of Cl permeability. *Biochim. Biophys. Acta.* 249-339, 1971.

15. C. O. Watlington. Regulation of Na Transport by alteration of chloride conductance. *Biochim. Biophys. Acta 288:* 482-485, 1972.

16. C. O. Watlington and L. B. Taylor. Humoral inhibition of Na transport by Cl permeability alteration. *Comp. Biochem. Physiol. 51*A: 733-736, 1975.

17. C. O. Watlington, G. Baldwin, R. King, S. Grossman, and H.
     Estep.   Chloride transport stimulatory factor in urine
     of chronically sodium-chloride loaded man. *The Lancet*,
     p. 169-171, July 23, 1977.

18. G. Baldwin, S. Jessee, and C. O. Watlington. Chloride trans-
     port stimulatory factor in human urine: Effects on Na[+]
     fluxes and preliminary characterization. *J. Clin. Endo-
     crinol. Metab. 49:* 930-936, 1979.

CHLORIDE TRANSPORT IN THE EHRLICH MOUSE ASCITES TUMOR CELL

Charles Levinson

Department of Physiology
The University of Texas Health Science Center
San Antonio, Texas

I. BRIEF HISTORY OF THE EHRLICH ASCITES TUMOR CELL

Since the data to be reviewed have been obtained from studies
utilizing the Ehrlich-Lettré ascites tumor (hyperdiploid), it is
worthwhile to briefly describe the origin and characteristics of
tumor cell.

Between 1896 and 1905, Paul Ehrlich and co-workers (1) at the
Frankfurt Institute established lines of over 60 transplantable
mouse mammary carcinomas. Many of these solid tumors arose spon-
taneously in stock mice and were identified simply as the Ehrlich
mouse mammary carcinoma. They were used by Ehrlich and others in
pioneering studies of cancer chemotherapy and tumor biology. In
1923, Loewenthal and Jahn (2) of the Robert Koch Institute, Berlin,
obtained an Ehrlich mouse carcinoma from the Frankfurt Institute
and maintained the tumor in solid form by passage from mouse to
mouse. Between 1927 and 1930, a number of experiments were
carried out in which "tumorbrei" or crude cell suspensions were

383

injected into the peritoneal cavity of mice. Only solid tumors
developed. However, in November 1930, a large number of viable
cells suspended in 0.20 ml of saline were injected. Again, in
most cases only solid tumors developed; however, in some mice a
different pattern was observed. Ten to fourteen days after the
innoculation a large volume of milky or bloody ascites formed
which contained large numbers of individual cells.   There was no
solid tumor formation, although a cellular infiltration of the
lymph vessels and fatty tissues was present. The ascites tumor
was readily transmissible to other mice and the frequency of
success was virtually 100%. The survival time of the innoculated
animals and the rapidity of the development of the ascites was a
function of the number of cells injected. At high dilution, solid
tumor formation without ascites was observed in a variable fraction
of the mice, and a certain minimal number of cells was necessary
to produce any tumor at all. Spontaneous regression of the ascites
tumor was never observed. Since its first description, the ascites
form (Ehrlich mouse ascites tumor cell) of the Ehrlich carcinoma
has been distributed to a large number of investigators.

For example, in the late 1930s Albert Fisher in Copenhagen
maintained the Ehrlich mammary carcinoma in three forms: solid,
ascites, and in tissue culture. The original tumor material
presumably was obtained from Loewenthal. At about this time, Hans
Lettré in Heidelberg also began his studies of potential cancer
chemotherapeutic agents using the ascitic form of the Ehrlich
mouse carcinoma cell as the cancer model. The source of the
Ehrlich-Lettré ascites tumor cell (Heidelberg strain) is obscure
but it most likely originated from Loewenthal's laboratory.

George Klein (3) of the Karolinska Institute obtained the
ascitic form, derived from tissue cultured cells, from A. Fisher in
1948, and showed that the aneuploid modal chromosome number is 80
(4). The cell was designated as tetraploid.

In 1950, T. S. Hauschka (5) received the tetraploid cell line
from Klein and between 1951 and 1954 a number of substrains, desig-

nated E_1-E_{11}, were cloned from the parent strain cell. Hauschka (6,7) obtained the Ehrlich-Lettré (Heidelberg) strain from Lettré in 1953 and noted as had Bayreuther (8), that the modal chromosome number was 45-46 or hyperdiploid. The relationship between these two distinct tumors (hypertetraploid and hyperdiploid), both iden- tified with Ehrlich's name and both presumably originating in Loewenthal's laboratory, is unclear. However, this should not be surprising since genetic drift in malignant cell populations, pro- pagated for many years in a variety of immunologically different host mice, is probably considerable. Since the direction of this drift is not predictable, the Ehrlich ascites tumor cells main- tained in numerous laboratories worldwide may still be similar, but far from identical.

It is instructive to end this brief history of the Ehrlich cell by pointing out that more than 45 separate listings under "Ehrlich ascites carcinoma" appear in the Imperial Cancer Research Fund (England) Crossed Referenced Bibliography Vol. I XVI (1964-1979) (9). With such a multiplicity of listings derived from so few stem lines, it is probable that careful evaluation would re- veal the existence of only a few varieties of the Ehrlich ascites carcinoma which have been widely disseminated.

II.   CHLORIDE TRANSPORT IN EHRLICH ASCITES TUMOR CELLS

Interest in the transport of anions across the cell membrane stems from the growth characteristics of the Ehrlich ascites tumor cell. This cell is concerned almost totally with growth and repli- cation which leads to a dramatic increase in the total cellular mass within the host mouse. For example, a mouse weighing 20 gm will yield 2-3 gm of single tumor cells suspended in 10-12 ml of ascitic fluid 7 days after innoculation with 10^7 cells. Conse- quently, the host supports the production of 10-15% of its body weight in new tissue. In order to accomplish this remarkable growth each cell must incorporate into its substance every 18-24 hr the complete complement of essential metabolites including water, electrolytes, and proteins. Since the cell membrane forms the boundary between the intracellular and extracellular fluids, it is important to establish the role of this structure in the regulation of the movement of materials both into and out of the cell.

A. Exchangeability

Several reports concerning the exchangeability of intracellular Cl^- have been published. Early studies suggested that intracellular Cl^- was not completely exchangeable with that of the extracellular medium (10-12). However, evidence has been presented that shows the nonexchangeable Cl^- found in earlier studies was due to an artifact in the measurement of total Cl^- from aqueous of the cell and does not represent a separate Cl^- compartment (13).

B. Transport

Since it is now well established that cellular Cl^- can completely exchange with that of the extracellular phase, we turn our attention to the nature of the exchange process. Evidence supporting the concept of simple passive diffusion of Cl^- has been presented (10,11,14). When extracellular Cl^- was replaced with isosmotic equivalents of NO_3^-, Cl^- transport varied linearly with the extracellular Cl^- concentration. Implicit in the conclusion of diffusional Cl^- transport was the assumption that NO_3^- in no way influenced the transmembrane movement of Cl^-.

This assumption has recently been reinvestigated. Both Aull et al. (15) and Hoffmann et al. (16) have shown that when NO_3^- is used to replace Cl^- in the medium, the cell water content and chloride distribution ratio remained constant and the relationship between Cl^- exchange and extracellular Cl^- concentration is concave upward rather than linear. Their data suggest that NO_3^-, rather than acting simply as an inert replacement anion, depresses Cl^- exchange. Levinson and Villereal (17). on the other hand, find a linear relationship between Cl^- exchange and extracellular Cl^- in agreement with the earlier studies (12,41,18). A constant distribution ratio of Cl^- (r_{Cl} = 0.39) even at high NO_3^- concentration (130 mM) was also observed.

Almost complete replacement of Cl^- by Br^- does not effect the cell water content or the Cl^- distribution ratio. However, the reduction of the extracellular Cl^- concentration by Br^- results in inhibition of steady-state Cl^- exchange (16,17).

Sulfate, but not inorganic phosphate, can also inhibit Cl^- exchange. For example, Levinson (19) measured the Cl^- efflux rate coefficient in cells incubated in 100 mM Cl^-, 100 mM sucrose medium. Replacement of sucrose with increasing concentrations of SO_4^{-2} led to progressive inhibition of Cl^- exchange.

The fact that NO_3^-, Br^-, and SO_4^{-2} alter the kinetics of Cl^- exchange (summarized in Table I) raises the possibility that Cl^- transport is not an example of simple passive diffusion but rather is a mediated process.

Evidence supporting this view has recently been presented. When the Cl^- concentration in the medium was varied either by substitution of lactate, sucrose, or gluconate, the steady-state Cl^- exchange, measured at 23^0 or 38^0C, exhibited saturation kinetics (16,20). In the case of lactate substitution, the relationship between flux and concentration is sigmoidal at low Cl^- concentrations while at concentrations greater than 100 mM the exchange process saturates. The apparent K_m is 50 mM. However, when sucrose replaced NaCl in the medium, transport reaches a maximum value at 50 mM and decreases with increasing concentration. The apparent K_m from this rising phase of the flux-concentration curve is 15 mM. It is not clear why differences exist between the kinetic parameters

TABLE I. *Chloride Transport in Ehrlich Tumor Cells*

Medium	$[Cl^-]_{cell}/[Cl^-]_{med}$	Flux vs. $[Cl^-]_{med}$	Reference
Cl^-/NO_3^-	0.39	Linear; $K_m = K_i$	12,14,17,18
	0.48	Concave upward	15,16
		$K_m > K_i$	
Cl^-/Br^-	0.37	Concave upward	17
	0.48	Concave upward	16
		$K_i > K_m$	
$Cl^-/Acetate$	0.39	Concave downward	17
		$K_m = 23$ mM Cl^-	
$Cl^-/Lactate$	0.33-0.55	Concave downward	16
		$K_m = 50$ mM Cl^-	
$Cl^-/Sucrose$	0.60-1.10	Concave downward	16
		$K_m = 15$ mM Cl^-	
$Cl^-/Gluconate$	0.30-0.57	Concave downward	20
		$K_m = 23.9$	

generated with lactate and sucrose substitution. They may be related to differences in ionic strength of the medium, the fact that
the Cl^- distribution ratio was not constant in either case, and the
possibility that lactate, particularly at high concentration, is an
inhibitor of Cl^- exchange. The observation that Cl^- transport is
depressed at concentrations greater than 50 mM is consistent with
the suggestion that Cl^- can inhibit its own transport perhaps by a
mechanism similar to that described for erythrocytes (21). In the
case of gluconate replacement, Cl^- exchange saturates at an extracellular Cl^- concentration of about 90 mM and the apparent K_m is
23 mM. No evidence for self-inhibition of Cl^- transport was
observed.

Additional evidence for the existence of carrier mediated Cl^-
transport comes from the studies of Levinson and Villereal (17).
The dependence of Cl^- exchange on the Cl^- concentration was determined when Cl^- was varied by isosmotic replacement with acetate.
Even though the Cl^- distribution ratio was unaffected by high concentration of acetate ($r_{Cl^-} = 0.394$), the cells lost K and gained
both Na and water. Aull et al. (15) have demonstrated similar
changes in the electrolyte content and also observed that a high
acetate concentrations (124 mM) relative to Cl^- (32 mM), the
efflux rate coefficient for Cl^- was depressed by 33%.

The relationship between Cl^- exchange and the Cl^- concentration (in acetate-Cl^- medium) suggests that Cl^- transport is composed of two components. One depends on a mechanism that saturates
at 50 mM Cl^- with an apparent K_m of 23 mM. The other resembles
free diffusion since with extracellular Cl^- concentrations above
50 mM, exchange is directly proportional to the Cl^- concentration.
The quantitative evaluation of the contribution of each pathway to
the steady state exchange process is complicated by the fact that
acetate has been shown to be a noncompetitive inhibitor of Cl^-
transport. However, the conclusion that Cl^- crosses the membrane
by both mediated and diffusional processes appears to be valid.

Hoffman et al. (16) have estimated the Cl^- conductance at

$38°C$ to be 14 µSiemens/cm^2, and from this calculated the diffusional component (J_D) as 4 pmols/cm^2 sec. or 4.9 mmols/Kg dry wgt $(min)^{-1}$. From J_D, the membrane potential (24 mV), and the extracellular Cl^- concentration (150 mM), the diffusional permeability coefficient ($P_{Cl}-$) was calculated and found to be 4 X 10^8 cm/sec. Total Cl^- exchange measured in 150 mM Cl^- and $38°C$ was 63 pmols/cm^2 sec, and consequently on 4/63 or 6% of the total exchange is diffusional while 94% occurs by an electrically silent transport system. Similarly Heinz *et al.* (22) found the net or diffusional permeability coefficient ($P_{Cl}-$) to be 0.057 ml/gm dry wgt $(min)^{-1}$ or 4.8 X 10^8 cm/sec and indicated this value to be about 20 times smaller than that derived from isotopic Cl^- exchange.

If we accept the contention that at least part of the trans-membrane movement of Cl^- is carrier mediated and obeys Michaelis-Menten kinetics, then the effects of Br^-, NO_3^-, and SO_4^{-2} on Cl^- exchange are consistent with assertion that these anions behave as competitive inhibitors (23). For example, a linear relationship between Cl^- exchange and extracellular Cl^- in NO_3^--containing media would result if the affinity of the transport system for Cl^- and NO_3^- were identical, that is, if $K_m (Cl^-) = K_i(NO_3^-)$. However, if the affinity of the transport system for NO_3^- or for Br^- exceeds that of Cl^- then a concave upward relationship between Cl^- exchange and extracellular Cl^- would result and $K_i (NO_3^-) < K_m(Cl^-)$ or $K_i (Br^-) < K_m(Cl^-)$.

In addition to mediated Cl^- self-exchange and net diffusional transport pathways, Geck and his colleagues have described a net cotransport system for Na^+, K^+, and Cl^- (24,25). This mode of transport is ouabain-insensitive but can be specifically inhibited by the diuretic, furosemide. Because the furosemide-sensitive fluxes of the Na^+, K^+, and Cl^- are not influenced by changes in the membrane potential, they appear to be electrically silent. Apparently a tight coupling exists between the furosemide-sensitive fluxes of Na^+, K^+, and Cl^- with a stoichiometry of 1:1:2. Chloride's participation in this transport system is obligatory since it cannot be replaced by NO_3^-, but can to some extent be re-placed by Br^-. Since this mode of transport has only been demon-strated under conditions of net uptake of Na^+, K^+, and Cl^-, it is not clear whether it is the same as Cl^--dependent, ouabain-insen-sitive K^+ self-exchange described by Bakker-Grunwald (26,27). In this system both furosemide and Cl^- replacement had similar inhi-bitory effects on K^+-self exchange.

C. Transport Inhibitors

Steady state Cl^- exchange is susceptible to inhibition by a variety of agents. Furosemide, a potent inhibitor of Cl^- transport in erythrocytes (28), inhibits steady-state Cl^- exchange by 86% when present at 10 mM and by 57% at 1 mM (15). Although transport was dramatically inhibited the cell Cl^- content did not change. These observations led to the conclusion that furosemide specifi-cally blocked an exchange diffusion system for Cl^-. However, it is doubtful that this agent is absolutely specific for the Cl^- transport system since furosemide can also inhibit net cotransport of Na^+, K^+, and Cl^-, as well as Cl^--dependent K^+ exchange. This may signify that furosemide nonspecifically alters the membrane in such a fashion as to modify both anion and cation transport pro-cesses. Alternatively, the action of furosemide may be much more specific in that it inhibits only those mediated transport pro-cesses that require Cl^- as a direct participant. Phloretin, like furosemide, also inhibits Cl^- exchange. However, this agent is effective at lower concentrations (e.g., 5×10^{-4} M) and inhibits the self-exchange of Cl^-, as well as that of SO_4^{-2} by 80% (17). In addition to inhibiting Cl^- exchange phloretin also blocks net Cl^- transport by 80% (16). This suggests that exchange and conductive pathways do not represent separate modes of transport but may involve common elements.

Anion transport in the Ehrlich ascites cell can also be inhi-

bited by SITS (4-acetamido, 4'-isothiocyano stilbene, 2,2'-disul-
fonic acid) and H_2DIDS (4,4'-diisothiocyano, 1,2-diphenyl ethane,
2,2'-disulfonic acid), two agents known to be specific inhibitors
of an anion transport in erythrocytes and other systems. These
agents are of particular interest because they do not enter the
erythrocyte but have the capacity to bind both reversibly and
irreversibly(covalently) to the membrane and in so doing inhibit
anion transport (29). The extent of inhibition is directly related to the number of molecules bound to a specific membrane protein with an apparent molecular weight of 95,000. SITS inhibits
the exchange of Cl^- but not that of K^+ (15). This occurs under
conditions where the intracellular Cl^- content remains constant.
Reduction of Cl^- transport was approximately 10% at 5×10^{-5} M
SITS. The inhibitory effect of SITS was completely reversed by
washing the cells in Ringer solution. When cells were pretreated
with 1×10^{-4} M SITS for 60 min and then washed, SO_4^{-2} transport
was inhibited by almost 90%, while Cl^- and inorganic phosphate
transport were unaffected (30). H_2DIDS, like SITS, reduces both
Cl^- and SO_4^{-2} self-exchange. However, H_2DIDS reduces SO_4^{-2} transport to a much greater degree than that of Cl^- (19). For example,
at 5×10^{-5} M, SO_4^{-2} exchange was reduced 90% when composed to
control while Cl^- exchange was inhibited only 22%. The inhibitory
affect of this agent could be almost completely reversed provided
the cells were washed following brief (less than 3 min) exposure
to H_2DIDS. Prolonged incubation with H_2DIDS (0.5 to 10×10^{-5} M;
30 min, 37^0C), on the other hand, led to an irreversible binding
which is manifest by progressive inhibition of SO_4^{-2} transport
with increasing initial concentration of H_2DIDS. For example, at
1×10^{-5} M, SO_4^{-2} transport was inhibited by 33% and at 5×10^{-5}
M inhibition increased to about 90%. However, even at the highest
concentrations tested, Cl^- transport was unaffected. The effects
of phloretin, SO_4^{-2}, and reversibly and irreversibly bound H_2DIDS
on the self exchange of Cl^- and SO_4^{-2} has led to the postulation
that Cl^- and SO_4^{-2} share a common transport mechanism which

possesses two reactive sites (19).

In summary, the available evidence indicates that in the Ehrlich ascites tumor cell Cl^- exchanges completely with that of the extracellular medium. This exchange process is apparently composed of two components. The first which exhibits saturation kinetics represents 90-95% of the total and has the characteristics of an equilibrating, electrically silent exchange diffusion. The second possesses the characteristics of a simple diffusion process and contributes 5-10% of the total exchange flux. Although kinetically dissimilar there is some evidence that these two processes may utilize the same transport component. In addition to these pathways Cl^- may also be cotransported with K^+ and Na^+ by an electrically silent ternary symport mechanism.

In red blood cells, convincing evidence has been obtained during the past few years that all inorganic anions are transported by a common carrier system. The fact that Br^-, NO_3^-, and SO_4^{-2} under the appropriate experimental conditions can inhibit Cl^- exchange may indicate that these anions are also transported by a common mechanism. The inhibitory effect of phloretin on the self-exchanges of both Cl^- and SO_4^{-2} is consistent with this idea. However, the differential effects of SITS and H_2DIDS on Cl^--exchange are more consistent with the existence of either two separate transport systems or a common system possessing two reactive sites.

REFERENCES

1. P. Ehrlich and A. Apolant. Beobachtungen uber maligne Mausetumoren. *Berlin Klin. Wsch.* 42: 871-874, 1905.

2. H. Loewenthal and G. Jahn. Ubertragungsversuche mit carcinomatoser Mause-Ascitesflussigkeit und ihr Verhalten gegen physikalische und chemische Einwirkungen. *Ztsch. Krebsforsch.* 37: 439-447, 1932.

3. E. Klein, N. B. Kurnick, and G. Klein. The effect of storage on the nucleic acid content and virulence of the mouse ascites tumor. *Exp. Cell. Res. 1:* 127-134, 1950.

4. G. Klein and L. Revez. Quantitative studies on the multiplication of neoplastic cells *in vitro*. I. Growth curves of the Ehrlich and MCIM ascites tumor. *J. Natl. Cancer Inst. 14:* 229-278, 1953.

5. T. S. Hauschka and A. Levan. Cytological and functional characteristics of single cell clones isolated from Krebs-2 and Ehrlich ascites tumors. *J. Natl. Cancer Inst. 21:* 77-136, 1958.

6. T. S. Hauschka. Cell population studies on mouse ascites tumors. *Trans. N. Y. Acad. Sci.* 15/15: 64-73, 1954.

7. T. S. Hauschka, S. T. Grinnell, L. Revez, and G. Klein. Quantitative studies on the multiplication of neoplastic cells *in vivo*. IV. Influence of doubled chromosome number on growth rate and final population size. *J. Natl. Cancer Inst. 19:* 12-32, 1957.

8. K. Bayreuther. Der chromosomen bestand des Ehrlich-Ascites-Tumors der Maus. *Z. Naturforschg. 7:* 554-557, 1952.

9. D. C. Roberts. Research Using Transplanted Tumours of Laboratory Animals: A Cross Referenced Bibliography, I-XVI, (1964-1979).

10. F. Aull. Measurement of the electrical potential difference across the membrane of the Ehrlich mouse ascites tumor cell. *J. Cell. Comp. Physiol. 69:* 21-32, 1967.

11. H. G. Hempling and H. Kromphardt. On the permeability of ascites tumor cells to chloride. *Fed. Proc. 24:* 709, 1965.

12. H. Kromphardt. Chloridtransport und Kationenpumpe in Ehrlich-Asciteszellen. *Eur. J. Bioch. 3:* 377-384, 1968.

13. L. O. Simonsen and A. -M. T. Nielsen. Exchangeability of chloride in Ehrlich ascites tumor cells. *Bioch. et Biophys. Acta. 241:* 522-527, 1971.

14. H. Grobecker, H. Kromphardt, H. Mariani, and E. Heinz. Unterschungen uber den Elektrolythaushalt der Ehrlich-Ascites-Tumorzelle. *Bioch. Zelt. 337:* 426-476, 1963.

15. F. Aull, M. S. Nachbar, and J. D. Oppenheim. Chloride self-exchange in Ehrlich ascites tumor cells. Inhibition by furosemide and 4-acetamido-4'-isothiocyanostilbene-2,2' disulfonic acid. *Bioch. et Biophys. Acta 471:* 341-347, 1977.

16. E. K. Hoffmann, L. O. Simonsen, and C. Sjøholm. Membrane potential, chloride exchange and chloride conductance in Ehrlich ascites tumor cells. *J. Physiol. (London) 296:* 61-84, 1979.

17. C. Levinson and M. L. Villereal. The transport of chloride in Ehrlich ascites tumor cells. *J. Cell. Physiol. 88:* 181-192, 1976.

18. F. Aull. The effect of external anions on steady state chloride exchange across ascites tumor cells. *J. Physiol. 221:* 755-771, 1972.

19. C. Levinson. Chloride and sulfate transport in Ehrlich ascites tumor cells: Evidence for a common mechanism. *J. Cell. Physiol. 95:* 23-32, 1978.

20. F. Aull. Saturation behavior of ascites tumor cell chloride exchange in the presence of gluconate. *Biochim. Biophys. Acta. 554:* 538, 1979.

21. M. Dalmark. Effects of halides and bicarbonate on chloride transport in human red blood cells. *J. Gen. Physiol. 67:* 223-234, 1976.

22. E. Heinz, P. Geck, and C. Pietrzyk. Driving forces of amino acids transport in animal cell. *Ann. N.Y. Acad. Sci. 264:* 428-441, 1975.

23. R. B. Gunn, M. Dalmark, D. C. Tosteson, and J. O. Wieth. Characteristics of chloride transport in red blood cells. *J. Gen. Physiol. 61:* 185-206, 1973.

24. P. Geck, C. Pietrzyk, B. C. Burckhardt, and E. Heinz. Elec-

trically silent cotransport of Na^+, K^+ and Cl^- in Ehrlich cells. *Bioch. et Biophys. Acta 600:* 432-447, 1980.

25. P. Geck and E. Heinz. Coupling of ion flows in cell suspension systems. *Ann. N.Y. Acad. Sci. 341:* 57-66, 1980.

26. T. Bakker-Grunwald. Effect of anions on potassium self-exchange in ascites tumor cells. *Bioch. et Biophys. Acta 513:* 292-295, 1978.

27. T. Bakker-Grunwald, J. S. Andre, and M. C. Neville. K^+ influx components in ascites cells: The effect of agents interacting with the $(Na^+ + K^+)$-pump. *J. Membr. Biol. 52:* 141-146, 1980.

28. P. C. Brazy and R. B. Gunn. Furosemide inhibition of chloride transport in human red blood cells. *J. Gen. Physiol. 68:* 583-599, 1976.

29. A. Rothstein, Z. I. Cabantchik, and P. Knauf. Mechanism of anion transport in red blood cells: Role of membrane proteins. *Fed. Proc. 35:* 3-10, 1976.

30. M. L. Villereal and C. Levinson. Chloride-stimulated sulfate efflux in Ehrlich ascites tumor cells: Evidence for 1:1 coupling. *J. Cell. Physiol. 90:* 553-564, 1977.

Index

U

Urine, human, stimulatory factor for chloride
transport in, 370–378

V

Vasoactive intestinal peptide (VIP)
cyclic AMP and, 286
in dogfish, 285–285
effect on rectal gland chloride secretion,
284–285
inhibition, 287–290